National Management Measures to Control Nonpoint Source Pollution from Forestry

United States Environmental Protection Agency
Office of Water
Washington, DC 20460
(4503F)

EPA-841-B-05-001

April 2005

National Management Measures to Control Nonpoint Source Pollution from Forestry

Nonpoint Source Control Branch
Office of Wetlands, Oceans and Watersheds
Office of Water
U.S. Environmental Protection Agency

April 2005

DISCLAIMER

This document provides guidance to States, Territories, authorized
Tribes, commercial and non-industrial private forest owners and
managers, and the public regarding management measures that may be used
to reduce nonpoint source pollution from forestry activities. At times
this document refers to statutory and regulatory provisions which
contain legally binding requirements. This document does not substitute
for those provisions or regulations, nor is it a regulation itself.
Thus, it does not impose legally-binding requirements on EPA, States,
Territories, authorized Tribes, or the public and may not apply to a
particular situation based upon the circumstances. EPA, State,
Territory, and authorized Tribe decision makers retain the discretion to
adopt approaches to control nonpoint source pollution from forestry
activities on a case-by-case basis that differ from this guidance where
appropriate. EPA may change this guidance in the future.

CONTENTS

FIGURES

TABLES

CHAPTER 1: INTRODUCTION

The Nation's aquatic resources are among its most valuable assets. Although environmental protection programs in the United States have successfully improved water quality during the past 25 years, many challenges remain. Significant strides have been made in reducing the effects of discrete pollutant sources, such as factories and sewage treatment plants (called point sources). But aquatic ecosystems remain impaired, mostly because of complex problems caused by polluted runoff, known as nonpoint source pollution.

Every 2 years the U.S. Environmental Protection Agency (EPA) reports to Congress on the status of the Nation's waters. The *1998 National Water Quality Inventory* (USEPA, 2000) reports that the most significant source of water quality impairment to rivers and streams and lakes, ponds, and reservoirs is agriculture, and the most significant source of impairment to estuaries is municipal point sources of pollution (Table 1-1). Other important sources of impairment or alterations that can impair water quality include hydrologic modifications like dams and channelization (a leading cause of impairment to rivers and streams and lakes, ponds, and reservoirs), urban runoff and storm sewer discharges (leading sources of impairment to all surface waters), and pollutants deposited from the atmosphere (a leading source of impairment to estuaries). The five leading pollutants impairing the Nation's waters are siltation, nutrients (from fertilizers and animal waste), bacteria, toxic metals, and organic enrichment that lowers dissolved oxygen (USEPA, 2000).[1] Siltation is the leading cause of water quality impairment to rivers and streams and the third leading cause of impairment to lakes, ponds, and reservoirs. Nine states list silviculture as a leading source of impairment to rivers and streams.[2]

> This guidance is designed to provide current information to state forestry program managers and foresters, commercial forest managers, private foresters and loggers, and nonindustrial private forest owners on nonpoint source pollution from forestry activities.

The Purpose and Scope of This Guidance

This guidance document is intended to provide technical assistance to state water quality and forestry program managers, nonindustrial private forest owners, industrial forest owners, and others involved with forest management on the best available, most economically achievable means of reducing the nonpoint source pollution of surface and groundwaters that can result from forestry activities. The guidance provides background information about nonpoint source pollution from forestry activities, including where it

---

[1] The term *pollutant* means dredged spoil, solid waste, incinerator residue, sewage, garbage, sewage sludge, munitions, chemical wastes, biological materials, radioactive materials, heat, wrecked or discarded equipment, rock, sand, cellar dirt and industrial, municipal, and agricultural waste discharged into water (Clean Water Act [Title 33, Chapter 26, Subchapter III, Section 1329]). The term *pollution* means the man-made or man-induced alteration of the chemical, physical, biological, and radiological integrity of water (Clean Water Act [Title 33, Chapter 26, Subchapter V, Sec. 1362(19)]).

[2] Nine states list silviculture as a major source of impairment to assessed rivers and streams: Arizona, California, Kentucky, Louisiana, Maine, New Mexico, Tennessee, Vermont, and West Virginia; 11 states/tribes list silviculture as a minor/moderate source of impairment to assessed rivers and streams: Coyote Valley Reservation, Florida, Hawaii, Minnesota, Mississippi, Ohio, Oklahoma, Oregon, South Carolina, Virginia, and Wisconsin; 6 states list silviculture as a source of impairment to assessed rivers and streams without specifying whether it is a major or minor/moderate source: Alaska, Colorado, Montana, North Carolina, Pennsylvania, and Washington. (Source: USEPA, 2000; National Water Quality Inventory, Appendix A-5.)

Table 1-1. Leading Pollutants and Sources Causing Impairment in Assessed Rivers, Lakes, and Estuaries (USEPA, 2000)

	Rivers and Streams[a]	Lakes, Ponds, and Reservoirs[b]	Estuaries[c]
Pollutants	Siltation	Nutrients	Pathogens (bacteria)
	Pathogens (bacteria)	Metals	Organic enrichment/ Low dissolved oxygen
	Nutrients	Siltation	Metals
Sources	Agriculture	Agriculture	Municipal Point Sources
	Hydromodification	Hydromodification	Urban runoff/ Storm sewers
	Urban runoff/ Storm sewers	Urban runoff/ Storm sewers	Atmospheric deposition

[a] Based on states' surveys of 23% of total river and stream miles.
[b] Based on states' surveys of 42% of total lake, reservoir, and pond acres.
[c] Based on states' surveys of 32% of total estuary square miles.

comes from and how it enters our waters. It presents the most current technical information about how to minimize and reduce nonpoint source pollution to forest waters, and it discusses the broad concept of assessing and addressing water quality problems on a watershed level. By assessing and addressing water quality problems at the watershed level, state program managers and others involved with forest management can integrate concerns about forestry activities with those of other resource management activities to identify conflicting requirements and provide balance between short-term impacts and long-term benefits (Table 1-2). This approach can maximize the potential for overall improvement and protection of watershed conditions and provide multiple environmental benefits.

The causes of nonpoint source pollution from forestry activities, the specific pollutants of concern, and general approaches to reducing the effect of such pollutants on aquatic resources are discussed in the Overview (Chapter 2). Also included in Chapter 2 is a general discussion of best management practices (BMPs) and the use of combinations of individual practices (BMP systems) to protect surface and groundwaters. Management measures for forest management and management practices that can be used to achieve the management measures are described in Chapter 3. Chapter 4 summarizes watershed planning principles and the application of management measures in a watershed context. Chapter 5 provides an overview of nonpoint source monitoring and tracking techniques.

This guidance does **not** replace the 1993 *Guidance Specifying Management Measures for Sources of Nonpoint Pollution in Coastal Waters.* The 1993 guidance still applies to coastal states.

Because this document is national in scope, it cannot address all practices or techniques specific to local or regional soils, climate, or forest types. Field research on management practices is ongoing in different parts of the country and under different harvesting circumstances to provide more guidance on how the practices mentioned in this guide and other management practices should be applied under specific circumstances. State laws and programs, or regional guidances published by the U.S. Forest Service, for instance, will have the criteria for site-specific management practice implementation. EPA encourages states to review their existing laws and programs for their relevance to forestry activities and to implement the management measures in this guidance within the context of state laws and programs wherever possible. In some cases very few adjustments to state laws and programs will be necessary to fully meet EPA's management measures. In other cases, major revisions or an entirely new program focus may be necessary. This guidance should prove useful in directing states toward those improvements that are necessary to protect water quality from forestry activities. Consult with

Table 1-2. Miles of Rivers and Streams Affected By Sources (USEPA, 2000).

SOURCE	MAJOR	MINOR	NOT SPECIFIED	TOTAL	TOTAL as Percent of Assessed Miles
Agriculture	21,856	102,264	46,630	170,750	20.3
Hydromodification	7,930	30,266	19,567	57,763	6.9
Nonirrigated Crop Production	2,551	34,747	9,186	46,484	5.5
Natural Sources	7,437	11,980	13,587	33,004	3.9
Urban Runoff/ Storm Sewers	5,747	20,060	6,504	32,310	3.8
Irrigated Crop Production	3,123	20,784	7,250	31,156	3.7
Municipal Point Sources	6,667	15,293	7,127	29,087	3.5
Animal Feeding Operations	2,736	24,908	108	27,751	3.3
Resource Extraction	5,948	9,771	9,612	25,231	3.0
Silviculture	717	14,884	4,420	20,020	2.4
Land Disposal	2,030	9,565	8,333	19,928	2.4
Range Grazing - Riparian and/or Upland	2,434	10,382	6,653	19,469	2.3
Habitat Modification (other than Hydro)	2,169	11,713	4,569	18,451	2.2
Channelization	3,024	9,677	4,802	17,503	2.1
Industrial Point Sources	3,409	7,335	3,051	13,795	1.6
Construction	1,653	6,331	4,452	12,436	1.5
Onsite Wastewater Systems (Septic Tanks)	874	3,123	7,834	11,831	1.4
Pasture Grazing - Riparian and/or Upland	1,262	9,335	0	10,597	1.3
Bank or Shoreline Modification	1,308	4,472	4,114	9,894	1.2
Other	768	4,375	2,495	7,638	0.9

state or local agencies, including the U.S. Department of Agriculture's Forest Service (USDA-FS), Natural Resources Conservation Service (NRCS), and Cooperative State, Research, Education, and Extension Service (CSREES); soil and water conservation districts; state forestry agencies; local cooperative extension services; and professional forestry organizations for additional information on nonpoint source pollution controls for forestry activities applicable to your local area. Resources and Internet sites related to forestry are listed in Appendices A and B.

This document provides guidance to states, territories, authorized tribes; commercial and nonindustrial private forest owners and managers; and the public regarding management measures that may be used to reduce nonpoint source pollution from forestry activities. At times this document refers to statutory and regulatory provisions that contain legally binding requirements. This document does not substitute for those provisions or regulations, nor is it a regulation itself. Thus, it does not impose legally binding requirements on EPA, states, territories, authorized tribes, or the public and may not apply to a particular situation based upon the circumstances. EPA, state, territory, and authorized tribe decision makers retain the discretion to adopt on a case-by-case basis approaches to control nonpoint source pollution from forestry activities that differ from this guidance where appropriate. EPA may change this guidance in the future.

Readers should note that this guidance is entirely consistent with the *Guidance Specifying Management Measures for Sources of Nonpoint Pollution in Coastal Waters* (USEPA, 1993), published under section 6217 of the Coastal Zone Act Reauthorization Amendments of 1990 (CZARA). This guidance, however, does not supplant or replace the 1993 coastal management measures guidance for the purpose of implementing programs under section 6217.

Under CZARA, states that participate in the Coastal Zone Management Program under the Coastal Zone Management Act are required to develop coastal nonpoint pollution control programs that ensure the implementation of EPA's management measures in their coastal management area. The 1993 guidance continues to apply to that program.

This document modifies and expands upon supplementary technical information contained in the 1993 coastal management measures guidance both to reflect circumstances relevant to differing inland conditions and to provide current technical information. It does not set new or additional standards for section 6217 or Clean Water Act section 319 programs. It does, however, provide information that government agencies, private sector groups, and individuals can use to understand and apply measures and practices to address sources of nonpoint source pollution from forestry.

What Is Nonpoint Source Pollution?

Nonpoint source pollution usually results from precipitation, atmospheric deposition, land runoff, infiltration, drainage, seepage, or hydrologic modification. As runoff from rainfall or snowmelt moves, it picks up and carries natural pollutants and pollutants resulting from human activity, ultimately dumping them into rivers, lakes, wetlands, coastal waters, and groundwater. Technically, the term *nonpoint source* is defined to mean any source of water pollution that does not meet the legal definition of *point source* in section 502(14) of the Clean Water Act of 1987:

> The term *point source* means any discernible, confined, and discrete conveyance, including but not limited to any pipe, ditch, channel, tunnel, conduit, well, discrete fissure, container, rolling stock, concentrated animal feeding operation, or vessel or other floating craft from which pollutants are or may be discharged. This term does not include agricultural storm water and return flows from irrigated agriculture.

Nonpoint sources, i.e., sources not defined by statute as point sources as described above, include return flow from irrigated agriculture, other agricultural runoff and infiltration, urban runoff from small or non-sewered urban areas, flow from abandoned mines, hydrologic modification, and runoff from forestry activities.

Although diffuse runoff is typically treated as nonpoint source pollution, runoff that enters and is discharged from conveyances such as those described above is treated as a point source discharge and therefore is subject to the permit requirements of the Clean Water Act. In contrast, nonpoint sources, including runoff from forestry activities, are not subject to federal permit requirements. Point source discharges usually enter receiving water bodies at some identifiable site and carry pollutants whose generation is controlled by some internal (e.g., industrial) process or activity, not by the weather. Point source discharges like municipal and industrial wastewaters, runoff or leachate from solid waste disposal sites, and storm sewer outfalls from large urban centers are regulated and permitted under the Clean Water Act.

Although water program managers understand and manage nonpoint sources in accordance with legal definitions and requirements, the nonlegal community often characterizes nonpoint sources in the following ways:

- Nonpoint source discharges enter surface and/or groundwaters in a diffuse manner at irregular intervals related mostly to weather.

- The pollutants arise over an extensive land area and move overland before they reach surface waters or infiltrate into groundwaters.

- The extent of nonpoint source pollution is related to uncontrollable climatic events and to geographic and geologic conditions and varies greatly from place to place and from year to year.

- Nonpoint sources are often more difficult or expensive to monitor at their point(s) of origin than point sources.

- Abatement of nonpoint sources is focused on land and runoff management practices, rather than on effluent treatment.

- Nonpoint source pollutants can be transported and deposited as airborne contaminants.

The nonpoint source pollutant of greatest concern with respect to forestry activities is sediment. The potential for sediment delivery to streams is a long-term (beyond 2 years) concern from almost all forestry harvesting activities and from forest roads regardless of their level of use or age (i.e., for the life of the road). Other pollutants of significance, including nutrients, temperature, toxic chemicals and metals, organic matter, pathogens, herbicides, and pesticides, are also of concern, and problems associated with these other pollutants (in the context of forestry activities) generally do not extend beyond 2 years from the time of harvest or are associated with a specific activity, such as an herbicide application. Nevertheless, all of these pollutants have the potential to affect water quality and aquatic habitat, and minimizing their delivery to surface waters and groundwater deserves serious consideration before and during forestry activities. Forest harvesting can also affect the hydrology of a watershed, and hydrologic alterations within a watershed have the potential to degrade water quality.

Programs to Control Nonpoint Source Pollution

During the first 15 years of the national program to abate and control water pollution (1972–1987), EPA and the states focused most of their water pollution control activities on traditional point sources. They regulated these point sources (and continue to regulate them) through the National Pollutant Discharge Elimination System (NPDES) permit program established by section 402 of the 1972 Federal Water Pollution Control Act (Clean Water Act). Under section 404 of the Clean Water Act, the U.S. Army Corps of Engineers and EPA also have regulated discharges of dredged and fill materials into wetlands.

As a result of the above activities, the United States has greatly reduced pollutant loads from point source discharges and has made considerable progress in restoring and maintaining water quality. However, the gains in controlling point sources have not solved all of our water quality problems. Studies and surveys conducted by EPA, other federal agencies, and state water quality agencies indicate that most of the remaining water quality impairments in our rivers, streams, lakes, estuaries, coastal waters, and wetlands result from nonpoint source pollution and other nontraditional sources, such as urban storm water discharges and overflows from combined sewers (sewers that carry both wastewater and storm water runoff). Summarized below are some legislative and programmatic efforts to control nonpoint source pollution from forestry activities.

Coastal Nonpoint Pollution Control Program

The Federal Coastal Nonpoint Pollution Control Program (6217) is designed to enhance state and local efforts to manage land use activities that degrade coastal habitats and waters.

In November 1990, Congress enacted the Coastal Zone Act Reauthorization Amendments (CZARA). These amendments were intended to address several concerns, including the effect of nonpoint source pollution on coastal waters.

To more specifically address the effects of nonpoint source pollution on coastal water quality, Congress enacted section 6217, *Protecting Coastal Waters* (codified as 16 U.S.C. section 1455b). Section 6217 requires that each state with an approved Coastal Zone Management Program develop a Coastal Nonpoint Pollution Control Program and submit it to EPA and the National Oceanic and Atmospheric Administration (NOAA) for approval. The purpose of the program is "to develop and implement management measures for nonpoint source pollution to restore and protect coastal waters, working in close conjunction with other state and local authorities."

Coastal Nonpoint Pollution Control Programs are not intended to replace existing coastal zone management programs and nonpoint source management programs. Rather, they are intended to serve as an update and expansion of existing programs and are to be coordinated closely with the coastal zone management programs that states and territories are already implementing in keeping with the Coastal Zone Management Act of 1972. The legislative history indicates that the central purpose of section 6217 is to strengthen the links between federal and state coastal zone management and water quality programs and to enhance state and local efforts to manage land use activities that degrade coastal waters and habitats.

Section 6217(g) of CZARA requires EPA to publish, in consultation with NOAA, the U.S. Fish and Wildlife Service, and other federal agencies, "guidance for specifying management measures for sources of nonpoint pollution in coastal waters." Section 6217(g)(5) defines management measures as

> economically achievable measures for the control of the addition of pollutants from existing and new categories and classes of nonpoint sources of pollution, which reflect the greatest degree of pollutant reduction achievable through the application of the best available nonpoint source control practices, technologies, processes, siting criteria, operating methods, and other alternatives.

EPA published *Guidance Specifying Management Measures for Sources of Nonpoint Pollution in Coastal Waters* (USEPA, 1993). In that document, management measures for urban areas; agricultural sources; forestry; marinas and recreational boating; hydromodification (channelization and channel modification, dams, and streambank and shoreline erosion); and wetlands, riparian areas, and vegetated treatment systems were defined and described. The management measures for controlling forestry nonpoint source pollution discussed in Chapter 3 of this document are based on those outlined by EPA in the coastal management measures guidance.

Nonpoint Source Program—Section 319 of the Clean Water Act

In 1987, in view of the progress achieved in controlling point sources and the growing national awareness of the increasingly dominant influence of nonpoint source pollution on water quality, Congress amended the Clean Water Act to focus greater national effort on nonpoint sources. Under this amended version, called the 1987 Water Quality Act,

Congress revised section 101, "Declaration of Goals and Policy," to add the following fundamental principle:

> It is the national policy that programs for the control of nonpoint sources of pollution be developed and implemented in an expeditious manner so as to enable the goals of this Act to be met through the control of both point and nonpoint sources of pollution.

More important, Congress enacted section 319 of the 1987 Water Quality Act, which established a national program to control nonpoint sources of water pollution. Under section 319, states, tribes, and territories address nonpoint source pollution by assessing the causes and sources of nonpoint source pollution and implementing management programs to control them. Section 319 authorizes EPA to issue grants to states, tribes, and territories to assist them in implementing management programs or portions of management programs that have been approved by EPA. In fiscal year 2001, Congress appropriated $237,476,800 for this purpose.

Section 319 nonpoint source pollution control programs are an important element of coastal states' efforts to comply with section 6217 Coastal Nonpoint Pollution Control Programs. Under section 6217, coastal states are directed to coordinate development of their coastal waters protection programs with their section 319 programs and related programs developed under other sections of the Clean Water Act, and two primary means of complying with section 6217 are through changes made to section 319 and Coastal Zone Management Programs.

> Section 319 requires states to assess nonpoint source pollution and implement management programs, and authorizes EPA to provide grants to assist state nonpoint source pollution control programs.

National Estuary Program—Section 320 of the Clean Water Act

EPA also administers the National Estuary Program under section 320 of the Clean Water Act. This program focuses on point source and nonpoint source pollution in geographically targeted, high-priority estuarine waters. In this program, EPA assists state, regional, and local governments in developing comprehensive conservation and management plans that recommend priority corrective actions to restore estuarine water quality, fish populations, and other designated uses of the waters.

Section 404 of the Clean Water Act

Section 404 of the Clean Water Act establishes a program to regulate the discharge of dredged and fill materials into waters of the United States, including wetlands. Activities regulated under this program include fills for development, water resource projects (such as dams and levees), infrastructure development (such as highways and airports), and conversion of wetlands to uplands for farming and forestry. The U.S. Army Corps of Engineers and EPA jointly administer the section 404 program. The Corps administers the day-to-day program, including permit decisions and jurisdictional determinations; develops policy and guidance; and enforces section 404 provisions. EPA develops and interprets environmental criteria used in evaluating permit applications; determines the scope of geographic jurisdiction; and approves and oversees state assumption. EPA also identifies activities that are exempt, enforces section 404 provisions, and has the authority to elevate or veto Corps permit decisions. In addition, the U.S. Fish and Wildlife Service, the National Marine Fisheries Service, and state resource agencies have important advisory roles.

Clean Water State Revolving Fund

The Water Quality Act of 1987, the last full reauthorization of the Clean Water Act, replaced the act's Clean Water Construction Grants Program with the Clean Water State Revolving Fund (CWSRF). The CWSRF is a state-based program to provide assistance to municipalities to construct wastewater treatment works, nonpoint source pollution control projects, and estuary protection. Congress insured that CWSRF could address all state water quality program priorities. CWSRF programs provided an average of $3.4 billion per year over the past 5 years, primarily in low-interest loans, to fund such water quality protection projects as well as watershed management projects. The CWSRF have provided more than $38.7 billion in funding over the life of the program.

Nationally, interest rates for CWSRF loans in 2002 averaged 2.5 percent, compared to market rates that averaged 5.1 percent. A CWSRF-funded project would therefore cost about 21 percent less than a project funded at the market rate. CWSRF loans can fund 100 percent of the project cost and provide flexible repayment terms up to 20 years.

States are required to match the federal funds received from CWSRF, but this match requirement is not passed on to loan recipients. Furthermore, the money received as a CWSRF loan can be leveraged as matching funds to obtain funding under other federal programs, such as 319 grants and USDA cost-share programs. This is because much of the CWSRF funds are recycled through loans, so fewer federal requirements apply to them compared to other federal funding sources.

CWSRF loans provide more than $200 million annually to control pollution from nonpoint sources and to protect estuaries, and total funding for these purposes has exceeded $1.6 billion. Some innovative funding examples follow.

❑ The Ohio EPA and Ohio Department of Natural Resources, Division of Forestry, are using Ohio's CWSRF to help Master Loggers and Certified Foresters purchase logging and tree planting equipment. Financed equipment includes bulldozers, tracked forwarders and hydro-bunchers, bridges, and mulching machines. Ohio hopes that this type of funding will support the successful use of BMPs on logging operations.

❑ The California CWSRF provided funds to landowners in the Tahoe Basin to assist them with the removal of dead and dying trees in a manner that minimized erosion and fully protected water quality. The area had a high risk of fire due to the large quantities of natural fuel for fires located on public and private lands throughout the basin.

❑ The Nature Conservancy of Ohio received three CWSRF loans totaling $264,000 for riparian zone conservation. The funds are used to protect 383 acres along Ohio's Brush Creek. The Nature Conservancy purchased 62 acres and obtained conservation easements on 321 acres. Protection measures include planting the riparian corridor with hardwood trees for streambank stabilization. "Restoring and preserving these riparian areas is an important part of controlling contaminated runoff that threatens water quality and stream habitat," said the director of Ohio EPA.

❑ Ohio EPA has worked to fund both point and nonpoint source projects through the newly developed Water Resource Restoration Sponsor Program (WRRSP). The WRRSP provides low-interest loans to communities for wastewater treatment plant improvements if the communities also sponsor water resource restoration projects. Provided that both projects qualify, CWSRF provides the financial support for both projects and reduces a community's interest rate on the total amount borrowed. As a result, the total amount repaid on the CWSRF loan for both projects is less than what would have been repaid on the wastewater treatment plant project alone. Ohio communities used $24 million of CWSRF loan funds to protect and restore 1,850 acres of riparian lands and wetlands and 38 miles of Ohio's stream corridors in 2000 and 2001. The WRRSP was designed to help prevent the loss of biodiversity and to maintain ecological health, and it has supported the acquisition of conservation easements, restoration of habitats, and modification of dams. The CWSRF program has assisted a variety of borrowers such as municipalities, communities of all sizes, farmers, homeowners, businesses, and nonprofit organizations. CWSRF recipients often partner with banks, nonprofits, local governments, and other federal and state agencies to leverage the maximum financing for their communities.

Sources: USEPA, undated a, undated b, 2002a, 2002b.

The basic premise of the program is that no discharge of dredged or fill material can be permitted if a practicable alternative exists that is less damaging to the aquatic environment or if the Nation's waters would be significantly degraded. In other words, an applicant for a permit is asked to show that

- Wetland effects have been avoided to the maximum extent practicable.

- Potential effects on wetlands have been minimized.

- Compensation has been provided for any remaining unavoidable effects through activities such as wetlands restoration and creation.

Regulated activities are controlled by a permit review process. An individual permit is required for potentially significant effects. However, for most discharges that will have only minimal adverse effects, the Army Corps of Engineers often grants general permits. These may be issued on a nationwide, regional, or state basis for particular categories of activities (for example, minor road crossings, utility line backfill and bedding) as a means to expedite the permitting process.

Section 404(f) exempts normal forestry activities that are part of an established, ongoing forestry operation. This exemption does not apply to activities that represent a new use of the wetland and that would result in a reduction in reach or impairment of flow or circulation of waters of the United States, including wetlands. In addition, section 404(f) provides an exemption of discharges of dredged or fill material for the purpose of constructing or maintaining forest roads, where such roads are constructed or maintained in accordance with BMPs to ensure that the flow and circulation patterns and chemical and biological characteristics of the navigable waters are not impaired, that the reach of the navigable waters is not reduced, and that any adverse effect on the aquatic environment will be otherwise minimized. (More information on wetlands and forestry, including a list of the aforementioned BMPs, is provided in Chapter 3, section J.)

Total Maximum Daily Loads—Section 303 of the Clean Water Act

A Total Maximum Daily Load (TMDL) is a statement of the total quantity of a pollutant that can be released to a water body or stretch of stream or river on a daily basis to maintain the water quality standard for the pollutant. A single water body might have many TMDLs, one for each pollutant of concern. A TMDL is the sum of the individual wasteload allocations for point sources, load allocations for nonpoint sources and natural background sources, plus a margin of safety for an individual body of water. TMDLs can be expressed in terms of mass of pollutant per unit time, to aquatic organisms toxicity, or other appropriate measures that relate to state water quality standards.

The process of creating TMDLs was established by Clean Water Act section 303(d) to guide the application of state standards to protect the designated "beneficial uses" (e.g. fishing, swimming, drinking water, fish habitat, aesthetics) of individual water bodies. Beginning in 1992, states, territories and authorized tribes were to submit lists of impaired waters (i.e., waters that do not meet water quality standards) to EPA every two years. Beginning in 1994, lists were due to EPA on April 1 of even-numbered years. States, territories, and authorized tribes rank the listed waters by priority, taking into account the severity of the pollution and the water body's designated uses.

A TMDL is established to identify reduction targets for two types of water pollution sources in rivers and streams:

- Point source pollution
- Nonpoint source pollution

While point sources of water pollution are regulated by discharge permits, nonpoint sources are controlled by the installation of BMPs, either voluntarily or by regulatory requirement, depending on the state.

A TMDL is a process as well as an outcome. The following are components of TMDL development:

- Problem identification
- Identification of water quality indicators and target values
- Source assessment
- Linkage between water quality targets and sources
- Allocations
- Follow-up monitoring and evaluation plan
- Assembling the TMDL

Forest harvesting; road construction, maintenance, and use; and abandoned roads in forests are the primary sources of sediment and other pollutants to water bodies from forestry activities. If a state determines that a priority water body is impaired by a pollutant that partially or wholly arises from forestry activities, the state develops a TMDL for the water body and in it determines the maximum allowable quantity of the pollutant that may be released from forestry activities. Some means of ensuring that no more than this quantity is released must then be implemented. BMPs are one method that could be used in conjunction with other methods chosen.

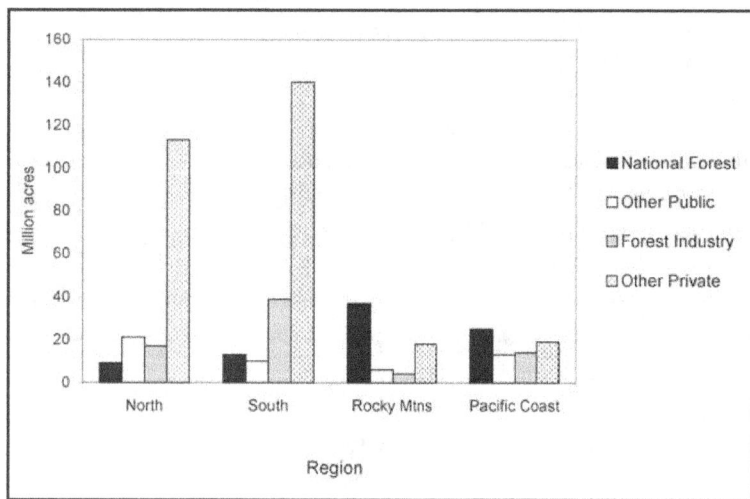

Figure 1-1.   Timberland ownership by region (Smith et al., 2001).

Forest Stewardship

Forest stewardship, including implementation of the management measures and BMPs in this guidance or similar ones (for instance, state-recommended BMPs) to minimize water quality impairment due to forest harvesting and associated activities, is the responsibility of those who own and harvest the land. In the United States, timberland ownership is divided among public agencies, the commercial forest industry, and other private timberland owners. On a national scale, 71 percent of timberland is owned privately and 29 percent publicly (Smith et al., 2001). The distribution of ownership among different public and private entities differs widely by region, as summarized in Figure 1-1. Figure 1-2 shows the distribution of forested land throughout the country.

This guidance is oriented toward the implementation of management measures and BMPs that will promote the protection of water quality, but it does not focus on assessing the quality of water that results from forestry activities. Other requirements, notably state water quality standards and designated uses, apply to all ownership categories and types of land-based activities. Thus, while different management measures and BMPs are recommended for forestry activities and agriculture, for instance, maintaining state water quality standards is the responsibility of those who undertake both activities.

Finally, it is important to mention that forests, especially well-managed forests, are a key element in any state, local, or federal water quality protection program. Forests and forested land, whether in a rural setting, along streams on agricultural land, intermixed with other land uses in suburban settings, or in urban locations, are natural filters for storm water runoff and one of the least expensive and most effective means of protecting water quality. It is the hope of EPA that the management measures and BMPs contained in this guidance, and the suggestions for their implementation, will help all persons involved with forestry activities and forest management to maintain the quality of the Nation's surface and groundwaters.

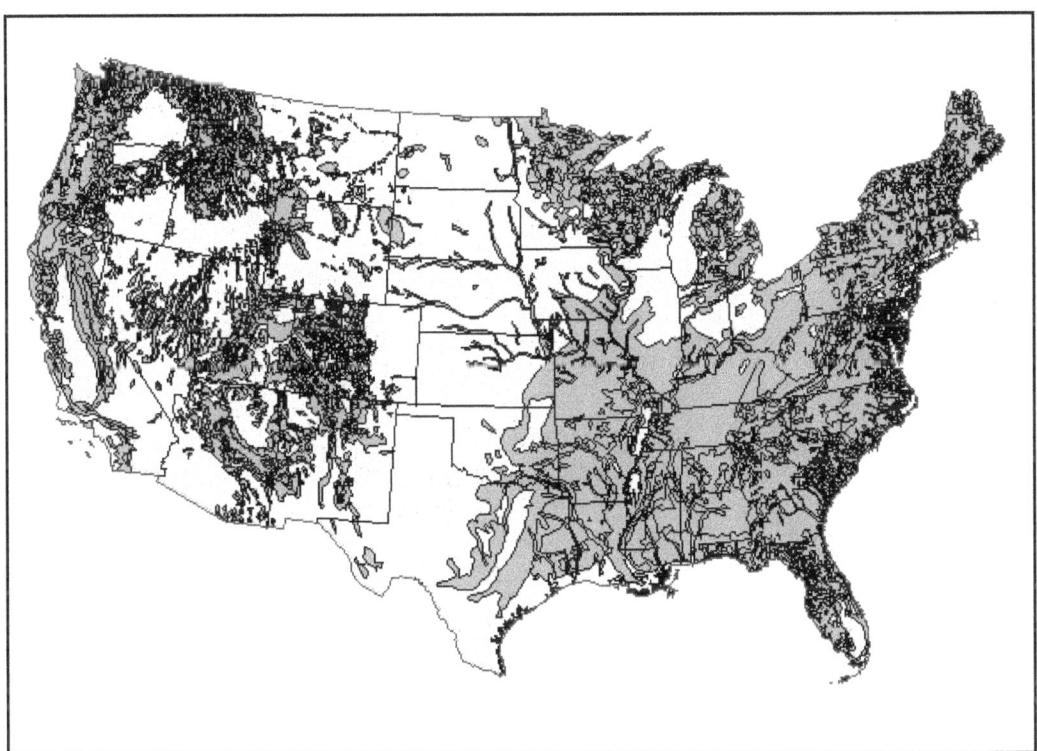

Figure 1-2. Forested lands of the United States.

CHAPTER 2: WATER QUALITY AND FORESTRY ACTIVITIES

Nonpoint source pollution remains a major challenge to meeting water quality standards and designated uses in much of the Nation. Chapter 1 defines and describes nonpoint source pollution. The potential for sediment delivery to streams is a long-term (beyond 2 years) concern from almost all forestry harvesting activities and from forest roads regardless of their level of use or age (i.e., for the life of the road). Other pollutants of significance, including nutrients, increased temperature, toxic chemicals and metals, organic matter, pathogens, herbicides, and pesticides, are also of concern, and problems associated with these other pollutants (in the context of forestry activities) generally do not extend beyond 2 years from the time of harvest or are associated with a specific activity, such as an herbicide application. Temperature effects might generally extend beyond 2 years because of the time necessary for regrowth to occur in harvested stream-side management areas (SMAs). Nevertheless, all of these pollutants have the potential to affect water quality and aquatic habitat and minimizing their delivery to surface waters and groundwater deserves serious consideration before and during forestry activities. Forest harvesting can also affect the hydrology of a watershed, and hydrologic alterations within a watershed also have the potential to degrade water quality. Forestry activities can also affect the habitats of aquatic species through physical disturbances caused by construction of stream crossings, equipment use within stream corridors, and placement of slash or other debris generated by forestry activities within streams. The effects of sediment and other pollutants on water quality in forested areas are discussed below.

The effects of forestry activities on surface waters are of concern to EPA and state and local authorities because healthy, clean waters are important for aquatic life, drinking water, and recreational use. Surface waters and their ecology can be affected by inputs of sediment, nutrients, and chemicals, and by alterations to stream flow that can result from forestry activities. The purpose of implementing management measures and best management practices (BMPs) to protect surface waters during and after forestry activities is to protect important ecological conditions and characteristics of the surface waters in roaded and logged forested areas. These conditions vary with water body type, but in general the ecological conditions that management measures and BMPs are intended to protect include the following:

- General water quality, by minimizing inputs of polluted runoff.

- Water temperature, by ensuring an adequate (but not excessive) and appropriate amount of shade along shorelines and streambanks.

- Nutrient balance, by providing for an adequate influx of carbon and nutrients that serve as the basis of aquatic food chains.

- Habitat diversity, by ensuring that inputs of large organic debris to the aquatic system are appropriate for the system.

- Hydrologic processes, by limiting disturbances to stream flow patterns, both seasonal and annual.

A great deal has been learned over the past 20 to 30 years about effective ways to reduce nonpoint source pollution from forestry activities. Developing more effective ways to control nonpoint source pollution in forested settings requires a basic understanding of forest hydrology and how forestry activities affect it. This chapter discusses the hydrologic processes of forested watersheds, the interaction of forestry activities with those processes, the general causes of nonpoint source pollution due to forestry activities, the specific pollutants and water quality concerns related to forestry activities, and general approaches to reducing the generation of pollutants. The information helps the reader understand how the management measures and BMPs discussed in Chapter 3 can minimize nonpoint source pollution and why proper implementation of BMPs is so critical to maintaining water quality in our forests.

Forested Watershed Hydrology

A watershed is an area that, due to its natural drainage pattern, collects precipitation and deposits it into a particular body of water. In western regions of the country these land areas are often called "drainages," and throughout the Nation they're sometimes referred to as river or stream "basins" (CWP, 2000). Streamflow is a critical element in understanding watershed processes and the effects of land use on those processes because it is the primary medium through which water, sediment, nutrients, organic material, thermal energy, and aquatic species move.

Streamflow is produced by vadose zone flow and groundwater seepage. Vadose zone flow is the flow that occurs between the ground surface and saturated soil, or the water table where groundwater lies. Rainfall and snowmelt supply and replenish both, but in a forested area only a portion of rainfall and snowmelt reaches surface waters. A portion is evaporated back to the atmosphere from the surface of leaves, other vegetative surfaces, and the ground. Some is absorbed by vegetation and either metabolized or transpired back to the atmosphere; and another portion is retained by the soil. Factors such as climate, soil type, topography, elapsed time since the last precipitation event, and amount of vegetation determine the portion of rainfall or snowmelt that actually reaches surface waters. The same factors, as well as soil structure (for instance, the presence of macropores created by animals or decayed roots, etc.) and geomorphology (e.g., depth to bedrock and type of underlying rock), determine how quickly moisture that infiltrates the soil reaches surface waters. If soil is already saturated or the quantity of rainfall or snowmelt is sufficient to exceed the soil's capacity to absorb moisture, surface runoff will occur, though it is not common in forested areas.

Surface runoff in a forested area is more likely to be caused by changes within a watershed than by excessive precipitation, and it is of concern because it has far more erosive power than subsurface flow. There is little storage of water that flows over a forest floor, whereas subsurface storage in soil can be substantial. For this reason, surface water flows down hillslopes more than 10 times faster than it flows through soil. Obstacles on the ground, such as leaf litter and woody debris, help slow surface runoff, but other factors can increase its velocity or volume. Such factors include a loss of vegetative cover that would contribute to evaporation and evapotranspiration, soil compaction, impervious surfaces, and cutslopes of roads or other soil disturbances where subsurface flow can be transformed into surface flow. Both the extent to which precipitation is delivered directly to the ground and prevented from infiltrating the soil and the amount of subsurface flow that is converted to overland flow are important factors that can affect the timing and volume of streamflow. When more water is delivered to streams faster than usual, stream flow peaks sooner and higher than normal, and instream erosion can occur.

Stormflow response in small basins depends primarily on hillslope processes, whereas that in large basins depends primarily on the geomorphology of the stream channel network. Consequently, land use changes and other site factors as mentioned above (e.g., soil compaction) affect streamflow in small basins more than in large basins. In any watershed, however, streamflow response to a given rain event largely depends on the capacity of the vegetation and soil to intercept rainfall or snowmelt. Saturated soil and little vegetative cover would tend to lead to a much faster streamflow response than dry soil and complete vegetative cover.

Streamflow during a season, the variability of streamflow within a season, and the variability of streamflow between seasons strongly influence channel form and processes. These factors also strongly affect aquatic and riparian species. In a stable stream—that is, one in equilibrium—each channel segment carries off sediment contributed from up-stream locations and from tributaries. When the sediment input rate is greater than the energy in the stream to carry off sediment, sediment accumulates and a channel aggrades. When a stream has more energy than what is necessary to carry the sediment the water is carrying, it can pick up extra sediment and incise the stream.

Forested riparian buffers can provide some measure of flow regulation under certain watershed conditions (Desbonnet et al., 1994). A primary way in which buffers reduce flow velocity is by slowing flow velocity and allowing absorption of water into soil. They also maintain streamside soils in a condition to absorb water by virtue of their extensive root systems that provide the soil structure necessary for a large quantity of absorption. Rainfall and runoff intensity, soil characteristics, hydrologic regime, and slope of the buffer and runoff source area are once again some of the factors that determine a forested riparian buffer's ability to regulate stream flow. A narrow forested buffer on a steep, nonvegetated slope has little ability to regulate flow, whereas a wide forested buffer on a gentle, vegetated slope could help reduce peak flow levels and provide for dry season flow.

Forestry Activities and Forest Hydrology

When one factor in a system changes, other factors may be affected as well. In a forested watershed, logging has the effect of both compacting and loosening soils due to the construction and use of roads, use of heavy machinery, logs being dragged over the ground or otherwise transported to yarding areas, and vegetation being removed. Roads and road ditches, ruts on the ground, and areas cleared of leaf litter or other soil coverings create opportunities for water channelling and flow diversion, which, if not properly controlled and directed, can generate erosive flows. Thus, the disturbances caused by logging in a forested watershed can lead to hydrologic changes within the same water-shed, which can in turn lead to nonpoint source pollution. Forestry activities and their potential effects on forest hydrology and water quality (through nonpoint source pollution) are discussed below.

A note on the concept of disturbance ecology is in order here. A forest is not an ecosystem that has been in perfect equilibrium from its beginning as a grassland to its mature state, modified only by the slow successional changes that occur naturally. Numerous disturbances occur along the way, ranging from those on a small scale (such as a treefall) to those on a large scale (such as a wildfire). Forests react to these disturbances in ways that can increase biodiversity and promote overall forest health. For many years people have managed forests—including protection from disturbance and unnatural disturbance (such as harvesting and altering land use)—without paying attention to the natural disturbance regime of the particular forest. An ecosystem approach to forest management is evolving as more is learned about natural disturbance, and forest management approaches are being developed that benefit both forests and people by creating disturbance in spatial and temporal patterns that closely resemble those of natural disturbances. Thus, forest management activities can be done such that the disturbances they cause benefit the forest ecosystem. Managing a forest this way, however, requires good knowledge of the forest ecosystem dynamics and consideration of all past, present, and future disturbance-creating activities within the forest ecosystem that could cumulatively create more disturbance—and thus unintended damage—than the project being considered, for instance road construction or a harvest.

Road Construction and Road Use

Roads are generally considered to be the major source of sediment to water bodies from harvested forest lands. They have been found to contribute up to 90 percent of the total sediment production from forestry activities (Megahan, 1980; Patric, 1976; Rothwell, 1983). There is some evidence that modern road building practices, such as locating roads on ridgetops instead of middle slopes, removing excavated material to an offsite location, and using full bench construction is reducing the amount of sediment delivered to streams from forest roads (Copstead, 1997). Erosion from roads can be disproportionately high because roads lack vegetative cover, are exposed to direct rainfall, have a tendency to channel water on their surfaces, and are disturbed repeatedly when used. Erosion from roads can be exacerbated by instability on cut-and-fill slopes, water flow over the road surface or through a roadside ditch, flow from surrounding areas becoming concentrated and channelled by a road surface, and lack of a protective surfacing. Much of the sediment load to streams that is associated with roads can be attributed to older roads, which may have been constructed with steep gradients and deep cut-and-fill sections and which may have poorly maintained drainage structures.

Numerous factors need to be considered to protect water quality from the potential effects of forest roads. Stream crossings of both older and modern forest roads and old forest roads that were placed near streams are the most troublesome source of sediment to streams. While roads contribute more to erosion on forested land on a per-area basis (e.g., quantity of eroded soil per acre of road versus per acre of undisturbed forest), they also occupy a disproportionately small amount of a forested area. Evidence indicates that the total amount of eroded soil from roads is not much if any greater than the total amount of soil eroded from the non-roaded surface of a forested area (Gucinski et al., 2001). A related factor is that a small percentage of road area may be responsible for most of the erosion from roads. Rice and Lewis (1986, cited in Gucinski et al., 2001) found that major erosional features of roads occupied only 0.6 percent of the length of roads. A final factor to consider is that soil loss from roads tends to be greatest during and immediately after road construction because of the unstabilized road prism and disturbance by passage of heavy trucks and equipment (Swift, 1984). Consideration of these factors to reduce water pollution from roads is provided in Chapter 3, section C, *Road Construction/Reconstruction*, and section D, *Road Management*.

Careful planning and proper road layout and design, however, can minimize erosion and substantially reduce the effects of roads on streams. The effect that a forest road network has on sediment input and flow changes in stream networks depends in part on how inter-connected the road and stream networks are. Roads generally are hydrologically connected to stream networks where subsurface groundwater flow is converted to channelled overland flow at road cuts, and road surface runoff drains directly to stream channels. Overland flow is delivered to streams much more quickly than subsurface flow, so the conversion of subsurface flow to overland flow and the connectivity of road networks to stream networks can have an effect on stormflow patterns in streams (Jones and Grant, 1996; Montgomery, 1994; Wemple et al., 1996). Careful road system planning, taking watershed processes, soil type, topography, and vegetative characteristics into account, and designing with natural drainage patterns to minimize hydrologic connections of the road network to streams and maximize opportunities for filtering surface drainage, can reduce these effects. Chapter 3, section A, *Preharvest Planning*, discusses these factors.

Timber Harvesting

Timber harvesting generally involves the use of forest roads (the effects of which are discussed separately above and in Chapter 3), skid trails (along which felled trees are dragged), yarding areas (where cut timber is collected for transport away from the harvest site), and machinery associated with harvesting, skidding, and yarding. Soil disturbance, soil compaction, and vegetation removal on the harvest site, skid trails, and yarding areas can contribute to water quality problems. Methods for minimizing the water quality effects of timber harvesting are discussed in Chapter 3, section E, *Timber Harvesting*.

The association between timber harvesting—especially clear-cut harvesting—and mass erosion events has been and continues to be controversial. Studies of landslides done up to the 1980s, primarily in the Pacific Northwest, found an association between clear-cutting and landslides, but the findings of the studies were inconclusive due to the way data were collected (Hockman-Wert, undated). Studies were often conducted using aerial photographs and concentrated on the steepest slopes. Aerial images cannot account for mass erosion that occurs under forest cover, and later research indicated that as much as 50 percent of mass erosion movements are unaccounted for on aerial photographs. While some studies found clear-cuts to lead to more landslides on steep slopes, when more gentle slopes were investigated the occurrence of landslides was found to be as common on forested sites as on clear-cut sites.

There is a general consensus that harvesting on steep slopes increases the landslide hazard for a period of time after the harvest. It is not clear, however, whether *more* or *larger* landslides occur due to harvesting. In an issue paper written for the Oregon Board of Forestry and to provide background information for policy decisions related to harvesting and public safety, Mills and Hinkle (2001) discuss the latest scientific evidence related to landslides and timber harvesting. They report that in three of four study areas higher landslide densities were found in stands that had been harvested within the previous nine years than in mature (i.e., more than 100 years old) forest stands, and that stands 30 to 100 years old had lower landslide densities than mature stands. They also report that the studies showed that average landslide volume was similar regardless of stand age.

Furthermore, landslides are known to be natural occurrences and important elements in stream ecology in that they are a primary means by which wood and gravel are delivered to streams to create fish habitat (Shaffer, undated). It may be, then, that landslides occur in steep areas regardless of land use history, but that harvesting may concentrate the occurrence of landslides into the 10 years after harvesting.

Geology, soil type, soil depth, and topography might have much more to do with determining whether a site is susceptible to landslides than land use history (Shaffer, nd). Underlying geology plays a role because porous bedrock drains water from soils quickly, while impermeable bedrock keeps water in the soil. Different types of bedrock, such as shales or granite, weather into different types of soils that will either promote or resist sliding. Soil type determines whether a soil binds well to itself and to bedrock to resist sliding or is easily dislodged to promote sliding. Soil depth determines how much soil volume there is above bedrock to absorb water before the soil becomes saturated and what the weight of soil available for sliding is. Water contributes to sliding not only by acting as a lubricant between soil and bedrock, but also by adding considerable weight to the soil. Two inches of rain in 24 hours adds 10 pounds of water in every square foot of soil. On flat topography, saturated soil will result in puddling or overland flow. On gently sloping topography, soil might "creep" downhill at the rate of a few inches a year. On steep topography, the combined weight of water and soil under saturated conditions can trigger a slide. Finally, vegetation provides soil binding to resist sliding, and root decay can make soils less cohesive. Root cohesion—the ability of roots to hold soil to a slope—is at its lowest about 10 years after a harvest (or some other event that kills trees, such as a wind storm after an ice storm). Depending on all of these factors—geology, soil type, soil depth, and topography, combined with the elements of precipitation and land use history—a landslide could occur before or after soil becomes saturated, before or after a harvest, and either slowly and progressively or suddenly and massively.

Finally, research on the effectiveness of different harvesting methods (e.g., clear-cutting or selective cutting) or logging practices to reduce landslide occurrence does not exist (Mills and Hinkle, 2001). The effectiveness of BMPs for minimizing the hazard of landslides from timber harvest sites is also not known.

Recent research in Canada has demonstrated that clear-cut harvesting can lead to increased mercury concentrations in runoff (McIlroy, 2001). Mercury is carried through the atmosphere from areas with sources such as coal combustion and incinerators, and can be deposited in forested areas. When those forested areas are clear-cut harvested, the additional runoff generated after the trees are removed might lead to increased mercury concentrations in the runoff. The Canadian study indicated that the effect is accentuated by heavy, clear-cut harvesting in large watersheds, and that the problem might be avoided by selective harvesting. Further study of the potential problem is needed to clearly portray the association, if any, between forest harvesting and mercury.

Another potential adverse effect of timber harvesting is an increase in stream water temperature—a water quality criterion for physical water quality—that can result if too much streamside vegetation is removed. Small streams are affected more by a loss of shade than are large streams. One reason that streamside buffer strips, or SMAs, are maintained is to minimize or prevent water temperature increases. Stream temperature maintenance is important for aquatic biota. For instance, stream temperature has been found to affect the time required for salmonid eggs to develop and hatch (Chamberlin et al., 1991). Fish and aquatic invertebrates are cold-blooded adapted to ranges of water temperature, and can be adversely affected by the water temperature exceeding the high temperature of the range for which they are adapted. Maintaining streamside vegetation in an amount sufficient to provide shade that maintains the stream temperature within the proper range is a key goal of the Streamside Management Area Management Measure (see Chapter 3, section B, *Streamside Management Areas*).

Timber harvesting along a stream can also affect stream ecology by removing overhanging trees and branches from which twigs, leaves, branches, and sometimes entire trees fall into the stream channel. Overhanging vegetation contributes organic material in the form of leaves and needles, and large woody debris, or LWD, to surface waters. These materials serve as a source of energy and provide nutrients for aquatic life and provide habitat diversity. They are a primary source of nutrients in small, low-order streams high in watersheds where aquatic vegetation might not be abundant and upstream sources of nutrients are limited. Farther downstream, instream sources of nutrients, such as aquatic plants and organic matter transported from upstream sources, are more abundant and organic debris from overhanging trees is a less important source of energy and nutrients. LWD is still important in these streams, however, for the habitat diversity it creates. LWD creates eddies, provides shelter and anchoring points for small aquatic animals, and forms areas of relatively calm water in flowing streams and rivers. SMAs protect these important ecological processes and benefits, without which stream waters might be prevented from attaining the water quality criterion of supporting aquatic life.

Site Preparation and Forest Regeneration

Site preparation is done to prepare a harvested site for regeneration. It can be accomplished mechanically using wheeled or tracked machinery, by the use of prescribed burning, or with applications of chemicals (herbicides, fertilizers, and pesticides). These techniques may be used alone or in combination. These operations can affect water quality if chemicals used and/or spilled during site preparation operations or soils disturbed during site preparation are transported to surface waters.

The chemicals associated with forestry operations that are of most concern from a water quality perspective are petroleum compounds, lubricants, and other machinery-related chemicals. Herbicides, pesticides, and fertilizers pose little threat to water quality if used and applied according to the specific directions for the chemical being applied and state and EPA guidelines. The herbicides and pesticides used in forestry operations are generally specific to the target vegetation and pose little threat to aquatic organisms, and they generally are short-lived in the environment. Fertilizers pose little threat to aquatic environments because they are used very infrequently for forestry operations, perhaps as little as two applications on a harvest site in 50 years.

Mechanical site preparation by large tractors that shear, disk, drum-chop, or root-rake a site can result in considerable soil disturbance over large areas (Beasley, 1979). Site preparation techniques can result in the removal of vegetation left after a harvest and forest litter, soil compaction and a loss of infiltration capacity, and soil exposure and disturbance. All of these effects can lead to increased erosion and sedimentation. They are most pronounced soon after a harvest and decrease over time, usually within 2 years, as vegetative cover returns to the harvested site.

Forest regeneration methods can be divided into two general types: (1) regeneration from sprouts and seedlings, either planted seedlings or those present naturally on a harvest site, and (2) regeneration from seed, which can be natural seed in the soil or seed from a broadcast application after a harvest. Loss of soil from a harvest site is obviously undesirable from a water quality perspective, and also because of the lowered soil productivity and tree regeneration that can result. Protecting a harvest site from undue disturbance during site preparation, therefore, is desirable both from water quality (reduced erosion) and site productivity perspectives. Means to protect soils from erosion and undue disturbance during site preparation and forest regeneration are discussed in Chapter 3, section F, *Site Preparation and Forest Regeneration*, and section H, *Revegetation of Disturbed Areas*.

Prescribed Burning

Prescribed burning is a method used to prepare a site for regeneration after a harvest, however because the methods for minimizing water quality effects due to fire are some-what specialized, it is treated separately in this document (see Chapter 3, section G, *Fire Management*). Prescribed burning of slash can increase erosion on some soils by elimi-nating protective cover and altering soil properties (Megahan, 1980). Burning can have the effect of making some soils water repellent, which will tend to increase runoff (Reid, 1993; Ziemer and Lisle, 1998). This effect can penetrate to a depth of 6 inches and persist for 6 or more years after a fire. Burning enhances infiltration in other soils. Which soils will be affected in what way cannot be consistently predicted, and the effect is evidently dependent on the type of vegetation in the area burned. Burning also releases nutrients, immediately increasing nitrogen available to plants, but produces an overall effect of decreasing nitrogen in the forest floor (Reid, 1993). Little effect occurs on soils not affected by fire.

The degree of erosion following a prescribed burn depends on soil erodibility; slope; timing, volume, and intensity of precipitation after a burn; fire severity; cover remaining on the soil; and speed of revegetation. Erosion resulting from prescribed burning is generally less than that resulting from roads and skid trails and from site preparation techniques that cause severe soil disturbance (Golden et al., 1984). However, serious erosion can occur following a prescribed burn if the slash being burned is collected or piled and soil on the harvest site is disturbed in the process of preparing for the burn.

The effects of fire on a watershed depend on burn severity and hydrologic events that follow a fire (Robichaud et al., 2000). Burn severity is related to the amount of vegetation loss and heat-related changes in soil chemistry due to a fire. In general, wildfire has a more severe effect on watershed processes than prescribed burning because it is more intense than a prescribed burn. Prescribed burns are generally set under conditions such that they can be controlled and the fire will burn lower and less intensely than would a wildfire. Given the potential effects that a severe burn can have on watershed processes, prescribed burning can be used effectively both for site preparation and to reduce the chances of wildfire—and the often more severe effects that the latter can have on water-shed processes.

Forestry Pollutants and Water Quality Effects

The discussion above focused on forestry activities, the potential they have for generating nonpoint source pollution and pollutants, and the watershed processes that can be affected by forestry activities. Below is a discussion of the pollutants that can be generated from forestry activities and the potential effects that these pollutants can have on water quality.

The nonpoint source pollution problem of greatest concern with respect to forestry activities is the addition of sediment to surface waters. Without adequate precautions, however, many water quality issues can arise from forestry operations:

- Sediment concentrations can increase because of accelerated erosion.
- Nutrients in water can increase after their release from decaying organic matter on the ground or in the water, or after a prescribed burn.
- Organic and inorganic chemical concentrations can increase because of harvesting and fertilizer and pesticide applications.
- Slash and other organic debris can accumulate in waterbodies, which can lead to dissolved oxygen depletion.
- Water temperatures can increase because of removal of riparian vegetation.
- Streamflow can increase because of reduced evapotranspiration and runoff channeling.

The discussions below of the individual pollutants that can be generated by forestry activities present the range of effects that might occur during and after road construction or use or a harvest. The particular effects of a forestry activities in a specific watershed will depend on the unique interaction of the characteristics of the area where the activities occur, time of year, harvesting method, and the BMPs used.

Sediment

Sediment deposited in surface waters is of concern in this guidance because of its potential to affect instream conditions and aquatic communities. Sediment is the pollutant most associated with forestry activities. Sediment is the solid material that is eroded from the land surface by water, ice, wind, or other processes and then transported or deposited away from its original location. Soil is lost from the forest floor by surface erosion or mass wasting (for example, landslides).

Surface erosion generally contributes minor quantities of sediment to streams in undisturbed forests, and the quantity of surface erosion depends on factors mentioned previously, such as soil type, topography, and amount of vegetative cover (Spence et al., 1996).

Rill erosion and channelized flow occur where rainwater and snowmelt are concentrated by landforms, including berms on roads and roadside ditches. They cause erosion most severely where water is permitted to travel for a long distance without interruption over steep slopes, because the combination of distance and slope tends to increase the volume and velocity of runoff. Sheet erosion, or overland flow, occurs occasionally on exposed soils where the conditions necessary for it, including saturated soil or a rainfall intensity that is greater than the ability of soil to absorb the water, but it is not common on forest soils.

Mass wasting—including slumps, earthflows, and landslides—occurs most often in mountainous regions where surface erosion is minor (Spence et al., 1996). It can contribute large quantities of sediment to streams—and stream ecology and fish populations may depend on this sediment; but it occurs episodically, usually following heavy rains. Clearcutting can promote landslides on steep slopes where other factors, such as type and depth of soil and type of bedrock, are favorable for landsliding. These other factors have a lot to do with whether a landslide will occur at a site, and tree removal increases the chance that a landslide will occur on a site that is prone to landsliding within a 10-year timeframe after a harvest (Mills and Hinkle, 2001). If topographic and geologic conditions at a site are favorable for landslides, then landslides are likely to occur at the site whether it is harvested or not, though harvesting may certainly affect the timing, volume, and composition of a slide. Many landslides occur on completely forested areas (Hockman-Wert, undated) and landslides are important to stream ecology in that they provide wood and gravel important to the creation of fish habitat (Shaffer, undated).

Gucinski and others (2000) reviewed the scientific information available on forest roads and forest road-related issues in a paper, *Forest Roads: A Synthesis of the Scientific Information*, for the U.S. Forest Service. The authors review information related to the direct physical and ecological effects, the indirect landscape effects, and the direct and indirect socioeconomic effects of forest roads. The reviewers conclude that forest roads can lead to mass failures if road fills and stream crossings are improperly located, culverts are too small to pass flood waters and debris, roads are sited poorly, surface and subsurface drainage is modified by a road, or water is diverted from a road to unstable soil areas. Furthermore, the reviewers emphasize that on most roads only a small percentage of a road's surface, as little as 1 percent or less, contributes to mass wasting. Many of the studies reviewed were conducted on roads that were constructed in the 1970s and 1980s. While studies of roads constructed with more modern road-building technologies, including technologies that incorporate the BMPs discussed in Chapter 3, *Road Construction/Reconstruction* (section C) and *Road Management* (section D), are not widely available yet, use of the modern technologies may lead to reduced mass wasting and water quality impacts from roads in general in the future.

Forest road stream crossings can be sites of sedimentation and hydrologic change if an inappropriate type and size of crossing is installed. A culvert that is too small will not permit the passage of debris and water during flood events, and can lead to instream erosion and culvert blowout. A culvert, ford, or bridge that is improperly installed can cause erosion at the site of the crossing. Problems associated with stream crossings can be avoided by proper planning (Wiest, 1998). Crossings can be located where gradient or channel alignment are relatively uniform and selected to be large enough for floodwaters and instream debris to pass through. The advantages and disadvantages of various stream crossing structures are summarized in Table 2-1. Management measures and BMPs for preventing problems at stream crossings associated with forestry activities are discussed in Chapter 3, sections C, *Road Construction/Reconstruction*, and D, *Road Management*.

An excessive quantity of sediment in a water body can cause or lead to a variety of problems. Sediment can reduce a water body's ability to support aquatic life when it fills the spaces between rocks and grains of sand where many organisms live, forage, and spawn, hindering these activities. Fine sediments, of the size that can be deposited between grains of sand, are most threatening to fish. If deposited on fish eggs, fine sediments can reduce egg-to-fry survival and fry quality by suffocating eggs and forming a physical barrier to emerging larvae. Different species have different tolerances to fine sediment due to the fry having different head diameters. Coarse sediment can cap a

Table 2-1.   Advantages and Disadvantages of Stream Crossing Structures

Stream Crossing Structure	Advantages	Disadvantages	Notes
Circular Pipe Ditch Relief Culvert	Stable and reliable for steep grades; less erosion and more economical than surface cross drains for high-traffic roads	Needs periodic maintenance and inspection to avoid plugging; if too small can plug and lead to erosion	Should be located far enough above stream crossings to avoid releasing ditch drainage water directly into streams
Bottomless or Log Culvert	Preserves natural streambed and gradient; no significant change in water velocity; maintains normal stream width	Vulnerable to erosion and downcutting; large logs might be required to achieve adequate flow with log culverts; expensive and can be difficult to install; not practical where footings cannot be placed in stable, nonerodible material	Generally spans the entire streambed and minimizes effects on the natural stream channel
Embedded Pipe Arch Culvert	When properly installed, maintains natural stream channel width, grade, and sediment transport characteristics	Complex and time-consuming installation; sizing must account for area lost to embedding; fitting with machinery possible only if the diameter is large enough to permit machine entry	Must be constructed on suitable bedding material; suitable on bedrock when concrete footings can be used
Ford	Useful for low-water crossings	Can be barriers to fish passage during low-flow conditions	Stream channel and slope must be suitable; useful where transportation requirements are seasonal
Bridge	Best option for maintaining natural stream channel	Expensive; requires special installation techniques; difficult to fit to tight road curves	Requires determination of 50- or 100-year flow

gravel streambed and restrict the emergence of alevins (Murphy and Miller, 1997). Murphy and Miller (1997) found that fine sediment deposited in spawning gravels after timber harvest contributed to a 25 percent reduction in chum salmon escapement.

High sediment concentrations in the water can cause pools—preferred by some salmon species such as coho—to fill with sediment and reduce or destroy essential rearing habitat. When streams are affected by high sediment deposition, these formerly productive low-gradient reaches become wide and shallow and recovery of fish habitat can take decades (Frissell, 1992).

Sediment suspended in water increases turbidity, limiting the depth to which light can penetrate if turbidity is increased to a sufficient degree and, thus, potentially reducing photosynthesis and oxygen replenishment. A quantity of suspended sediment far in excess of that normally present in a water body can suffocate aquatic animals and severely limit the ability of sight-feeding fish to find and obtain food.

Increased Temperature

Temperature increases in streams are of concern because of the potential effects on aquatic species. The water quality criterion for temperature is set for waters to protect aquatic biota, and the temperature tolerance limits of fish are used to indicate whether a water body's temperature has been adversely affected. When streamside vegetation is removed, any increase in solar radiation reaching the stream can increase the water temperature. The temperature increase can be dramatic in smaller (lower order) streams and can heat the water to beyond the tolerance limits of some aquatic species. Increased water temperatures can also accelerate the chemical processes that occur in the water, decrease the ability of a water body to hold oxygen, and lower the concentration of dissolved oxygen.

Because streams in forests are shaded, fish species in forested streams tend to be cooler-water species, such as salmon and trout, than fish species in non-forested streams. The duration of an elevated temperature and the availability of cool pools of water are among the factors that determine how severe an effect a temperature increase has on fish and other biota. An elevated water temperature can retard growth, reduce reproductive success, increase susceptibility to disease, decrease the ability to avoid predators, and decrease the ability to compete for food (Spence et al., 1996).

Riparian forested buffers, as discussed above and in Chapter 3, section B (*Streamside Management Areas*) are a primary means of minimizing temperature increases due to timber harvesting. The role of riparian forested buffers in regulating ambient stream temperature, however, varies with stream width and vegetation type, as well as other factors such as stream depth, orientation to the sun, and surrounding topography. A narrow stream with a complete riparian forested buffer might receive as little as 1 to 3 percent of the total incoming solar radiation, whereas a wide mid-order stream might receive as much as 10 to 25 percent. Riparian vegetation, therefore, has less ability to regulate water temperature as stream width increases (Spence et al., 1996).

Nutrients

Nutrients, such as nitrogen and phosphorus from fertilizers, soil, and plant material, are primary chemical water quality constituents. They can enter water bodies attached to sediments, dissolved in the water, or transported by air. Forest harvesting can increase nutrient leaching from the soil, though the effect generally subsides to near precutting levels within two years of a harvest. Low to moderate increases in nutrient levels may have no or a beneficial effect on an aquatic environment, but excessive amounts of nutrients can stimulate algal blooms or an overgrowth of other types of aquatic vegetation. This can in turn lead to an increase in the amount of decomposing plant material in an aquatic system and, in turn, increased turbidity and biological oxygen demand. The latter effect can decrease dissolved oxygen concentrations, with potentially detrimental effects to aquatic biota. Chapter 3, section I, *Forest Chemical Management*, discusses methods for minimizing the adverse effects of forestry activities on nutrient balances.

Organic debris, discussed below, can be an important source of nutrients in an aquatic environment, and SMAs play an important role in organic debris inputs and maintaining nutrient balances in aquatic forest ecosystems.

Organic Debris

Organic debris—primarily composed of leaves, twigs, branches, and fallen trees—is an important element of water quality in that it provides nutrients and stream structure that are important to supporting aquatic life. It ranges in size from suspended organic matter in water to fallen trees. Large woody debris, or LWD, can be whole trees or tree limbs that have fallen into streams. It creates the physical habitat diversity essential to supporting aquatic life. As a structural element, it influences the movement and storage of sediment and gravel in streams and stabilizes streambeds and banks (Spence et al., 1996). Small organic litter—primarily leaves in deciduous forests and cones and needles in coniferous forests—is an important source of nutrients for aquatic communities. It usually decomposes over a year or more, depending on forest type.

When streamside vegetation is removed, inputs of organic debris decrease and the amount of sunlight reaching the water increases. A stream that might previously have relied primarily on sources of nutrients external to the stream (fallen debris) can be forced to rely primarily on instream sources (such as algal growth and instream vegetation). The latter may not be present in high-order streams.

Organic debris generated during forestry activities includes residual logs, slash, litter, and soil organic matter. These materials can perform some of the same positive functions as naturally occurring LWD and organic litter. If their abundance in a stream is substantially greater than normal, however, they can also block or redirect streamflow, alter nutrient balances, and decrease the concentration of dissolved oxygen as they decompose and consume oxygen. Observing management guidelines for streamside management areas, discussed in Chapter 3, section B, *Streamside Management Areas*, is a key means to minimize ecological and water quality effects due to organic debris.

Forest Chemicals

Chemicals that enter surface waters can be toxic to aquatic biota, make it difficult to attain drinking water quality criteria, and degrade the aesthetics of streams. The most harmful substances considered under the general category of "forest chemicals" and used during forestry operations are fuel, oil, and lubricants; coolants; and others used for harvesting and road-building equipment. Simple precautions can prevent water quality deterioration, whereas improper use and management of chemicals used during forestry operations can result in degraded water quality.

Fertilizers, herbicides, and pesticides are used to prepare a site for regeneration and to protect forests from disease and pests. Adverse effects on water quality due to forest chemical applications typically result from not following the specific application instructions for the chemical being used, such as specifications for the quantity to apply and the distance to maintain around watercourses (Norris and Moore, 1971). Generally, the water quality and aquatic biota threats due to fertilizers, herbicides, and pesticides are small because the chemicals are applied at most only one to three times at a harvest site and they specifically target biochemical pathways present only in plants, rendering them of little danger to aquatic animals. Furthermore, the half-lives of forestry herbicides are on the order of less than 100 days, so bioaccumulation in aquatic species is rarely of concern. Precautions for minimizing water quality effects due to forest chemical use are discussed in Chapter 3, section I, *Forest Chemical Management*.

Hydrologic Modifications

Streamflow is a concern because of the instream changes that can occur if the quantity of streamflow or the timing of streamflow is changed substantially as a result of a forest harvest or repeated forest harvesting. The dynamics of forest harvesting and streamflow response are discussed above under *Forested Watershed Hydrology*. Methods of minimizing the streamflow effects of forest roads and timber

---

harvesting are discussed in Chapter 3, and particularly in sections C, *Road Construction/Reconstruction*, D, *Road Management*, and E, *Timber Harvesting*.

If forest roads or timber harvesting result in a more rapid delivery of runoff to streams than before roads were present or timber was harvested, peak flows can be increased. This can lead to increases in channel scouring, streambank erosion, downstream sedimentation, and flooding. The magnitude of changes in peak flows after logging depends on the size of the watershed and the amount of land harvested, and to a lesser extent on road building. Changes are usually greatest in small watersheds and where a large percentage of the surrounding watershed is logged at one time. Streamflow can be increased as a result of forest road building alone, but this usually occurs only in small, upland watersheds where streams and streamflow are small and the amount of impervious or heavily compacted surface from the harvest and associated activities is large in proportion to the areal extent of the watershed. Downstream flooding is rarely a consequence of logging in small, upstream watersheds (Adams and Ringer, 1994).

Normally, when only a small portion (e.g., less than 15 percent) of a watershed is harvested, flow is not altered in associated streams. Where more than 15 to 20 percent of the forest canopy is removed, streamflow typically increases. Any increase is greatest in the first years after harvest and typically becomes smaller with time as vegetation grows on harvested sites. Streamflow generally returns to the original level within 20 to 60 years, depending on forest and land type (Adams and Ringer, 1994).

Physical Barriers

Forest road stream crossings can be sites of hydrologic change, sedimentation, and debris buildup if the appropriate type and size of crossing are not selected. Improperly installed culverts at stream crossings can lead to erosion around the culvert and of the road surface when the design storm is exceeded or if debris inhibits or redirects flow. This can result in excessive sedimentation and channel alterations downstream. Culverts installed above the grade of a stream can create a barrier to upstream fish migration. Any of the following conditions associated with culverts can block fish passage: water velocity at the culvert is too fast, water depth at the culvert is too shallow, there is no resting pool below the culvert, the culvert is too high for a fish to jump, or the culvert is clogged because of lack of maintenance.

Problems associated with stream crossings can be avoided by proper planning (Wiest, 1998). Crossings can be located where they do not cause large increases in water velocity and there are not large changes in gradient or channel alignment. Doing so can minimize effects on sedimentation and fish passage. Planning for safe fish passage involves determining the type and extent of fish habitat, the species of fish present in the stream, and the window during which instream work can occur without harming fish habitat or interfering with fish migration. Adequate fish passage is that which conserves the free movement of fish in and about streams, lakes, and rivers in order that they can complete critical phases of their life cycles. It permits adult fish to migrate to spawning areas and juvenile fish to accompany adult fish or make local moves to rearing or overwintering areas. The advantages and disadvantages of various stream crossing structures are summarized in Table 2-1.

Fords, bridges, and culverts of various sizes, shapes, and materials can be installed to avoid hydrologic and habitat changes and to provide adequate fish passage. Road crossings and culverts also need to be installed to fail when the design storm is exceeded to prevent substantial sedimentation. Management measures and BMPs for preventing physical barriers in streams associated with forestry activities are discussed in Chapter 3, sections C, *Road Construction/Reconstruction*, and D, *Road Management*.

Cumulative Effects

Cumulative effects occur when two or more activities cause the same response within a watershed (e.g., lead to increased stream flow at a given time of year), when multiple responses disturb the same resource (e.g., increased stream flow and sediment yield both affect the same stream reach), when one response provokes another (e.g., increased stream flow induces scouring around culverts), or when responses interact to produce another (e.g., road construction on a steep slope and unusually heavy rains produce a mass soil movement) (Reid, 1993). Cumulative effects can occur spatially, when numerous activities conducted at different locations within a watershed contribute to instream responses, or temporally, when a single activity repeated in the same place or different activities conducted in different places at different times have an additive effect. Most land use activities affect only one of four environmental parameters—vegetation, soils, topography, or chemicals—and other watershed changes result from initial effects on these factors. If a change in vegetation or another one of these four factors is persistent or affects watershed transport processes or rates, cumulative effects can result.

Cumulative effects are of concern with respect to forest roads; forest road construction, use, and maintenance; and forest harvesting because the changes that can occur in watershed processes following these activities can persist for many years. This persistence increases the potential for cumulative effects to occur. Examples of potential persistent effects due to forestry activities include the delivery of sediment to streams from a forest road used repeatedly over a period of years and increased subsurface flow and decreased evapotranspiration due to a reduced amount of vegetation at a harvest site.

Forest roads and timber harvesting can cause changes to a landscape or stream on a temporal scale far different from that associated with the life of the road or duration of the harvest. A road may be constructed and used for many years, and its effect on a landscape can continue for years after it is no longer needed. Cafferata and Spittler (1998) found that "legacy" roads can be significant sources of sediment for decades after their construction. Reid (1998) also found that sedimentation rates may increase 25 years or more after logging roads are abandoned as they begin to fail and erode. A harvest might occur in one season, or numerous harvests in a watershed might occur over a number of years, and during the months or years afterward temporary roads and stream crossings might be removed and the ground or streambeds rehabilitated. In contrast, recovery of a forest, instream recovery from channel erosion, habitat recovery, and aquatic community recovery occur on time scales much longer than the harvest. The long-term recovery times provide ample opportunity for other disturbances to contribute to cumulative effects.

Consider the following study of cumulative effects, modeled using Monte Carlo simulations of four hypothetical watersheds (Ziemer et al., 1991). Each watershed was a 10,000-ha, fifth-order watershed typical of one that might be located in coastal Oregon or California at 300 to 500 meters of elevation and 30 kilometers inland from the coast. Annual rainfall was simulated at 1500 millimeters. The four watersheds were simulated to have the following treatments:

- One watershed was simulated as undisturbed.

- One watershed was simulated as clear-cut and roaded within 10 years of the commencement of harvesting, with harvesting beginning at the upper reaches of the watershed and progressing toward the mouth.

- One watershed was simulated as harvested at the rate of 1 percent per year, beginning at the mouth and progressing upstream.

- The fourth watershed was again simulated as harvested at a rate of 1 percent per year, but with the harvests widely dispersed throughout the watershed.

These harvesting patterns were simulated as being repeated each 100 years, and in each watershed (except the unharvested one) one-third of the road network was simulated to be rebuilt each 100 years. The greatest differences between the treatments were noticed in the first 100 years, and they related most to the rate of treatment. That is, to whether the harvests were concentrated or dispersed temporally. By the second 100 years, the primary difference between the treatments was in the timing of the impacts. Interestingly, the simulation indicated that temporally dispersing the harvest units did not reduce cumulative effects.

The conclusion reached by the authors was that current estimates of cumulative effects due to logging *underestimate* the effects because they accumulate over much longer periods than previously thought, but they *overestimate* the benefits of temporally dispersing harvests in a watershed. Concentrating the treatments (over 10 years instead of 100 years) increased the chances of cumulative effects on the affected resources.

A more detailed discussion of issues related to cumulative effect assessment is provided in Chapter 4, *Using Management Measures to Prevent and Solve Nonpoint Source Pollution Problems in Watersheds.*

Mechanisms to Control Forestry Nonpoint Source Pollution

Nonpoint source pollution control practices for forestry activities are referred to as *best management practices* (BMPs), *management practices, accepted forestry practices, management measures, BMP systems, management practice systems,* and the like. Some of these terms have specific uses in legislation and regulations, whereas other terms are found in technical manuals, journal articles, and informational materials. Forestry management practices have been developed by all states, though they may not exist as a separate program or set of rules or guidelines. In some states, forest protection guidelines are contained within watershed protection or water quality protection programs, in some they are incorporated into erosion and sedimentation control programs, while in others a separate program of forestry rules or guidelines governs harvesting activities. Links to all state forestry programs, with information on the agencies that are involved in protecting forests in the states, can be found at the Web site www.usabmp.net.

BMPs are individual practices (such as leaving a streamside management area) that serve specific functions (such as protecting streams from temperature increases and filtering sediment and nutrients from runoff). *Management measures*, as the term is used in this guidance, are environmental goals to be attained by using one or more BMPs. For instance, minimizing sediment delivery to streams (part of the overall goal of the Management Measure for Streamside Management Areas [see Chapter 3, section B]) from harvest sites might be accomplished with the following BMPs: maintaining a riparian buffer; locating roads, yarding areas, and skid trails away from streams; and not using machinery in streams.

BMPs are the building blocks for BMP systems and management measures, and the implementation of the forestry management measures in this guidance, as appropriate to the situation, can result in comprehensive water quality protection for most harvesting operations.

Management Measures

The management measures in this guidance contain technology-based performance expectations and, in many cases, specific actions to be taken to prevent or minimize nonpoint source pollution. Management measures are means to control the entry of pollutants into surface waters. Management measures achieve nonpoint source pollutant control goals through the application of nonpoint pollution control BMPs, which may be technologies, processes, siting criteria, operating methods, or other alternatives. Chapter 3 contains the management measures and recommended BMPs controlling nonpoint source pollution from forestry activities.

For example, the Management Measure for Site Preparation and Forest Regeneration (see section F) contains the performance expectation *Confine on-site potential nonpoint source pollution and erosion resulting from site preparation and the regeneration of forest stands*. Statements of BMPs or actions that can be taken to achieve this performance expectation (e.g., *Conduct mechanical tree planting and ground-disturbing site preparation activities on the contour of sloping terrain*) are generally included in the management measure statement. Even so, in most cases there is considerable flexibility to determine how to best achieve the performance expectations for the management measures. EPA's management measures for forestry and BMPs recommended to be used to achieve them are described in Chapter 3.

Best Management Practices

BMPs can be structural (e.g., culverts, broad-based dips, windrows) or managerial (e.g., preharvest planning, forest chemical management, fire management). Both types are used to control the delivery of nonpoint source pollutants to receiving waters in one of three ways:

- They minimize the quantity of pollutants released (pollution prevention).
- They retard the transport or delivery of pollutants, either by reducing the amount of water (and thus the amount of the pollutant) transported or by improving deposition of the pollutant (delivery reduction).

- They render the pollutant harmless or less harmful before or after it is delivered to a water body through chemical or biological transformation.

BMPs are usually designed to control a particular type of pollutant from a specific land use or activity. For example, stream crossings are specified and designed to control erosion from stream banks where roads cross them and sediment delivery from roads to streams. BMPs might also provide secondary benefits. Streamside management areas, for instance, reduce sediment delivery to streams and protect streams from temperature increases, and they also provide a source of large organic debris to streams and habitat for wildlife.

Sometimes, however, a BMP might increase the generation, transport, or delivery of a pollutant and is best used in combination with other BMPs. Site preparation, for example, is generally performed for commercial timber regeneration, but can temporarily expose soil to erosive forces. Therefore, sedimentation control BMPs, such as establishing SMAs of widths suitable to retain the anticipated quantity of eroded soil and not conducting mechanical site preparation on steep slopes, are recommended to be combined with site preparation techniques.

Which BMP is *best* for in a given situation depends on many factors. Criteria for determining which BMP is best for a particular forestry activity might include the harvesting technique, frequency of road use, topography, soil type, climate, amount of maintenance feasible BMPs will require, the willingness of landowners to implement BMPs (in a program of voluntary implementation, for instance), and BMP cost and cost-effectiveness. The relative importance assigned to these and other criteria in judging what is best varies among states, within states, and among landowners, often for very good reasons. For example, erosion control considerations are very different in mountainous western regions versus relatively flat southeastern coastal plain regions. Some BMPs that can be used to achieve the forestry management measures are described in Chapter 3.

Best Management Practice Systems

The distinction between BMPs selected for particular areas or aspects (e.g., roads, yarding areas, skid trails, stream crossings) of a harvest activity and a *BMP system* is similar to the difference between controlling pollutant sources individually and controlling them based on a TMDL. Pollutant sources, especially point sources, controlled on an individual basis are analyzed independently relative to a standard for a type of industry and water quality criteria for the receiving water body. A TMDL incorporates all pollutant sources affecting a water body and limits loads for individual sources relative to the assimilative capacity of the water body. Similarly, BMPs selected for individual aspects of a harvest activity views those activities or areas independently of other activities and areas to control water pollution, while approaching water quality considerations from the point of view of a BMP system would involve considering the harvest and all of its activities and affected areas from a hydrologic perspective, examining the flow of surface water and groundwater over the entire site, and determining the best locations for sediment, nutrient, and other pollutant interception. As an example, consider a harvest operation that involves road repairs, a stream crossing, creation of a yarding area, and site preparation. Individual BMPs can be selected for each aspect of the harvest operation. That is, BMPs for sediment retention (for example) could be chosen for the road segment, others selected for the stream crossing, and still others placed on the yarding area.

Each set of BMPs for these separate areas would be selected to control sediment runoff from that area alone. Alternatively, the spatial relationship of the three areas from a water flow or hydrologic perspective could be considered to understand how BMPs selected for the site preparation work might be altered somewhat to capture sediment from the yarding area, thus eliminating the need for separate BMPs for the yarding area. Also, it might be noticed that a different type or orientation of BMPs along the road segment could significantly reduce the potential for sediment delivery along the road to the stream crossing, thus permitting a change in the stream crossing to better ensure retaining the natural stream shape. The BMP system approach might reduce the total number of BMPs required and increase the efficiency of the BMPs for protecting water quality, and thus reduce the cost of the operation.

Structural and managerial BMPs used as part of a BMP system can be selected, designed, implemented, and maintained in accordance with site-specific considerations (e.g., slope, soil type, proximity to streams, and layout of the harvest) so they work effectively together. Planning BMP use as part of a system also helps to ensure that design standards and specifications for the individual BMPs are compatible so they will achieve the greatest amount nonpoint source pollution control possible with the least cost.

Cost Estimates for Forest Practice Implementation

Estimates of the per acre cost of implementing BMPs for timber harvests were arrived at based on information obtained from published reports on regional studies of the cost of BMP implementation and cost estimates based on the regulatory structure of forestry practice programs. Studies have been conducted on the cost of implementing forestry practices for water quality and soil protection in the Southeast and some western states (Aust et al., 1996; Dissmeyer and Foster, 1987; Dubois et al., 1991; Henly, 1992; Lickwar, 1989; Olsen et al., 1987). Costs associated with complying with forest practices in states where their implementation is either voluntary or regulated, with differing numbers and types of requirements depending on the state, have also been estimated (Table 2-2) (Ellefson et al., 1995).

Some cost information for forest practice implementation is based on the average increased cost of conducting a harvest when management measures, i.e., a suite of practices, are used versus when they are not used (Table 2-3). Costs provided in this way emphasize the difficulty in separating the costs of implementing individual forest practices. This difficulty is due to incorporating the cost of using numerous BMPs into the accomplishment of a single harvesting or road construction activity, and spreading the cost for individual practices across the accomplishment of multiple activities. For example, the cost of adhering to a state regulation for stream crossings might be spread among the costs of planning a harvest to minimize the number of stream crossings, designing and constructing forest roads to accommodate the plan and minimize instream effect to water quality and fish, and the actual construction of the stream crossings. Furthermore, these costs differ with each harvest because the terrain, soils, location of harvest site relative to streams, and hydrology are different at each harvest site. Therefore, all costs presented here are best regarded as rough estimates.

Table 2-2.   Estimations of Overall Cost of Compliance with State Forestry BMP Programs by Program Type

Applicability	Cost Estimation	Reference

Virginia and southeastern states (applicable to central and northern states)	Voluntary-to-mandatory implementation ($)
	  Coastal plain region: = $11.70 per acre
	  Piedmont region: = $30.40 per acre
	  Mountain region: = $44.50 per acre
	
	Stringent/Enforceable implementation ($)
	  Coastal plain region: = $21.40 per acre
	  Piedmont region: = $38.00 per acre
	  Mountain region: = $49.10 per acre
		Aust et al., 1996

California	Average cost = $250 per acre Inland areas = $81 - $414 per acre[a] Coastal areas = $460 per acre[a]	Henly, 1992
Oregon, Washington, Alaska	Average cost = $175 - $373 per acre Noncoastal areas = $175 per acre Coastal areas = $373 per acre	Ellefson et al., 1995(Division between coastal and noncoastal based on California model)
Nevada, New Mexico, Idaho	Other western states with forest practice regulation. Cost per acre is estimated as the average of costs in western states without forest practice regulation and the low-end cost given for Oregon noncoastal forests: ($125 + $175)/2 = $150 per acre	
Arizona, Colorado, Montana, Utah, Wyoming, Hawaii	Western states without forest practice regulation. Cost per acre is estimated as one-half of California's noncoastal cost: $250/2 = $125 per acre	

Note: All costs in 1998 dollars.
[a] Excluding most costly scenario.

The costs of implementing state forest practices arise from conducting timber surveys, preparing management plans, constructing roads, and implementing practices specifically designed to protect water quality. Many of these costs are borne whether or not a stream or other surface water is located on or near a harvest site, though additional costs (e.g., designing and flagging an SMA, constructing stream crossings) are incurred where streams are present. Costs also take the form of lost revenue from trees that are not harvested to ensure compliance with forest practices. Revenue might be reduced if merchantable trees are left standing in SMAs or when selective cutting is called for rather than clear-cutting. Although the loss of revenue is a real "cost" to landowners, it is very market- and species-dependent and is generally not included in the cost estimates provided here. The overall costs of complying with regulatory forestry BMP programs might be borne by forest landowners alone or shared among landowners, timber operators, and others (Figure 2-1).

Factors that typically affect the cost of implementing forest practices include the type of terrain on which a harvest occurs (with costs for harvesting on steeper

Table 2-3.   Estimations of Implementation Costs by Management Measure in the Southeast and Midwest

Practice	Average Cost	Cost Range	Comments	Reference
Planning			Savings were associated with avoiding problem soils, wet areas, and unstable slopes. Maintenance savings resulted from revegetating cut and fill slopes, which reduced erosion. Southern states.	Dissmeyer and Foster, 1987
Savings from road design/ location	($385/mi)			
Savings in maintenance	($231/mi)			
SMA	$3,996		Costs for average tract size of 1,361 ac; include marking and foregone timber value. Southern states.	Lickwar, 1989
Road Construction		$5,301/mi - $42,393	Lower end for no gravel and few culverts; upper end for complete graveling and more culverts. West Virginia.	Kockenderfer and Wendel, 1980
		$14,801 -$42,393	Lower end for 1,832-ac forest with slopes <3%; upper end for 1,148-ac forest with slopes >9%. Southern states.	Lickwar, 1989
		$229/mi - $11,604/mi	Lower end for grass surfacing; upper end for large stone surfacing. Appalachia.	Swift, 1984
Construction Phase (as percent of total cost)		10% 20 - 25% 20 - 25% 10% 30 - 40%	Equipment and Material Clearing, grubbing, and slash disposal Excavation Culvert installation Rock surfacing	USDA-SCS, cited in Weaver and Hagans, 1994
Road Maintenance	$2,205-$3,941		Lower end for roads constructed without BMPs; upper end for roads constructed with BMPs. Costs over 20 years discounted at 4%.	Dissmeyer and Frandsen, 1988
Mechanical Site Preparation	$140/ac	$77/ac -$281/ac	Lower end for disking only; upper end for shear-rake-pile-disk. Southern states.	Dubois et al., 1991
		$75/ac -$180/ac	Lower end for light preparation, including hand; upper end for chemical-mechanical site preparation.	Minnesota, 1991
Regeneration		$84/ac - $355/ac	Lower end for direct seeding; upper end for tree planting with purchased planting stock.	Illinois, 1990
	$50/ac	$48/ac - $60/ac	Lower end for machine planting; upper end for hand planting. Southern states.	Dubois et al., 1991
Revegetation	$22,741		Cost for average sized tract of 1,361 ac; includes seed, fertilizer, mulch. Southern states.	Lickwar, 1989
		$132/ac - $239/ac	Lower end for introduced grasses; upper end for native grasses. Includes seedbed preparation, fertilizer, chemical application, seed, seedlings.	Minnesota, 1991
Prescribed burning	$13/ac	$10/ac - $19/ac	Lower end for windrow burning; upper end for burning after chemical site preparation. Southern states.	Dubois et al., 1991
Pesticide application	$102/ac	$56/ac - $138/ac	Lower end for ground application; upper end for aerial application. Southern states.	Dubois et al., 1991
Fertilizer application	$63/ac	$43/ac - $73/ac	Lower end for ground application; upper end for aerial application. Southern states.	Dubois et al., 1991

Note: All costs in 1998 dollars.

terrain typically being higher than costs for harvesting on flatter terrain) and the regulatory structure of forest practice rules. Compliance in states that have numerous and stringent forest practice regulatory requirements generally costs more than compliance in states where regulatory requirements are fewer or less stringent, or are voluntary. Some states have single regulations that can add significantly to the cost of forest harvesting. An example is the requirement for a detailed forest harvest plan in California. This alone places compliance with forest practices in California in a category by itself.

Table 2-2 summarizes estimations of the overall per-harvest cost of complying with forest practice regulations in different regions and states. Table 2-3 provides cost estimates for implementation of individual management measures in the Southeast and Midwest. The costs, updated to 1998 dollars, have been verified with state and federal forest management agencies and have been found to be representative of actual expenditures. Although most of the cost information came from case studies in the southeastern United States, they are representative of costs incurred nationwide. Costs vary depending on the site-specific nature of the timber harvesting area. Table 2-4 provides estimates of costs for installing individual road construction and erosion control BMPs. Costs are provided by region. Factors that affect implementation costs are mentioned in the *Comments* column.

Other costs, where available, are provided for individual management measures or BMPs within the appropriate discussions in Chapter 3.

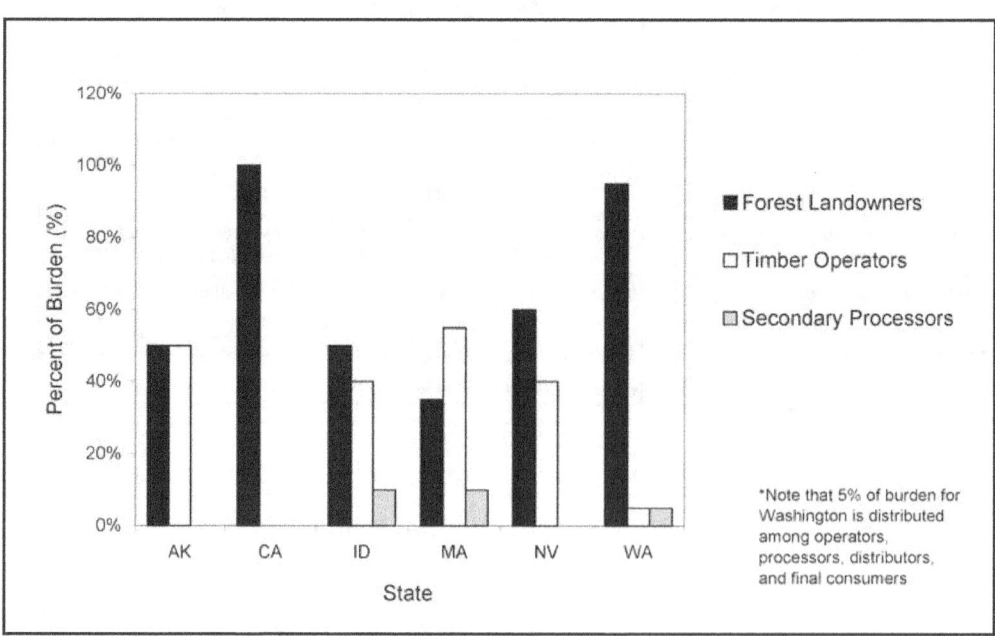

Figure 2-1. Distribution of the cost of regulatory programs among different groups in representative states (Ellefson et al., 1995).

Table 2-4. Estimations of Construction and Implementation Costs for Individual BMPs, by Region

BMP	Approximate Construction and Implementation Costs per BMP Installed, by Region							Comments
	Northeast[1]	Southeast[2]	Midwest[3]	Rocky Mountains[4]	Northwest[5]	Southwest[6]	Alaska[7]	
Broad-based dip		$40	$40 - $90	$50 - 60	$25 - 35	$100 - $130	$30 - $40	Depends on the cost of labor, equipment, and terrain (Northwest costs include profit and overhead).
Waterbar		$20 (not including labor)	$60 - $75 (on skid trails)	n/a	$100	$45 - $60	$25 - $35	Cost varies with size and construction material.
Mulch		$71 (ton)	$20 - $80 (ton)	n/a	$1,500/(ac) (hydro-mulch)	$400 - $500 (ac)	$80 - $90 (ton)	Cost varies with regional market price and haul distance.
Seed	$1,000 (ac) (hydro-seed)	$1 - $6 (lb)	$0.50 - $10 (lb)	$6 (lb)	$400 - $450 (ac)	$200 - $400 (ac)	$7 - $10 (lb)	Cost varies with species of seed, regional market price, and terrain.
Riprap		n/a	$5 - $10 (yd³)	$2 (yd³)	$15 - $30 (yd³)	n/a	$19 - $37 (yd³)	Price varies with size of rock used.
Gravel		$6 - $10 (ton)		$35,000-$40,000 (mile, 14" W x 4" D)	$16 - $26 (yd³)	$30 (yd³)	$18 - $22 (yd³)	Cost varies with the size of rock and haul distance.
Culvert		$420	$500 - $2,000	$19 (ft, 18" pipe)	$26 (ft, 24" pipe) $100 (ft, 72" pipe)	$24 (ft, 18" pipe)	$23 (ft, 18" pipe)	Cost varies with size and length of culvert. Costs provided reflect base cost for installation.
Straw Matting		$56 (roll, 7.5' x 120')		n/a	$2 (yd²)	$1 - $3 (yd²)	$2.50 (yd²)	Cost varies with size of matting.
Geotextiles		$378 (700 yd²)	$2 - $6 (ft)	$8 - $12 (ft)	$1 - $2 (ft)	n/a	$14 (ft)	Woven geotextiles are the only geotextile recommended for road-stream crossings.
Hardwood Mats (pallets)	$120 - $200	$120 - $200	$170 (10' x 12')	$120 - $200	$120 - $200	$120 - 200	$155 (10' x 12')	Cost varies with size.
Turn-outs	$40 - $50	$50 - $70	$50 - $70	$50	$50	$40 - $50	$71	Cost varies with equipment and labor costs.
Silt Fence		$24 (24" H x 100' L)	not commonly used	not commonly used	$1.50 (yd²)	$4 (ft)	$2 (yd²)	Cost varies with regional prices and length.
Dust Control	$1,000 (mile, using calcium chloride)				$1,000 - $3,000 (mile, annually)	$190 (ton)		Varies widely with traffic level.
Temporary Bridge		$500 - $20,000	$500 - $15,000	$200 - $25,000	$1,000 - $2,000 (ft)	n/a	$1,250 - $2,500 (ft)	Cost varies widely with quality of materials used, width, and span.
Barge (Alaska)	–	–	–	–	–	–	$1,000 (hr)	Barge transport in southeastern Alaska (Tongass Natl. Forest) is the most common means to deliver material to a site.

Note: All costs are per unit provided (ac = acre; ft = linear foot; hr = hour; lb = pound; yd² = square yard; yd³ = cubic yard; D = depth; H = height; L = length; W = width). Where units are not provided, costs per BMP installed.
[1] Schmid, 2000
[2] Holburg, 2000; Marzac, 2000
[3] Hansit, 2000; Gambles, 2000
[4] Taylor, 2000
[5] Dorn, 2000; Hullet, 2000; Wilbrecht, 2000; Yoder, 2000
[6] Leyba, 2000
[7] Jenson, 2000

CHAPTER 3: MANAGEMENT MEASURES

Scope of This Chapter

For the purposes of this guidance, EPA has addressed the activities associated with forestry activities that could affect water quality through nine management measures. A separate management measure is applicable specifically to forested wetlands. The management measures are stated as steps to be taken, guidelines for operations, or goals to be achieved for protecting water quality during the related phases or activities. The following are EPA's forestry management measures:

- Preharvest planning
- Streamside management areas
- Road construction/reconstruction
- Road management
- Timber harvesting
- Site preparation and forest regeneration
- Fire management
- Revegetation of disturbed areas
- Forest chemical management
- Wetland forest management

Numerous BMPs are associated with each management measure. BMPs are specific actions, processes, or technologies that can be used to achieve a management measure. These BMPs are very similar to those recommended by most states. Because of the national scope of this guidance, however, some of the particulars of implementation (such as prescriptions for sizes of pipes, lengths of road at particular slopes, and other such site- or region-specific details) are not included as part of the descriptions of BMPs. Implementation of one or more BMPs is usually necessary to achieve the level of pollution control intended by a single management measure.

Each management measure is addressed in a separate section of this chapter. Each section contains the wording of the management measure, which has not been changed from that in the 1993 CZARA guidance; a description of the management measure's purpose or how it can be used effectively to protect water quality; and information on BMPs that are suitable, either alone or in combination with other BMPs, to achieve the management measure. Where new or improved versions of BMPs have been developed, they are discussed in this guidance. Many of the BMPs were in the 1993 CZARA guidance, and most can be found in state forest practices manuals. For recommendations on widths of streamside management areas, slopes and lengths of culverts, and other criteria for your specific area, consult a state forest practices manual or contact your local forester.

Since the forestry management measures developed for CZARA are for the most part a system of BMPs commonly used and recommended by states and the U.S. Forest Service, many BMPs are already being implemented at many harvest sites and on many forest roads. Where the BMPs in place are inadequate to protect water quality, augmenting them with additional or complementary BMPs might be all that is necessary. Where measures are lacking and water quality is or might become impaired, this guidance can assist in the choice of BMPs suitable to the source of water quality impairment.

Management Measure Effectiveness

States have used a number of approaches for assessing the effectiveness of management measures and BMPs. Florida and South Carolina have assessed their effectiveness using bioassessment techniques and stream habitat assessment. Florida has compared sites adjacent to harvests with non-logged reference sites, and South Carolina has also compared sites upstream from harvests to those downstream from harvests and conditions at the same site before harvests to those after harvests. Maine and Virginia have placed in-stream water quality samplers in streams near forest harvest operations. South Carolina and Washington have used a weight-of-evidence approach, in which a variety of different assessment approaches are used and the conclusion about effectiveness arrived at most by the different approaches is accepted as the overall conclusion. South Carolina has concluded from its weight-of-evidence assessments that on sites with perennial streams, BMP compliance checks, stream habitat assessment, and benthic macroinvertebrate assessments can be used effectively to assess BMP effectiveness.

All of the approaches have produced valuable information about BMP effectiveness. The conclusions from these studies are many:

- BMP assessment monitoring is important for determining that the standards for design and implementation of BMPs are appropriate for the soils and topography where they are to be used.

- One or more BMP assessment approaches, including BMP compliance and an in-stream habitat or macroinvertebrate approach, can help determine whether BMP implementation standards are adequate.

- Once adequate implementation standards have been developed, rigorous BMP compliance checks generally suffice as an indicator of BMP effectiveness. The compliance checks are used to verify that BMPs are being installed properly and in a timely manner, and that they are maintained adequately.

- It is important to assess the effectiveness of BMPs under a variety of site conditions and to tailor implementation standards to different types of soils, slopes, and regional site characteristics if the BMPs are to be effectively applied.

- Application of BMPs per implementation standards during forest harvesting protects water quality in adjacent streams. BMPs protect stream ecology and stream temperature, and they prevent sedimentation.

- When BMPs are not properly applied, they do not adequately protect water quality. Improperly applied BMPs can result in stream sedimentation, changes in stream morphology, increased average water temperatures, wider water temperature fluctuations, and changes to stream ecology.

- Many water quality problems that arise from forest harvesting are associated with improperly applied BMPs or not having used BMPs. The most frequently misapplied or missing BMPs are those for road surface drainage control, erosion control prior to the harvest, stream crossings, and SMAs.

- Some states do not adequately address some water quality problems associated with forest harvesting. BMPs for ephemeral drainages need to be developed and the circumstances under which ephemeral drainages require BMPs needs to be determined. Ephemeral drainages can produce or deliver large quantities of sediment to other streams if left unprotected after a harvest.

- The most important BMPs for protecting stream water quality are properly sized SMAs, properly designed BMPs for erosion control implemented prior to the commencement of road construction and harvesting, properly designed stream crossings, and comprehensive preharvest plans.

Examples of Management Measure Effectiveness

Examples of how BMPs can operate as a system to control nonpoint source pollution are given in a paper that summarizes a national effort by USDA's Forest Service to develop analysis procedures for estimating the economic benefits of soil and water resource management (Dissmeyer and Foster, 1990). The paper focuses on benefits in five areas—timber, forage, fish, enhanced water quality, and road construction and maintenance. The benefits noted from the use of resource management systems are expressed as increased timber production, increased forage on the harvest site, and benefits to other resources from improved soil and water resource management. The following are the examples of the proper implementation of resource management systems provided in Dissmeyer and Foster (1990) and Dissmeyer and Frandsen (1988). Each example begins with a hypothetical situation and then describes how BMPs apply to the situation.

Example 1 focuses on soil and water resource management in road construction and maintenance. In this example, a main haul road is built across problem soils, cutbanks yield excessive surface runoff and erode easily, the runoff volume from the site is sufficient to erode through the road surface and road subgrade, road maintenance (without BMPs installed) is needed every 3 years, and the road is assumed to be used for 20 years. Applying a resource management system to this situation, the following solution was devised: construct the road with midslope terraces in the cutbanks; install water diversions above the cutbanks; and seed, fertilize, and mulch the cutbanks. The total estimated repair costs over 20 years were calculated at $2,137 for materials, labor, and cost of technical assistance. The one-time installation of BMPs, which would eliminate the need for maintenance every 3 years, would cost $1,200. The resulting net present value, or economic benefit to the property owner, of installing the BMPs in this example was calculated as $937 (all cost figures in 1990 dollars).

Example 2 relates to recouping timber growth and yield losses through skid trail rehabilitation. Skid trails and skid roads in harvest areas are areas where sediment is lost, and as a result the timber yield in primary skid trails and on skid roads is in general severely reduced. Soils in skid trails can become severely compacted, limiting water infiltration and thus soil moisture availability and tree root development. Finally, soil nutrients are removed during skidding and during road construction. A resource management system solution to this problem involves using the following BMPs: ripping and tilling the soil, waterbarring, seeding, fertilizing, and mulching. Using these practices as a system, the net present value of timber volume recovered (based on estimations provided in published studies) would be $210 per acre based on a harvest of shortleaf pine stands and $237 per acre in hardwood stands. Note that the economic returns are positive in high-value shortleaf pine stands and negative in low-value hardwood stands. The study notes, however, that the herbaceous growth from applying a system of resource BMPs in hardwood stands would have positive value for hunting and environmental protection.

Example 3 relates to the effect of site preparation, which can affect sediment production, soil productivity, and timber growth and yields. Poor site preparation practices that compact the soil, remove litter, and remove nutrients adversely affect soil productivity and sediment retention. The study, based on modeling data from independent studies of BMPs used for site preparation, found that site preparation results in economic benefits. Specifically, investing $50 *more* per acre in preparing a site with shearing and windrowing *reduced* future maintenance costs by $129 per acre, compared to chopping and burning.

These examples highlight the economic and ecological advantages of using management measures and BMPs as a system to reduce effects on surface waters and to ensure more rapid site regeneration and healthier timber stands.

3A: Preharvest Planning

Management Measure for Preharvest Planning

Perform advance planning for forest harvesting that includes the following elements where appropriate:

(1) Identify the area to be harvested including location of water bodies and sensitive areas such as wetlands, threatened or endangered aquatic species habitat areas, or high-erosion-hazard areas (landslide-prone areas) within the harvest unit.

(2) Clearly mark these sensitive areas with paint or flagging tape, or in another highly visible manner, prior to harvest or road construction.

(3) Time the activity for the season or moisture conditions when the least effect occurs.

(4) Consider potential water quality effects and erosion and sedimentation control in the selection of silvicultural and regeneration systems, especially for harvesting and site preparation.

(5) Reduce the risk of occurrence of landslides and severe erosion by identifying high-erosion-hazard areas and avoiding harvesting in such areas to the extent practicable.

(6) Consider additional contributions from harvesting or roads to any known existing water quality impairments or problems in watersheds of concern.

Perform advance planning for forest road systems that includes the following elements where appropriate:

(1) Locate and design road systems to minimize, to the extent practicable, potential sediment generation and delivery to surface waters. Key components are:

- locate roads, landings, and skid trails to avoid to the extent practicable steep grades and steep hillslope areas, and to decrease the number of stream crossings;

- avoid to the extent practicable locating new roads and landings in Streamside Management Areas (SMAs); and

- determine road usage and select the appropriate road standard.

(2) Locate and design temporary and permanent stream crossings to prevent failure and control effects from the road system. Key components are:

- size and site crossing structures to prevent failure;

- for fish-bearing streams, design crossings to facilitate fish passage.

(3) Ensure that the design of road prism and the road surface drainage are appropriate to the terrain and that road surface design is consistent with the road drainage structures.

(4) Identify and plan to use road surfacing materials suitable to the intended vehicle use for roads that are planned for all-weather use.

(5) Design road systems to avoid high erosion or landslide hazard areas. Identify these areas and consult a qualified specialist for design of any roads that must be constructed through these areas.

Each state should develop a process (or utilize an existing process) that ensures that the management measures in this chapter are implemented. Such a process should include appropriate notification, compliance audits, or other mechanisms for forestry activities with the potential for significant adverse nonpoint source effects based on the type and size of operation and the presence of stream crossings or SMAs.

Management Measure Description

The objective of this management measure is to ensure that forestry activities, including timber harvesting, site preparation, and associated road construction, are planned with water quality considerations in mind and conducted without significant nonpoint source pollutant delivery to streams or other surface waters. Road system planning is an essential part of this management measure because road construction is the main soil destabilizing activity carried out in forestry, and avoidance is the most cost-effective means of dealing with unstable terrain (Weaver and Hagans, 1994).

A basic tenet of road planning is to minimize the number of road miles constructed in a watershed through basin-wide planning. A second tenet is to locate roads to minimize the risk of water quality impacts. Good road location and design can greatly reduce the sources and transport of sediment. Road systems can be designed to minimize the length and surface area of roads and skid trails, the size and number of landings, and the number of stream crossings, and to locate all of these road system elements as far from surface waters as feasible. Minimizing stream crossings is especially important in sensitive watersheds.

Preharvest planning includes consideration of the potential water quality and habitat effects of the component parts of the harvest, including the harvesting system (e.g., clear-cut or selective cut); the yarding system (e.g., skyline cable or ground skidding); the road system; and postharvest activities such as site preparation. Water quality considerations can most effectively be incorporated into preharvest planning by determining which pollutants are likely to be generated during each of the phases of the harvest and how best to ensure that they are kept out of surface waters. Reviewing Section 2 can help with the task of identifying the pollutants, and Section 3 provides information on the BMPs that will minimize their entry into surface waters.

The water quality effects of yarding can be reduced with thoughtful preharvest planning. Yarding done with ground skidding equipment can cause much more soil disturbance than cable yarding. McMinn (1984) compared a skidder logging system and a cable yarder for their relative effects on soil disturbance (Table 3-1). With the cable yarder, 99 percent of the soil remained undisturbed (the original litter still covered the mineral soil), whereas the amount of soil remaining undisturbed after logging by skidder was only 63 percent. Whether cable yarding, ground skidding, or skyline yarding is best for the particular harvest is based on whether the stand is even-aged or uneven-aged, the terrain, cost, and other factors. Among these other factors should be the need and means to protect water quality.

Table 3-1.   Comparison of the Effect of Conventional Logging System and Cable Miniyarder on Soil in Georgia (McMinn, 1964)

Disturbance Class[a]	Cable Skidder	Miniyarder
Undisturbed	63%	99%
Soil exposed	12%	1%
Soil disturbed	25%	0%

[a]   Undisturbed = original duff or litter still covering the mineral soil.
    Exposed = litter and duff scraped away, exposing mineral soil, but no scarification.
    Disturbed = Mineral soil exposed and scarified or dislocated.

Preharvest planning is the time to consider how harvested areas are to be replanted or regenerated to prevent erosion and effects on water bodies after the harvest has occurred. At the same time, it is important to consider other activities that have occurred recently, will coincide with the harvesting, or are scheduled to occur in the watershed where harvesting is to take place, as well as the overall soil, habitat, and water quality conditions of the watershed. Other activities within the watershed that can also stress water systems include land use changes from forest to agriculture, residential development or other construction, and applications of pesticides or herbicides. Cumulative effects on soils, water quality, and habitats from other activities and the proposed forest practices can result in excessive erosion and pollutant transport, and detrimental receiving water effects (Sidle, 1989). Cumulative effects are influenced by forest management activities, natural ecosystem processes, and the distribution of other land uses within a watershed. Forestry operations such as timber harvesting, road construction, and chemical use can increase runoff of nonpoint source pollutants and thereby contribute to preexisting impairments to water quality.

A previously completed cumulative assessment might exist for the area to be harvested, in which case it can be determined whether water quality problems, if any, in the watershed are attributable to the types of pollutants that might be generated by the planned forestry activity. If more pollutants of the same types are likely to be generated as a result of the harvesting activity, adjustments to the harvest plan or use of management practices beyond those normally used might be necessary. For instance, consider selecting harvest units with low sedimentation risk, such as flat ridges or broad valleys; postponing harvesting until existing erosion sources are stabilized; or selecting limited harvest areas using existing roads. The need for additional measures, as well as the appropriate type and extent, is best considered and addressed during the preharvest planning process.

During preharvest planning, it is also particularly important to plan implementation of management practices to be used to control sediment delivery from sources that are characteristically erosion-prone and lead to water quality impairment at stream crossings, landings, road fills on steep slopes, road drainage structures, and roads located close to streams. Constructing roads through high-erosion-hazard areas can lead to serious water quality degradation and should be avoided when possible. Some geographical areas (e.g., the Pacific coast states) tend to have more serious erosion problems (landslides, major gullies, etc.) after road construction than other areas. Factors such as climate, slope steepness, soil and rock characteristics, and local hydrology influence this potential. A person trained to recognize high-erosion hazard areas should be involved with preharvest planning.

Erosion hazard areas are often mapped by public agencies, and these maps are one tool to use in identifying high-erosion-hazard sites. The U.S. Geological Survey has produced geologic hazard maps for some areas. The USDA Natural Resources Conservation Service (NRCS) and Agricultural Farm Service Agency (FSA), as well as state and local agencies, might also have erosion-hazard-area maps.

Benefits of Preharvest Planning

The Virginia Department of Forestry found that preharvest planning is one of the three BMPs that are crucial to water quality protection. The other two are the establishment and use of streamside management areas (SMAs) and properly designed and constructed stream crossings. Although all BMPs are considered to be important, these three were found to be the most important to preventing water quality degradation.

In a study conducted by Black and Clark (no date), sediment concentrations were compared from stream waters in an unlogged watershed, a watershed where a harvesting operation with thorough preharvest planning had been conducted, and a watershed where a harvesting operation with no preharvest planning had been conducted. Sediment concentrations in the water from the unlogged watershed averaged 4 parts per million (ppm), those in the water from the watershed with the planned logging operation averaged 5 ppm, and those from the watershed with the unplanned harvest averaged 31 ppm (Figure 3-1). Preharvest planning in this study took into consideration road siting and construction techniques, landing siting, yarding techniques, and other BMPs intended to minimize erosion and sediment loss.

Of course, BMPs are effective only when properly designed, constructed, implemented, and maintained. Too often, BMPs are not installed early enough in the process to effectively control nonpoint source pollution, or they are not maintained properly, which can lead to their failure and to sedimentation or other forms of pollution. In general, poor BMP effectiveness can be attributed to one or more of the following:

- A lack of time or willingness to plan timber harvests carefully before cutting begins.
- A lack of skill in or knowledge of designing effective BMPs.
- A lack of equipment needed to implement effective BMPs.
- The belief that BMPs are not an integral part of the timber harvesting process and can be engineered and fitted to a logging site after timber harvesting has been completed.

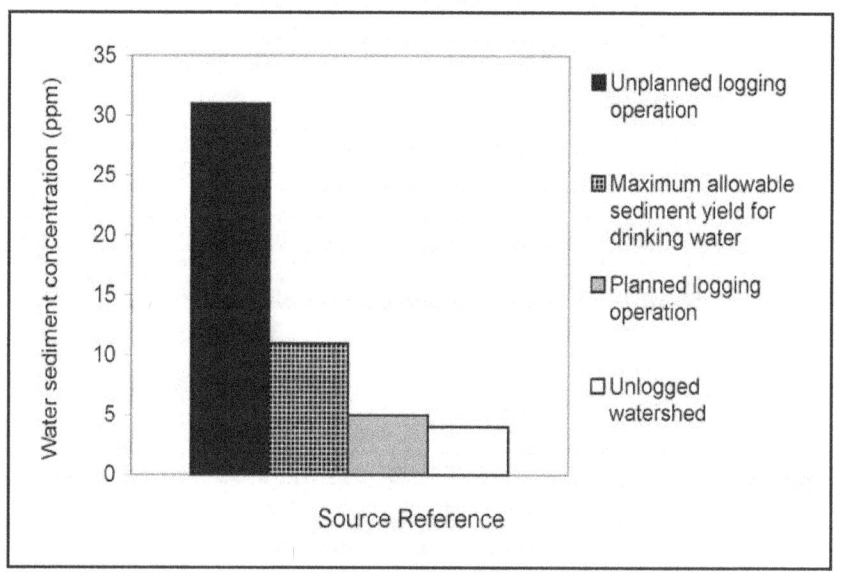

Figure 3-1. **Comparison of sediment concentrations in runoff from various forest conditions to drinking water standard (after Black and Clark, nd).**

Best Management Practices

Harvest Planning Practices

◆ *Use topographic maps, aerial photographs, soil surveys, geologic maps, and seasonal precipitation information—as slow long duration precipitation can be as limiting as high intensity short duration rainfall—to augment site reconnaissance to lay out and map harvest units. Identify and mark, as appropriate:*

- Sensitive habitats that need special protection, such as threatened and endangered species nesting areas.
- Streamside management areas.
- Steep slopes, high-erosion-hazard areas, and landslide-prone areas.
- Wetlands.

◆ *In warmer regions, schedule harvest and construction operations during dry periods or seasons. Where weather permits, schedule harvest and construction operations during the winter to take advantage of snow cover and frozen ground conditions.*

◆ *Consider potential water quality and habitat effects when selecting the silvicultural system as even-aged (clear-cut, seed tree, or shelterwood) or uneven-aged (group or individual selection). The yarding system, site preparation method, and any pesticides that will be used can also be considered during preharvest planning. As part of this practice, consider the potential effects from and extent of roads needed for each silvicultural system.*

◆ *In high-erosion-hazard areas, trained specialists (geologist, soil scientist, geotechnical engineer, wild land hydrologist) can identify sites that have high risk of landslides or that might become unstable after harvest. These specialists can recommend specific practices to reduce the likelihood of erosion hazards and protect water quality.*

◆ *Determine what other harvesting activities, chemical applications, or other potentially polluting activities are scheduled to occur in the watershed and, where appropriate, conduct the harvest at a time and in such a manner as to minimize potential cumulative effects.*

Road System Planning Practices

Road Location Practices

◆ *Preplan skid trail and landing locations on stable soils and avoid steep gradients, landslide-prone areas, high-erosion-hazard areas, and poor-drainage areas.*

- Plan to minimize roads, stream crossings, landings, skid trails, and activities on unstable soils and steep slopes.
- Locate landings outside of SMAs and ephemeral drainage areas.
- Locate new roads and skid trails outside of SMAs, except where necessary to cross drainages.

- Locate roads away from stream channels where road fill extends within 50 to 100 horizontal feet of the annual high water level. (Bankfull stage is also used as a reference point for this.)

◆ *Systematically design transportation systems to minimize total mileage.*

- Compare layouts for roads, skid trails, landings, and yarding plans, and determine which will result in the least soil disturbance and erosion.
- Locate landings to minimize skid trail and haul road mileage and disturbance of unstable soils.

◆ *Identify areas that would need the least modification for use as log landings and use them to reduce the potential for soil disturbance. Avoid using areas, such as ephemeral drainages, that could contribute considerably to nonpoint source pollution if high precipitation occurs during the harvest. Use topographic maps and aerial photographs to locate these areas.*

◆ *Plot feasible routes and locations on aerial photographs or topographic maps to assist in the final determination of road locations. Compare the possible road location on-the-ground and proof the layout to ensure that the road follows the contours. Design roads and skid trails to follow the natural topography and contour, minimizing alteration of natural features.*

Proper design can reduce the area of soil exposed by construction activities. Figure 3-2 presents a comparison of road systems. Following the natural topography and contours can reduce the amount of cut and fill needed and consequently reduce both road failure potential and cost. Ridge routes and hillside routes are good locations for ensuring stream protection because they are removed from stream channels and the intervening undisturbed vegetation acts as a sediment barrier. Wide valley bottoms are good routes if stream crossings are few and roads are located outside SMAs.

◆ *Plan the management of existing and future roads and road systems to minimize environmental problems arising from them.*

Roads analysis is an integrated ecological, social, and economic approach to transportation planning addressing both existing and future road systems. The U.S. Forest Service's Roads Analysis procedure, developed by a team of Forest Service scientists and managers, is designed to help national forest managers bring their road systems into balance with current social, economic, and environmental needs. The top priority is to provide road systems that are safe for the public, responsive to public needs, environmentally sound, affordable, and efficient to manage. A roads analysis provides scientific information used to inform decision makers about effects, consequences, options, priorities, and other factors. This information is essential to plan efficiently and manage the forest transportation crisis. The iterative procedure for conducting the roads analysis consists of six steps aimed at producing needed information and maps (USDA Forest Service, 1999):

- **Step 1**: Set up the analysis. The analysis is designed to produce an overview of the road system. An interdisciplinary team develops a list of information needs and a plan for the analysis.
- **Step 2**: Describe the situation. The interdisciplinary team describes the existing road system in relation to current forest management plans. Products from this step include a map of the existing road system, descriptions of access needs, and

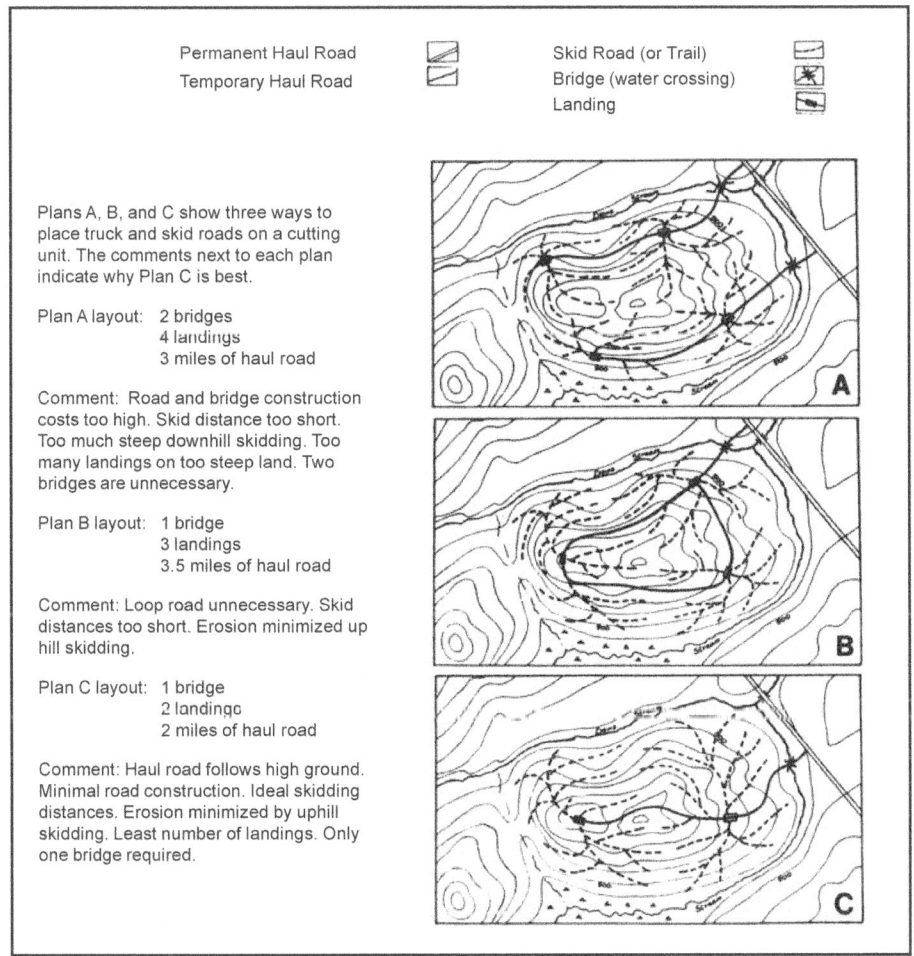

Permanent Haul Road
Temporary Haul Road

Skid Road (or Trail)
Bridge (water crossing)
Landing

Plans A, B, and C show three ways to place truck and skid roads on a cutting unit. The comments next to each plan indicate why Plan C is best.

Plan A layout: 2 bridges
                 4 landings
                 3 miles of haul road

Comment: Road and bridge construction costs too high. Skid distance too short. Too much steep downhill skidding. Too many landings on too steep land. Two bridges are unnecessary.

Plan B layout: 1 bridge
                 3 landings
                 3.5 miles of haul road

Comment: Loop road unnecessary. Skid distances too short. Erosion minimized up hill skidding.

Plan C layout: 1 bridge
                 2 landings
                 2 miles of haul road

Comment: Haul road follows high ground. Minimal road construction. Ideal skidding distances. Erosion minimized by uphill skidding. Least number of landings. Only one bridge required.

Figure 3-2.   An example of laying out sample road systems for comparison purposes (Hynson et al., 1982).

information about physical, biological, social, cultural, economic, and political conditions associated with the road system.

- **Step 3**: Identify issues. The interdisciplinary team, in conjunction with the public, identifies important road-related issues and the information needed to address them. The interdisciplinary team also determines data needs associated with analyzing the road system in the context of the important issues, for both existing and future roads. The output from this step includes a summary of key road-related issues, a list of screening questions to evaluate them, a description of the status of relevant available data, and a list of additional data needed to conduct the analysis.

- **Step 4**: Assess benefits, problems, and risks. After identifying the important issues and associated analytical questions, the interdisciplinary team systematically examines the major uses and effects of the road system, including the environmental, social, and economic effects of the existing road system and the values and sensitivities associated with unroaded areas. The output from this step is a synthesis of the benefits, problems, and risks of the current road system and the risks and benefits of building roads into unroaded areas.

- **Step 5**: Describe opportunities and set priorities. The interdisciplinary team identifies management opportunities, establishes priorities, and formulates technical recommendations that respond to the issues and effects. The output from this step includes a map and a descriptive ranking of management options and technical recommendations.

- **Step 6**: Report. The interdisciplinary team then produces a report and maps that portray management opportunities and provide supporting information important for making decisions about the future characteristics of the road system. This information sets the context for the development of proposed actions to improve the road system and for future amendment and revision of forest plans.

◆ *Consider using or upgrading existing roads to minimize the total amount of road construction necessary whenever practical and when less adverse environmental impact would be caused.*

Existing roads should be used where they are in good condition or can be feasibly upgraded, unless using the roads would cause more water quality impacts than building a new road elsewhere (Weaver and Hagans, 1994). When an existing road is available on the side of a drainage opposite the harvest site, consider using it instead of constructing a new road to minimize the amount of soil disturbance due to new road construction. Avoid using existing or previously-used roads, however, if they are likely to create water quality problems, such as if they were constructed next to streams in valleys.

Road Design Practices

◆ *In moderately sloping terrain, plan for road grades of less than 10 percent, with an optimal grade of between 3 percent and 5 percent. In steep terrain, short sections of road at steeper grades can be used if the grade is broken at regular intervals. On steep grades, vary road grades frequently to reduce culvert and road drainage ditch flows, road surface erosion, and concentrated culvert discharges.*

Gentle grades are desirable for proper drainage and economical construction. Steeper grades are acceptable for short distances (200-300 feet), but an increased number of drainage structures might be needed above, on, and below the steeper grade to reduce runoff potential and minimize erosion. Heavy traffic on steep grades can result in surface rutting that renders crowning, outsloping, and insloping ineffective. On sloping terrain, no-grade road sections are difficult to drain properly and are best avoided when possible.

◆ *Design skid trail grades to be 15 percent or less, with steeper grades only for short distances.*

◆ *In designing roads for steep terrain, avoid the use of switchbacks through the use of more favorable locations. Avoid stacking roads above one another in steep terrain by using longer span cable harvest techniques.*

◆ *Avoid locating roads where they will need fills on slopes greater than 60 percent. When necessary to construct roads across slopes that exceed the angle of repose, use full-bench construction and/or engineered bin walls or other stabilizing techniques.*

◆ *Plan to use full-bench construction and remove fill material to a suitable location where constructing road prisms on side slopes greater than 60 percent.*

◆ *Design cut-and-fill slopes to be at stable angles, or less than the normal angle of repose, to minimize erosion and slope failure potential.*

The degree of steepness that can be obtained is determined by the stability of the soil. Figure 3-3 presents recommended stable backslope and fill slope angles for different soil materials.

• Use retaining walls, with properly designed drainage, to reduce and contain excavation and embankment quantities. Vertical banks can be used without retaining walls if the soil is stable and water control structures are adequate.

• Balance excavation and embankments to minimize the need for supplemental building material and to maximize road stability.

• Avoid the use of road fills at drainage crossings as water impoundments unless they have been designed as an earthfill dam (in which case they might be subject to section 404 requirements). These earthfill embankments need outlet controls to allow draining prior to runoff periods and a design that permits flood flows to pass.

◆ *Try to avoid springs wherever possible. However, where they must be crossed, provide drainage structures for springs that flow to roads and that flow continuously for longer than 1 month, rather than allowing road ditches to carry the flow to a drainage culvert.*

Avoiding springs will limit disruptions to the natural hydrology of an area and limit the extent to which roads can become integrated into an area's drainage system. Unmanaged springs can compromise sections of roads and contribute to erosion and sedimentation.

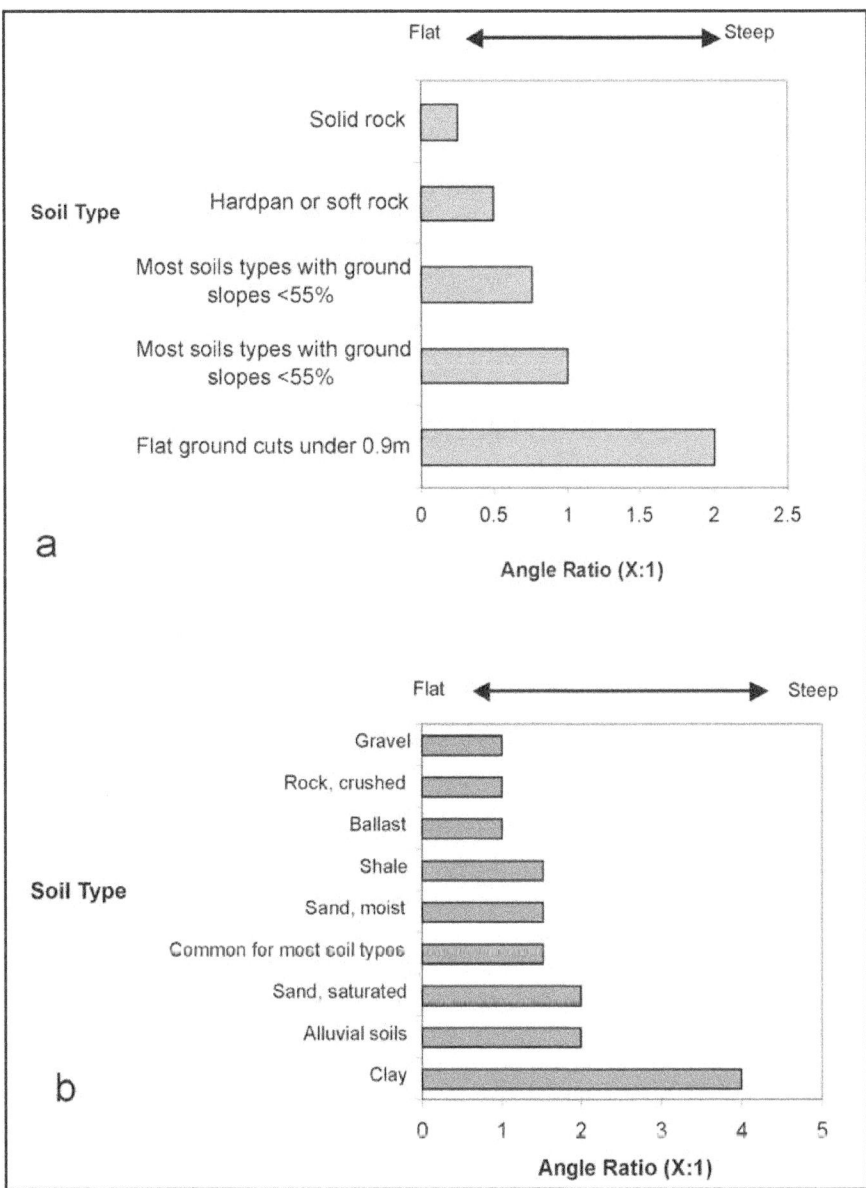

Figure 3-3.   Maximum recommended stable angles for (a) backslopes and (b) fill slopes (after Rothwell, 1978).

◆ *Design roads crossing low-lying areas so that water does not pond on the upslope side of the road.*

   • Use overlay construction techniques with suitable nonhazardous materials for roads crossing muskegs.

   • Provide cross drains at short intervals to ensure free drainage and avoid ponding, especially in sloping areas.

   • Provide adequate cross drainage to maintain natural dispersed hydrologic flows through wet areas.

◆ *Plan water source developments, used for wetting and compacting roadbeds and surfaces, to prevent channel bank and stream bed effects.*

◆ *Design access roads such that they do not provide sediment to the water source.*

Road Surfacing Practices

◆ *Select a road surface material suitable for the intended road use and likelihood of water quality effects.*

The volume and composition of traffic, the desired service life, and the stability and strength of the road foundation (subgrade) material will determine the type of road surfacing needed. Roads that are closer to streams or other surface waters should be considered for a durable, non-erosive surface.

◆ *Where grades increase the potential for surface erosion, design roads with a surface of gravel, grass, wood chips, or crushed rocks.*

◆ *Where a road is to be surfaced, select an appropriately sized aggregate, appropriate percentage of fines, and suitable particle hardness to protect road surfaces from rutting and erosion under heavy truck traffic during wet periods.*

When a road is to be used for only a short time period, consider not surfacing it, and closing it and returning the surface to natural vegetation after use.

Road Stream Crossing Practices

◆ *Lay out roads, skid trails, and harvest units to minimize the number of stream crossings.*

◆ *Design and site stream crossings to cross drainages perpendicular to the streamflow. Design road segments with water turn-outs and broad-based dips to minimize runoff directly entering the stream at the crossing.*

◆ *Locate stream crossings to avoid channel changes and minimize the amount of excavation or fill needed at the crossing. Apply the following criteria to determine the locations of stream crossings:*

   • Construct crossings at locations where the streambed has a straight and uniform profile above, at, and below the crossing.

   • Locate the crossing so the stream and road alignment are straight in all four directions.

   • Cross where the stream is relatively narrow with low banks and firm, rocky soil.

   • Avoid deeply cut streambanks and soft, muddy soil.

◆ *Choose stream-crossing structures (bridges, culverts, or fords) with the structural capacity to safely handle expected vehicle loads with the least disturbance to the watercourse.*

◆ *Design culverts and bridges for minimal effect on water quality. Install culverts of a size that is appropriate to pass a design storm. Opening size varies depending on climate, the drainage area upstream of where the stream-crossing structure is to be placed, and the likelihood of plugging with debris.*

Consider the following guidelines for culvert sizing, but consult the state forestry agency and local hydrologists: a 50-year design storm for small diameter culverts and a 100-year design storm for large diameter culverts and bridges. Bridges or arch culverts, which retain the natural stream bottom and slope, are preferred over pipe culverts for streams used for fish migrating or spawning areas (Figure 3-4). The FishXing Web site (http://www.stream.fs.fed.us/fishxing/index.html) provides software and learning systems for fish passage through culverts.

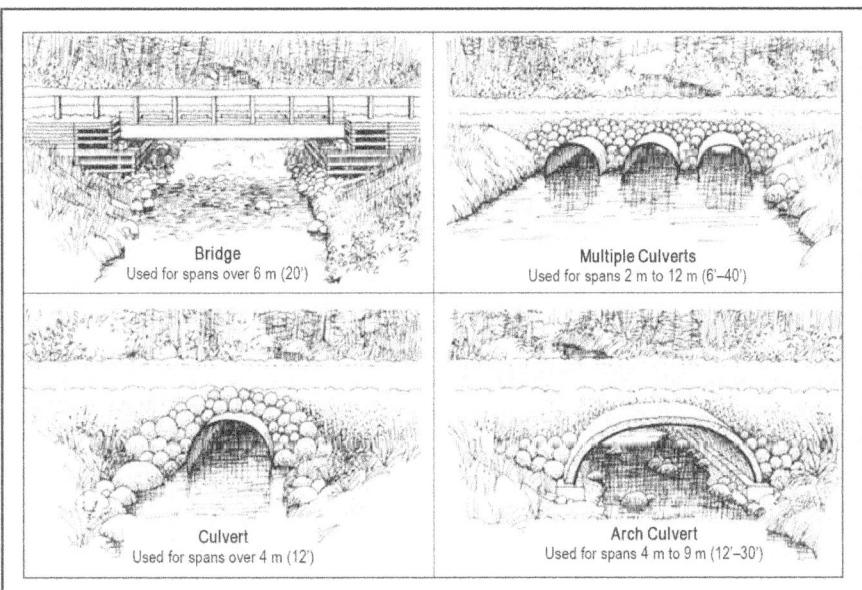

Figure 3-4.   Alternative water crossing structures (Ontario Ministry of Natural Resources, 1988).

◆ *The use of fords is best limited to areas where the stream bed has a firm rock or gravel bottom (or where the bottom has been armored with stable material), where the approaches are both low and stable enough to support traffic, where fish are not present during low flow, and where the water depth is no more than 3 feet.*

◆ *Design small stream crossings on temporary roads using temporary bridges.*

Temporary bridges usually consist of logs bound together and suspended above the stream, with no part in contact with the stream itself. This prevents stream bank erosion, disturbance of stream bottoms, and excessive turbidity. Provide additional capacity to accommodate debris loading that might lodge in the structure opening and reduce its capacity.

Scheduling Practices

◆ *Plan road construction or improvement to allow sufficient time afterward for disturbed soil and fill material to stabilize prior to use of the road.*

Compact and stabilize roads prior to use. This reduces the amount of maintenance needed during and after harvesting activities.

◆ *To minimize soil disturbance and road damage, plan to suspend operations when soils are highly saturated. This will reduce sediment runoff potential and creation of ruts in the haul road, landings, skid trails, and loading areas, which in turn will prevent possible damage to vehicles. Damage to forested slopes can also be minimized by not operating logging equipment when soils are wet, during wet weather, or when the ground is thawing.*

Preharvest Notification Practices

◆ *Encourage timberland owners and harvesters to submit a preharvest plan to the state for review prior to performing any road work or harvesting.*

States are encouraged to adopt notification mechanisms for harvest planning that integrate and avoid duplicating existing requirements or recommendations for notification, including severance taxes, stream crossing permits, erosion control permits, labor permits, forest practice acts, plans, and so forth. For example, states might recommend that a preharvest plan be submitted by the landowner to a single state or local office. The appropriate state agency might encourage forest landowners to develop a preharvest plan. The plan would address the components of this management measure, including the area to be harvested, any forest roads to be constructed, and the timing of the activity.

Many states currently use some process to ensure implementation of management practices. These processes are typically related to the planning phase of forestry operations and commonly involve some type of notification process. Some states have one or more processes in place that serve as notification mechanisms used to ensure implementation. These state processes are usually associated with forest practices acts, erosion control acts, state dredge and fill or CWA section 404 requirements, timber tax requirements, or state and federal incentive and cost share programs. Some state education and training programs are discussed in Section 2.

It is suggested that notification be encouraged prior to:

- Timber harvesting or commercial timber cutting.
- Road construction or road improvement.
- Stream crossing construction or any work within 50 feet of a watercourse or water body.
- Reforestation.
- Pesticide, herbicide, or fertilizer applications.
- Any work in a wetland.
- Conversion of forestland to a non-forest use.

3B: STREAMSIDE MANAGEMENT AREAS

Management Measure for Streamside Management Areas

Establish and maintain a streamside management area along surface waters, which is sufficiently wide and which includes a sufficient number of canopy species to buffer against detrimental changes in the temperature regime of the water body, to provide bank stability, and to withstand wind damage. Manage the SMA in such a way as to protect against soil disturbance in the SMA and delivery to the stream of sediments and nutrients generated by forestry activities, including harvesting. Manage the SMA canopy species to provide a sustainable source of large woody debris needed for in-stream channel structure and aquatic species habitat.

Management Measure Description

Streamside management areas (SMAs), also commonly referred to as streamside management zones or riparian management areas or zones, are areas of riparian vegetation along streams that receive special management attention because of their value in protecting water quality and habitat. Riparian vegetation is highly beneficial to water quality and aquatic habitat. Riparian areas reduce runoff and trap sediment from upslope areas and may reduce nutrients in runoff (Belt et al., 1992). Canopy species shade surface waters, moderating water temperature and providing detritus that serves as an energy source for streams. Trees in riparian areas are a source of large woody debris (LWD) to surface waters. Riparian areas provide important habitat for aquatic organisms and terrestrial species.

The width of SMAs is determined in one of two ways: (1) a fixed minimum width is recommended or prescribed, or (2) a variable width is determined based on site conditions such as slope (Phillips et al., 2000) (Figure 3-5). SMAs need to be of sufficient width to protect the adjacent water body. A minimum width of 35 to 50 feet is generally recommended for SMAs to be effective. Areas such as intermittent channels, ephemeral channels, and depressions need to be given special consideration when determining SMA boundaries. Channels should be disturbed as little as possible to maximize the effectiveness of an SMA, as disturbance in and adjacent to a SMA can contribute considerably to pollutant runoff volumes. SMAs also need to be able to withstand wind damage or blowdown. For example, a single rank of canopy trees is not likely to withstand blowdown and maintain the functions of an SMA.

Table 3-2 presents North Carolina's recommendations for SMA widths for various types of water bodies dependent on adjacent upland slope. Maine's recommended filter strip widths are dependent on the land slope between the road and the water body (Table 3-3). SMA widths might vary along a stream's course and on opposite sides of the same stream. SMA width is measured along the ground from the streambank on each side of the stream and not from the centerline of the watercourse (Georgia Forestry Commission, 1999).

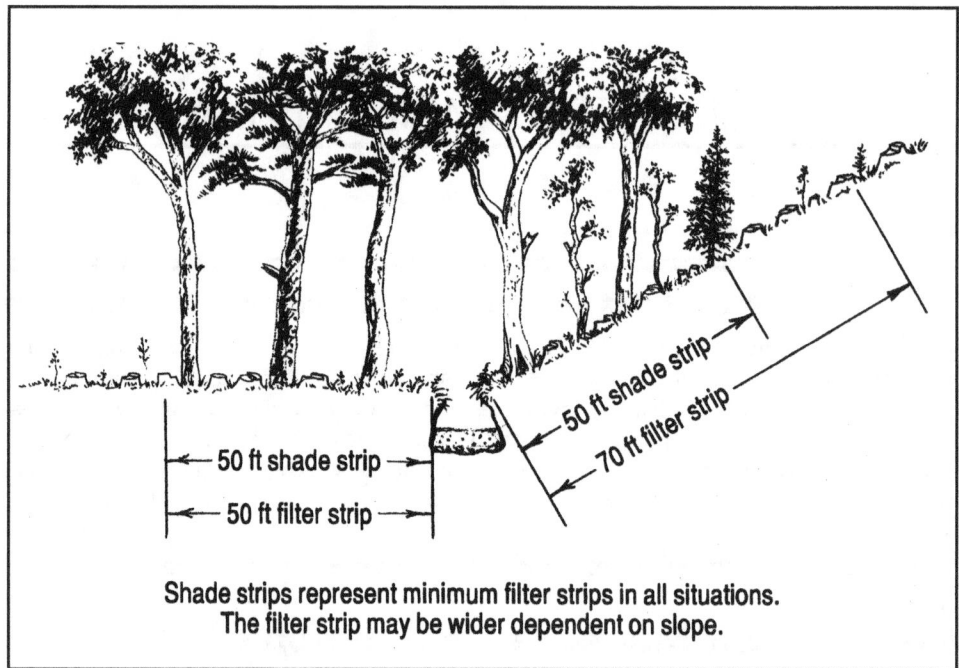

Shade strips represent minimum filter strips in all situations.
The filter strip may be wider dependent on slope.

Figure 3-5. Calculation of slope—an important step in determining SMA width (Georgia Forestry Commission, 1999).

Table 3-2. Recommended Minimum SMZ Widths (North Carolina Division of Forest Resources, 1989)

Type of Stream or Water Body	Percent Slope of Adjacent Lands				
	0–5	6–10	11–20	21–45	46+
	SMZ Width Each Side (feet)				
Intermittent	50	50	50	50	50
Perennial	50	50	50	50	50
Perennial trout waters	50	66	75	100	125
Public water supplies (Streams and reservoirs)	50	100	150	150	200

Table 3-3. Recommendations for Filter Strip Widths (Maine Forest Service, 1991)

Slope of Land (%)	Width of Strip (ft along ground)
0	25
10	45
20	65
30	85
40	105
50	125
60	145
70	165

A sufficient number of large trees in an SMA provide for bank stability and a sustainable source of large woody debris. LWD consists of naturally occurring dead and downed woody materials, not to be confused with logging slash or debris. Trees to be maintained or managed in the SMA can provide large woody debris to the stream at a rate that maintains beneficial uses associated with fish habitat and stream structure. Woody debris is added at the site and downstream at a rate that is sustainable over a long time period.

A sufficient number of canopy species are maintained in an SMA also to provide shading to the stream water surface to prevent changes in the temperature regime of the water body and to prevent harmful temperature- or sunlight-related effects on the aquatic biota. If the existing shading conditions for the water body prior to activity are known to be less than optimal for the stream, SMAs can be managed to increase shading of the water body.

Lakeside management areas, or LMAs—the lake and pond equivalent of SMAs—should also be left around lakes and ponds on harvest sites (Minnesota Forest Resources Council, 1999; Wisconsin Department of Natural Resources, 2003). The width of LMAs varies depending on site conditions, as do the recommended widths of SMAs. Topography, hydrology, size of water body, size of adjacent harvest area, harvest method, forest management objectives (e.g., timber production, wildlife), whether the water body contains sensitive fish species, and tree species composition all influence the size and leave-tree recommendations for LMAs.

Generally, LMAs should be as wide as SMAs, or generally between 50 and 100 feet wide, though where sensitive fish species are present in the water body, a wider LMA— up to 200 feet—may be necessary to fully protect water quality.

Other considerations for timber harvesting near lakes and ponds include ensuring that some trees are left on all areas surrounding water bodies all the way to the top of the adjacent slope, and using an extended rotation period within LMAs (as should be done for SMAs) to minimize soil and riparian area disturbance.

To preserve SMA integrity for water quality protection, some states limit the type of harvesting, timing of operations, amount harvested, or reforestation methods used in them. SMAs are managed to use only harvest and forestry methods that prevent soil disturbance in the SMA. Additional operational considerations for SMAs are addressed in subsequent management measures. Practices for SMA applications to wetlands are described in the *Wetlands Forest Management Measure* (Chapter 3, section J).

Benefits of Streamside Management Areas

The effectiveness of SMAs in regulating water temperature depends on the interrelationship between vegetative and stream characteristics. Specifying leave tree and stream shade quantities is an effective way to prevent detrimental temperature changes. An example of a leave tree specification might be Leave trees that provide midsummer and midday shade to the water surface, and preferably a quantity of trees that provide a minimum of 50 percent of the summer midday shade. Shade cover is preferably left distributed evenly within the SMA. If a threat of blowdown exists, leave trees may be clumped and clustered as long as sufficient shade at the reach scale is provided.

Lynch and others (1985) studied the effectiveness of SMAs in controlling suspended sediment and turbidity levels (Table 3-4). A combination of practices were applied,

Table 3-4.   Storm Water Suspended Sediment Delivery for Treatments (Pennsylvania) (Lynch et al., 1985)

Water Year and Treatment	Annual Average Suspended Sediment in mg/L (Range)
1977	
Forested control	1.7 (0.2–8.6)
Clear-cut-herbicide	10.4 (2.3–30.5)
Commercial clear-cut with BMPs[a]	5.9 (0.3–20.9)
1978	
Forested control	5.1 (0.3–33.5)
Clear-cut-herbicide	— [b](1.8–38.0)
Commercial clear-cut with BMPs[a]	9.3 (0.2–76.0)

[a]Buffer strips, skidding in streams prohibited, slash disposal away from streams, skid trail and road layout away from streams.
[b]Data not available

including SMAs and prohibitions on skidding, slash disposal, and roads located in or near streams. Average storm water-suspended sediment and turbidity levels in the area without these practices were very high compared to those of the control and SMA/BMP sites. Table 3-5 presents data on how effective different cutting practices and buffer strips are in preventing debris from entering the stream channel (Froehlich, 1973).

Hall and others (1987) studied the effectiveness of SMAs in protecting streams from temperature increases, large increases in sediment load, and reduced dissolved oxygen (Table 3-6). The value of SMAs for protecting streams from water temperature changes is clear from the 30 °F maximum daily increase in stream temperature observed during the study. The study also showed that not leaving a SMA can cause sediment increases streams, and more recent research has demonstrated that SMAs might be effective in

Table 3-5.   Average Changes in Total Coarse and Fine Debris of a Stream Channel After Harvesting (Oregon) (Froehlich, 1973)

Cutting Practice	Natural Debris	Material Added in Felling	% Increase
	(tons per hundred feet of channel)		
Conventional tree-felling	8.1	47	570
Cable-assisted directional felling	16	14	112
Conventional tree-felling with buffer strip[a]	12	1.3	14

[a]Buffer strips ranged from 20 to 130 feet wide for different channel segments.

Table 3-6.   Comparison of Effects of Two Methods of Harvesting on Water Quality (Oregon) (Hall et al., 1987)

Watershed	Method	Streamflow	Water Temperature	Sediment	Dissolved Oxygen
Deer Creek	Patch cut with buffer strips (750 acres)	No increase in peak flow	No change	Increases for one year due to periodic road failure	No change
Needle Branch	Clearcut with no stream protection (175 acres)	Small increases	Large changes, daily maximum increase by 30 °F, returning to pre-log temp. within 7 years	Five-fold increase during first winter, returning to near normal the fourth year after harvest	Reduced by logging slash to near zero in some reaches; returned to normal when slash removed

intercepting overland flow and some sediment it contains, but not in intercepting sediment contained in channelized flow (Belt et al., 1992; Keim and Schoenholtz, 1999). Keim and Schoenholtz (1999), in a study on highly erodible soils in Mississippi, found that the primary means by which SMAs reduce sediment delivery to streams is by preventing soil disturbance next to the stream and not by intercepting sediment from upland sources. Finally, the study demonstrated the effect that logging slash placed in streams has in depleting dissolved oxygen as it decomposes.

Hartman and others (1987) compared the physical changes associated with logging using three streamside treatments—leaving a variable-width strip of vegetation along a stream (least intensive); clear cutting to the margin of a stream, but with virtually no instream disturbance (intensive); and clear-cutting to the stream bank with some yarding near the stream and pulling merchantable timber from the stream (most intensive). They performed their study to observe the effect of different SMAs on the supply of woody debris. The volume and stability of large woody debris decreased immediately in the most intensive treatment area, decreased a few years after logging in the careful logging area, and remained stable where streamside trees and other vegetation remained.

The costs associated with SMAs vary according to site conditions. SMAs can be more difficult to lay out on rough terrain or along a stream or river that meanders a lot due to the need to adjust the SMA width appropriately. Also, harvesters or landowners take into account the value of merchantable timber left unharvested because of SMA restrictions. No single SMA width or layout is preferable for all sites in terms of cost. Dykstra and Froelich (1976a) concluded in one study that a 55-foot buffer strip was the least costly on a million-board-foot (mfb) basis, but they cautioned that cost is not the only factor to consider when deciding what type of stream protection to use (Table 3-7).

There are several research papers that focus on the costs of SMA implementation. Lickwar (1989) examined the costs of SMAs as determined by varying slope steepness (Table 3-8) in different regions in the Southeast and compared them to road construction and revegetation practice costs. He found that SMAs are the least expensive practice, in general, and that their cost is approximately the same regardless of slope. The costs associated with use of alternative buffer and filter strips were also analyzed in an Oregon study (Olsen, 1987) (Table 3-9). In that study, increasing the SMA width from 35 feet on each side of a stream to 50 feet reduced the value per acre by $75 (discounted cost) to $103 (undiscounted cost), or an approximate 2 percent increase in harvesting cost per acre (from $3,163 discounted to $5,163 undiscounted). Doubling the SMA width from

Table 3-7.   Average Estimated Logging and Stream Protection Costs per MBF (Oregon) (Dykstra and Froehlich, 1976a)

Cutting Practice	Total Cost		Volume Foregone
	Average	Range	
Conventional felling	$70.98	$62.74–85.74	None
Cable-assisted directional felling (1.43% breakage saved within 200-foot stream)	$74.62	$61.19–89.49	—
Cable-assisted felling (10% breakage saved)	$70.59	$56.00–85.42	—
Buffer strip (55 feet wide)	$66.86	$56.84–79.55	0 - 6 percent
Buffer strip (150 feet wide)	$77.78	$69.70–86.74	6 - 17 percent

Note: All costs updated to 1998 dollars.
[a]Cost estimates for each of 10 areas studied by Dykstra and Froehlich were averaged for this table.

Table 3-8. Cost Estimates (and Cost as a Percent of Gross Revenues) for Streamside Management Areas (Lickwar, 1989)

Practice Component	Steep Sites[a]	Moderate Sites[b]	Flat Sites[c]
Streamside Management Zones	$2,958 (0.52%)	$3,441 (0.51%)	$3,363 (0.26%)

Note: All costs updated to 1998 dollars.
[a] Based on a 1,148-acre forest and gross harvest revenues of $573,485. Slopes average over 9 percent.
[b] Based on a 1,104-acre forest and gross harvest revenues of $678,947. Slopes ranged from 4 percent to 8 percent.
[c] Based on a 1,832-acre forest and gross harvest revenues of $1,290,641. Slopes ranged from 0 percent to 3 percent.

Table 3-9. Cost Effects of Three Alternative Buffer Strips (Oregon): Case Study Results with 640-acre Base (36 mbf/acre) (Olsen, 1987)

	Scenario		
	I	II	III
Average buffer width (feet on each side)	35	50	70
Percent conifers removed	100	60	25
Percent reclassified Class II streams[a]	0	20	80
Harvesting restrictions	Current	New	New
Road Construction			
New miles	2.09	2.14	3.06
Road and landing acres	10.9	11.1	15.9
Cost total (1000's)	$96.00	$102.00	$197.00
Cost/acre	$149.00	$160.00	$307.00
Harvesting Activities[b]			
mmbf harvested	22.681	22.265	20.277
Acres harvested	638.3	635.5	633.1
Cost total (1000's)	$3,104.00	$3,101.00	$2,842.00
Cost/acre	$4,841.00	$4,835.00	$4,432.00
Cost/mbf	$136.87	$139.26	$140.17
Inaccessible Area and Volume			
Percent area in buffers	1.3	3.9	14.0
mmbf left in buffers	0.000	0.313	2.214
Acres unloggable	1.44	4.32	6.72
mmbf lost to roads and landings	0.202	0.205	0.295
Undiscounted Costs (1000's)			
Road cost	$96.00	$102.00	$197.00
Harvesting cost	$3,104.00	$3,101.00	$2,842.00
Value of volume foregone[c]	$38.00	$101.00	$413.00
Total	$3,238.00	$3,304.00	$3,451.00
Cost/acre	$5,060.00	$5,163.00	$5,393.00
Reduced dollar value/acre	—	$103.00	$323.00
Discounted Costs			
Cost with 4% discount rate (1000's)	$2,023.00	$2,071.00	$2,195.00
Cost/acre	$3,162.00	$3,237.00	$3,431.00
Reduced value/acre	—	$75.00	$269.00

Note: mmbf = millon board feet; mbf = thousand board feet.
1986 dollars.
[a] Generally, only Class I streams are buffered.
[b] Includes felling, landing construction and setup, yarding, loading, and hauling.
[c] Volume foregone x net revenue ($150/mbf).

35 to 70 feet on each side of a stream reduced the dollar value per acre by approximately 3 times, adding approximately 8 percent to the discounted harvesting costs.

According to the Vermont Agency of Natural Resources, adequately sized SMAs are the best means to protect water quality (VANR, 1998). The agency conducted habitat assessments and bioassessments on stream segments above and below harvest sites and before and after harvesting and determined that SMAs are particularly important for protecting small headwater streams and ephemeral stream channels. The Virginia Department of Forestry also monitored BMP implementation and effectiveness and determined that although improvement was needed in meeting minimum standards of implementation, properly implemented SMAs (together with stream crossings and preharvest plans) are crucial to protecting water quality.

The Oregon Department of Forestry similarly found that application of a riparian rule (passed in 1987) results in stream protection that generally maintains pre-operation vegetative conditions.

Where SMAs were found to be ineffective or less effective than possible, the Virginia Department of Forestry discovered that in some cases this was the result of careless timber harvesting in the SMAs, a lack of adequately sized SMAs on adjacent intermittent streams, or gaps in SMAs caused by cutting in them.

Of course, BMPs are effective only when properly designed and constructed. In general, poor BMP effectiveness can be attributed to one or more of the following:

- A lack of time or willingness to plan timber harvests carefully before cutting begins.

- A lack of skill in or knowledge of designing effective BMPs.

- A lack of equipment needed to implement BMPs effectively.

- The belief that BMPs are not an integral part of the timber harvesting process and can be engineered and fitted to a logging site after timber harvesting has been completed.

- A lack of timely implementation and maintenance of BMPs.

Best Management Practices

◆ *Minimize disturbances that would expose the mineral soil of the SMA forest floor. Do not operate skidders or other heavy machinery in the SMA.*

◆ *Locate all landings, portable sawmills, and roads outside the SMA.*

◆ *Restrict mechanical site preparation in the SMA, and encourage natural revegetation, seeding, and hand planting.*

◆ *Limit pesticide and fertilizer usage in the SMA. Establish buffers for pesticide application for all flowing streams.*

◆ *Directionally fell trees away from streams to prevent excessive quantities of logging slash and organic debris from entering the water body. Remove slash and debris unless consultation with a fisheries biologist indicates that it should be left in the stream for large woody debris.*

There is no "correct" amount of organic debris that streams should have. Streams have natural amounts of organic debris (e.g., fallen leaves, twigs, limbs, and trees), but the amount varies with season, tree falls, storms, and so forth. Aquatic organisms are adapted to the annual (and longer) range of the quantities of organic debris in the stream. As discussed in Chapter 2, large woody debris, or LWD, alters sediment and water routing and, thereby, affects channel morphology, provides structure and complexity to aquatic and terrestrial organism habitats, and is a source of nutrients for aquatic organisms. Periodic variations in the influx of sediment and LWD also contribute to habitat heterogeneity that is reflected in diverse aquatic communities. When areas upslope from a stream are changed enough that the quantity of organic debris that reaches a stream is significantly changed (i.e., so much that it is too little or too much for the stream's dynamics and the aquatic organisms), it can be detrimental to the aquatic system and be considered a water quality problem. Removing trees from near the stream edge, harvesting older trees on upslope areas, and burning that removes forest floor litter could all reduce inputs of organic debris to the aquatic system and adversely affect stream ecology.

Retaining SMAs along streams is one step to take to ensure that the streams are provided with sufficient inputs of organic debris. Leaving slash and other logging debris in a stream could exceed the natural high limit of organic debris inputs for the stream's ecology and adversely affecting the stream. Removing felled material from streams on a site where changes have occurred that will reduce inputs of organic debris in the future could leave the stream with less organic debris than the stream ecology is adapted to. Maintaining stream water quality—which includes habitat diversity for aquatic life support—does not necessarily imply reducing inputs of woody debris to a stream, therefore, but rather means not altering the aquatic system to a degree in either direction (too much or too little) that stream ecology is adversely affected. A fisheries biologist will be able to help with decisions on what sizes and quantities of woody debris, if any, should be left in a stream to mimic natural conditions. Table 3-10 compares the goals of two types of LWD projects. Further information on the role and importance of LWD in streams and on placing LWD in streams can be obtained from the U.S. Army Corps of Engineers' Ecosystem Management and Restoration Research Program (EMRRP). A paper issued under the program, *Streambank habitat enhancement with large woody debris* (Fischenich and Morrow, 2000), can be found on the Web at http://el.erdc.usace.army.mil/elpubs/pdf/sr13.pdf.

♦ *Apply harvesting restrictions in the SMA to maintain its integrity.*

Vegetation, including trees, should be left in the SMA to achieve the desired objective for the area, such as maintain shading and bank stability and to provide adequate woody debris to create habitat diversity and provide nutrients to surface waters. This provision for leaving residual trees might be specified in various ways. For example, the Maine Forestry Service specifies that no more than 40 percent of the total volume of timber 6 inches diameter breast height (DBH) and greater be removed in a 10-year period, and that the trees removed be reasonably distributed within the SMA. Florida recommends leaving a volume equal to or exceeding one-half the volume of a fully stocked stand. The number of residual trees varies inversely with their average diameter. A shading specification that is independent of the volume of timber might be necessary for streams where temperature changes could alter aquatic habitat.

Table 3-10. Goals of Two Main Types of LWD Projects (Fischenich and Morrow, 2000)

	Category 1	Category 2
LWD Project Goals	Improve habitat by increasing LWD quantities in a stream	Alter flows to improve aquatic habitat

3C: ROAD CONSTRUCTION/RECONSTRUCTION

Management Measure for Road Construction/Reconstruction

(1) Follow preharvest planning (as described under the Management Measure for Preharvest Planning) when constructing or reconstructing the roadway.

(2) Follow designs planned under the Management Measure for Preharvest Planning for road surfacing and shaping.

(3) Install road drainage structures according to designs planned under the Management Measure for Preharvest Planning and regional storm return period and installation specifications. Match these drainage structures with terrain features and with road surface and prism designs.

(4) Guard against the production of sediment when installing stream crossings.

(5) Protect surface waters from slash and debris material from roadway clearing.

(6) Use straw bales, silt fences, mulching, or other favorable practices on disturbed soils on unstable cuts, fills, etc.

(7) Avoid constructing new roads in streamside management areas to the extent practicable.

Management Measure Description

Road construction is one of the largest potential sources of forest activity-produced sediment (Megahan, 1980), and road and drainage crossing construction practices that minimize sediment delivery to surface waters are essential for protecting water quality. Water quality degradation resulting from forest roads is mostly attributable to sediment loss during road construction, erosion that occurs within a few years after road construction, soil loss from heavy road use, and road failure during storm events that exceed the road's design capacity. An early study of erosion from road construction concluded that the amount of sediment produced by road construction is directly related to the percent of area occupied by roads, whether a road is given a protective surface, and the amount of protection provided to loose soils on back slopes and fill slopes (King, 1984) (Table 3-11). Best management practices related to these aspects of road construction, and for stream crossing construction, are the subject of this management measure. Erosion and water quality degradation are also problems associated with older, unmaintained roads, and BMPs for road maintenance are the subject of the next management measure.

General Road Construction Considerations

Road design and construction that are tailored to the topography and soils and that take into consideration the overall drainage pattern in the watershed where the road is being constructed can prevent road-related water quality problems. Lack of adequate consideration of watershed and site characteristics, road system design, and construction techniques appropriate to site circumstances can result in mass soil movements, extensive surface erosion, and severe sedimentation in nearby water bodies. The effect that a forest

Table 3-11. Effects of Several Road Construction Treatments on Sediment Yield in Idaho (King, 1984)

Watershed Area (acres)	Area in Roads (percent)	Treatment	Increase of Annual Sediment Yield[a] (percent)
207	3.9	Unsurfaced roads; Untreated cut slope; Untreated fill slope	156
161	2.6	Unsurfaced roads; Untreated cut slope dry seeded	130
364	3.7	Surfaced roads; Cut and fill slopes straw mulched and seeded	93
154	1.8	Surfaced roads; Filter windrowed; Cut and fill slopes straw mulched and seeded	53
70	3.0	Surfaced roads; Filter windrowed; Cut and fill slopes hydro-mulched and seeded	25
213	4.3	Surfaced roads; Filter windrowed; Cut and fill slopes hydro-mulched and seeded	19

[a] Measured in debris basins.

road network has on stream networks largely depends on the extent to which the road and stream networks are interconnected. Road networks can be hydrologically connected to stream networks where road surface runoff is delivered directly to stream channels at stream crossings or via ditches or gullies that direct flow off of the road and then to a stream, and where road cuts transform subsurface flow into surface flow in road ditches or on road surfaces that delivers sediment and water to streams much more quickly than without a road present and increases the risk of mass wasting (Jones and Grant, 1996; Montgomery, 1994; Wemple et al., 1996). The combined effects of these drainage network connections are increased sedimentation and peak flows that are higher and arrive more quickly after storms. This in turn can lead to increased instream erosion and stream channel changes. This effect is strongest in small watersheds (Jones et al., In press).

Site characteristics are first considered during preharvest planning, and it is important to review the harvesting plan at the harvest site before construction begins to verify assumptions made during planning. On-site verification of information from topographic maps, soil maps, and aerial photos is necessary to ensure that locations where roads are to be cut into slopes or built on steep slopes or where skid trails, landings, and equipment maintenance areas are to be located are appropriate to the use. If an on-site visit indicates that changes to road, skid trail, or landing locations can reduce the risk of erosion, the project manager can make these changes prior to construction, and in some cases as the project progresses.

Road drainage features tailored to the site and its conditions prevent water from pooling or collecting on road surfaces and thereby prevent saturation of the road surface, which can lead to rutting, road slumping, and channel washout. It is especially important to ensure that road drainage structures are well constructed and designed for use during logging operations because the heavy vehicle use during harvesting creates a high potential for the contribution of large quantities of sediment to runoff.

Some roads are temporary or seasonal use roads, and their construction should not generally involve the high level of disturbance generated by the construction of permanent, high-standard roads. However, temporary or low-standard roads still need to be constructed and maintained to prevent erosion and sedimentation, and many of the BMPs discussed for this management measure are applicable to temporary road construction.

In a study in three headwater watersheds in the mountains of central Idaho, 70 percent of sediment deposition from roads constructed on the watersheds, where the slope ranged from 15 to 40 percent, occurred during the first year after construction, and one-fourth of this deposition occurred during road construction (Ketcheson and Megahan, 1996). In this study, sediment usually traveled less than 100 meters (m) from its source. The distance that sediment traveled varied depending on its source: the distance traveled from fills, rock drains, berm drains, and landings was between 4 m and 20 m, while that from cross drains was 50 m. The maximum travel distance from some cross drains was more than 250 m. Cross drains have a larger source area from which runoff is collected, including the road prism and upslope watershed area, and this accounted for more sediment being deposited than from all other sources combined. These findings highlight the importance of road placement, design, and construction in relation to watercourse location and the installation of BMPs to control runoff sedimentation from roads.

Based on the findings of studies such as this, it is clear that erosion control practices need to be applied while a road is being constructed, when soils are most susceptible to erosion, to minimize soil loss to water bodies. Since sedimentation from roads often does not occur incrementally and continuously, but in pulses during large rainstorms, it is important that road, drainage structure, and stream crossing design take into consideration a sufficiently large design storm that has a good chance of occurring during the life of the project. Such a storm might be the 10-year, 25-year, 50-year, or even 100-year, 12- to 24-hour return period storm. Sedimentation cannot be completely prevented during or after road construction, but the process is certainly exacerbated if the road construction and design are inappropriate for the site conditions or if the road drainage or stream crossing structures are insufficient.

Several common practices minimize erosion during road construction. In general, it is recommended that forest roads be constructed as a single lane for minimum width and outsloped with minimal cut-and-fill, where conditions are suitable (Weaver and Hagans, 1984). These roads should cause the least disturbance and have lower maintenance costs. Figure 3-6 illustrates various erosion and sediment control practices. Aspects of road construction addressed by the BMPs discussed under this management measure are introduced below. Further information is provided in the discussions of the individual BMPs.

Road Surface Shape and Composition

The shape of a road is an important component of runoff control. Terminology related to road construction and road shape is illustrated in Figure 3-7. Road drainage and runoff control are obtained by shaping the road surface to be insloping, outsloping, or crowned (Figure 3-8). Road surfaces need to have and maintain one of these shapes at all points to ensure good drainage (Moll et al., 1997). Insloping roads can be particularly effective where soils are highly erodible and directing runoff directly to the fill slope would be detrimental. Outsloped roads tend to dissipate runoff more than insloped roads, which concentrate runoff at cross drain locations, and are useful where erosion of the backfill or

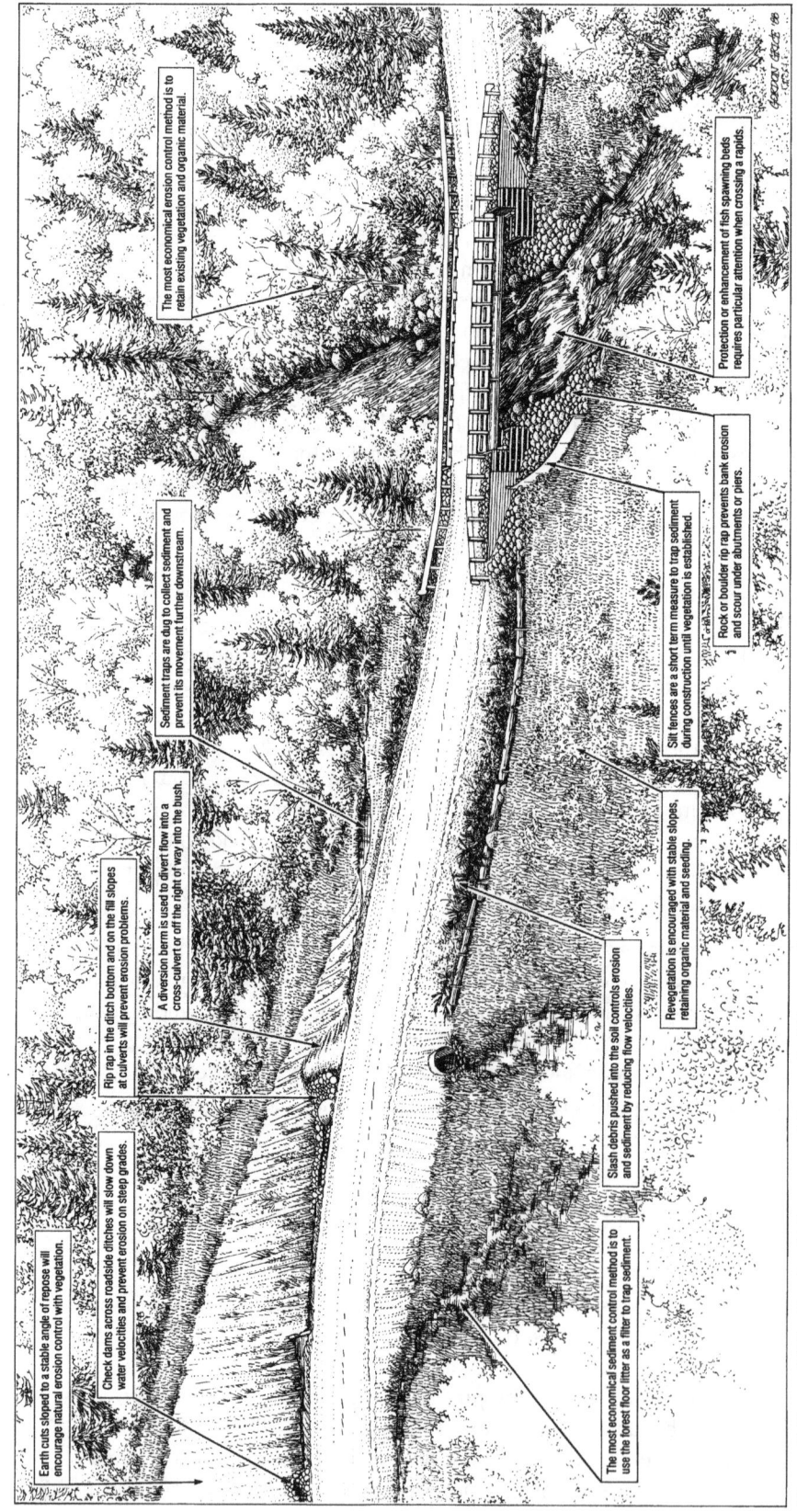

Figure 3-6. Mitigation techniques used for controlling erosion and sediment to protect water quality and fish habitat (Ontario MNR, 1988).

National Management Measures to Control Nonpoint Source Pollution from Forestry

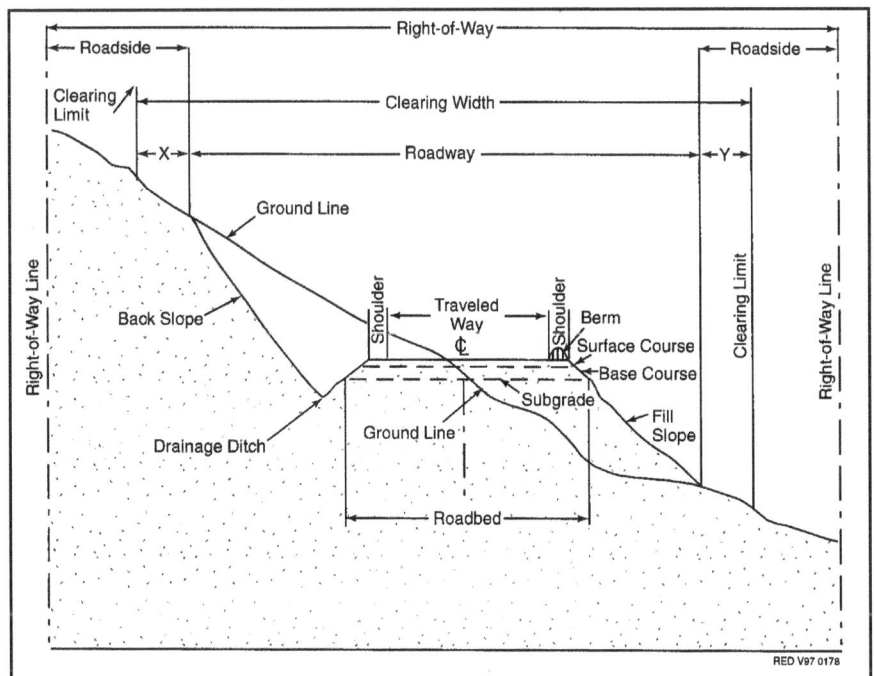

Figure 3-7.   Illustration of road structure terms (Moll et al., 1987).

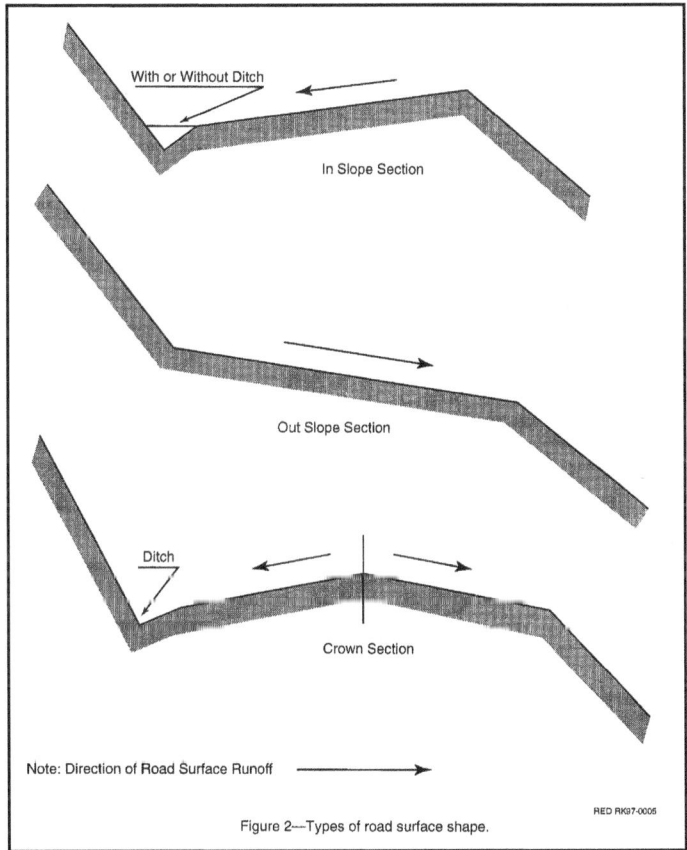

Figure 2—Types of road surface shape.

Figure 3-8.   Types of road surface shape (Moll et al., 1997).

Sediment Runoff Distance and Quantity Vary with Source

Seventy percent of sediment deposition from roads constructed on three headwater watersheds in the mountains of central Idaho, where the slope ranged from 15 to 40 percent, occurred during the first year after construction, and on-fourth of this occurred during road construction.

Sediment generally traveled less than 100 m from its source. Average sediment travel distances from fills, rock drains, berm drains, and landings were between 4 m and 20 m, while that from cross drains was 50 m. The maximum travel distance from some cross drains was more than 250 m.

The larger source area for runoff from cross drains, including he road prism and upslope watershed areas, accounts for more sediment deposited form them and for the sediment from them traveling farther than from other sources.

(Source: Ketcheson and Megahan, 1996)

ditch soil might be a problem. Crowned roads are particularly suited to two-lane roads and to steep single-lane roads that have frequent cross drains or ditches and ditch relief culverts (Moll et al., 1997). Crowns, inslopes, and outslopes will quickly lose effectiveness if not maintained frequently, due to micro-ruts created by traffic when the road surface is damp or wet.

The composition of a road surface can be chosen to effectively control erosion from the road surface and slopes. It is important to choose a road surface that is suitable to the topography, slope, aspect, soils, and intended use. Small, temporary, dry season roads can be left unsurfaced and decommissioned after use to minimize their impact to water quality. Roads that will be used more intensively or for long periods can have road surfaces formed from native material, aggregates, asphalt, or other suitable materials. Any of these surface compositions can be shaped in one of the ways discussed above. Surface protection of the roadbed and cut-and-fill slopes with a suitable material can

- Minimize soil losses during storms
- Reduce frost heave erosion production
- Restrain downslope movement of soil slumps
- Minimize erosion from softened roadbeds

Numerous studies have been conducted and have demonstrated the potential of a suitable road surface composition to control erosion and sedimentation from forest roads. Swift (1985) found that applying 20 centimeters (cm) of crushed rock to forest roads in the southern Appalachian mountains yielded sediment runoff of 0.06 ton/acre/inch of rainfall, a significant reduction from the 1.475 ton/acre/inch of rainfall yielded by a road surface covered by only 5 cm of crushed rock (Figure 3-9). In another study in the Appalachian mountains, Kochenderfer and Helvey (1984) demonstrated that using 1-inch crusher-run gravel or 3-inch clean gravel reduced erosion from road surfaces to less than one-half of that from 3-inch crusher-run gravel, and to only 12 percent of the erosion rate measured from an ungraveled road surface (Table 3-12). In a more recent study (Johnson and Bronsdon, 1995), a surface of bituminous oil or 15 to 20 cm of gravel reduced erosion rates by as much as 96 percent below that measured from unsurfaced roads (Figure 3-10). In the same study, logging slash left on roads was also found to provide a protective layer and reduced erosion by 75 to 87 percent compared to unsurfaced roads.

Properly shaping a road surface (i.e., insloped, outsloped, or crowned) might not suffice to control drainage adequately, and drainage structures in addition to the relief culverts on insloped and crowned roads might be necessary for drainage control (Moll et al., 1997). Structures such as broad-based dips, turnouts, and cross drains can be used under such conditions, and these BMPs are further discussed below. The proper choice of drainage structure, in combination with the chosen surface shape, and effective installation of the

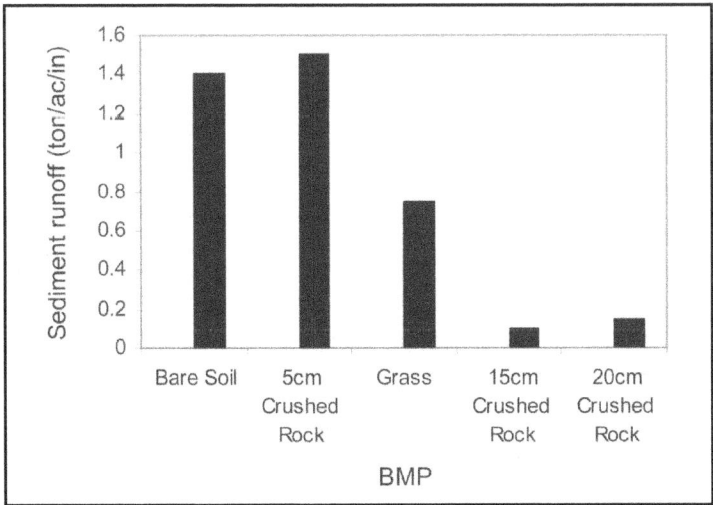

Figure 3-9. Comparison of sedimentation rates (as tons of sediment in runoff per acre per inch of rainfall) from different forest road surfaces (after Swift, 1984).

Table 3-12. Effectiveness of Road Surface Treatments in Controlling Soil Losses in West Virginia (adapted from Kcohcndcrfcr and Hclvoy, 1984)

Surface Treatment	Average Annual Soil Losses (tons/acre)[a]
Ungraveled	44.4
3-inch crusher-run gravel	11.4
1-inch crusher-run gravel	5.5
3-inch clean gravel	5.4

[a] Six measurements taken over a 2-year period.

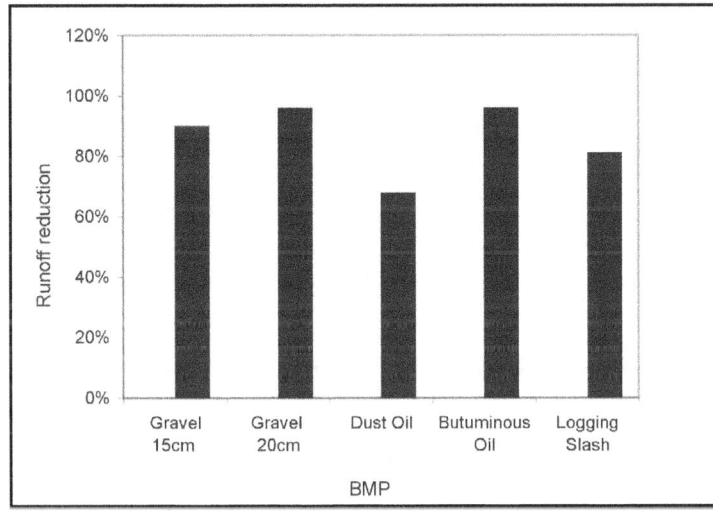

Figure 3-10. Percent of reduction in sediment runoff from a forest road surface with different treatments. Percent reduction in erosion is the amount below that observed on an untreated road (after Johnson and Bronsdon, 1995).

drainage structures is crucial to minimizing erosion from roads and sedimentation in water bodies. Improper or insufficient installation of road drainage structures is the cause of many road failures, whereas proper installation of the correct structure can reduce erosion potential, extend the useful life of a road, and decrease the need for road maintenance.

Slope Stabilization

Road cuts and fills can be a large source of sediment once a logging road is constructed. Stabilizing back slopes and fill slopes as they are constructed is an important process in minimizing erosion from these areas. Combined with graveling or otherwise surfacing the road, establishing grass or using another form of slope stabilization can significantly reduce soil loss from road construction. If constructing on an unstable slope is necessary, as it sometimes is, consider consulting with an engineering geologist or geotechnical engineer for recommended construction methods and to develop plans for the specific road segment. Unstable slopes that threaten water quality should always be considered unsuitable for road building (Weaver and Hagans, 1984).

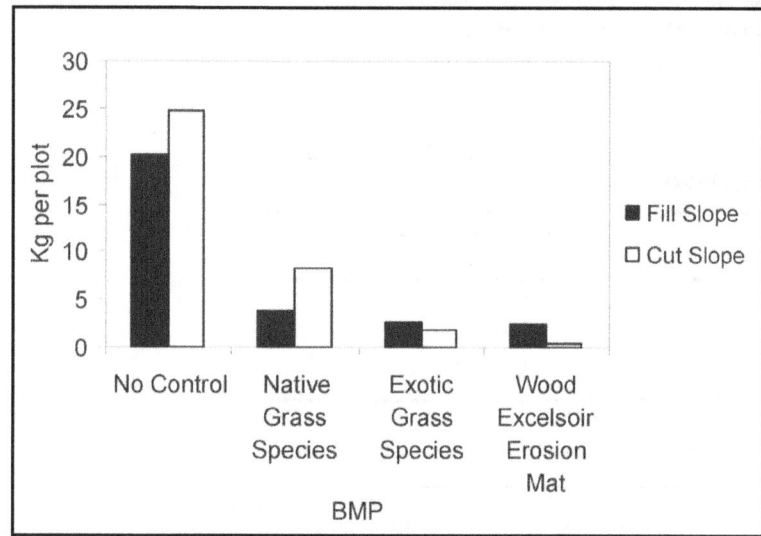

Figure 3-11. **Sediment yield from plots using various forms of ground covering. Sediment yield is per plot area over a 6-month period; plots measured 1.5 m x 3.1 m (after Grace et al., 1998).**

Planting grass on cut-and-fill slopes of new roads can effectively reduce erosion, and placing forest floor litter or brush barriers on downslopes in combination with establishing grass is also an effective means to reduce downslope sediment transport (Tables 3-13 and 3-14). Grass-covered fill is generally more effective than mulched fill in reducing soil erosion from newly constructed roads because of the roots that hold the soil in place, which are lacking with any other covering placed on the soil. Because grass needs some time to establish itself, a combination of straw mulch with netting to hold it in place can be used to cover a seeded area and effectively reduce erosion during the period while grass is growing. The mulch and netting provide immediate erosion control and promote growth of the grass. Figure 3-11 shows the results of a study conducted by Grace and others (1998) to demonstrate the erosion control capacities of different cut-and-fill slope stabilization BMPs on forest roads. The results of several studies on different types of slope stabilization BMPs are summarized in Table 3-15.

Table 3-13. **Reduction in the Number of Sediment Deposits More Than 20 Feet Long by Grass and Forest Debris (Swift, 1986)**

Type of Soil Protection	Degree of Soil Protection	Number of Deposits per 1,000 Feet of Road
Grassed fill, litter and brush burned	Low	13.9
Bare fill, forest litter		9.9
Mulched fill, forest litter	↕	8.1
Grassed fill, forest litter, no brush barrier		6.9
Grassed fill, forest litter, brush barrier	High	4.5

Table 3-14. Comparison of Downslope Movement of Sediment from Roads for Various Roadway and Slope Conditions (Swift, 1986)

Comparisons	Sites (no.)	Mean Slope (%)	Distance (feet)		
			Mean	Max	Min
All sites	88	46	71	314	2
Barrier[a]					
Brush barriers	26	46	47	156	3
No brush barrier	62	47	81	314	2
Drainage[b]					
Culvert	21	40	80	314	30
Outsloped without culvert	56	47	63	287	2
Unfinished roadbed with berm	11	57	95	310	25
Grass fill and forest litter[c]	46	40	45	148	2
With brush barrier	16	39	34	78	3
With culvert	4	20	37	43	30
Without culvert	12	45	32	78	3
Without brush barrier	30	41	51	148	2
With culvert	7	37	58	87	30
Without culvert	23	42	49	148	2

[a] Examined the effectiveness of leaving brush barriers in place below road fills, rather than removing brush barriers.
[b] Compared roads where storm water was concentrated at a culvert pipe to outsloped roads without a culvert. The berm was constructed on an unfinished roadbed to prevent downslope drainage.
[c] Compared effectiveness of brush barriers versus drainage (culvert) systems.

Table 3-15. Effectiveness of Surface Erosion Control on Forest Roads (adapted from Megahan, 1980, 1987)

Stabilization Measure	Portion of Road Treated	Percent Decrease in Erosion[a]	Reference
Hydro-mulch, straw mulch, and dry seeding[b]	Fill slope	24 to 58	King, 1984
Tree planting	Fill slope	50	Megahan, 1974b
Wood chip mulch	Fill slope	61	Ohlander, 1964
Straw mulch	Fill slope	72	Bethlahmy and Kidd, 1966
Excelsior mulch	Fill slope	92	Burroughs and King, 1985
Paper netting	Fill slope	93	Ohlander, 1964
Asphalt-straw mulch	Fill slope	97	Ohlander, 1964
Straw mulch, netting, and planted trees	Fill slope	98	Megahan, 1974b
Straw mulch and netting	Fill slope	99	Bethlahmy and Kidd, 1966
Straw mulch	Cut slope	32 to 47	King, 1984
Terracing	Cut slope	86	Unpublished data[c]
Straw mulch	Cut slope	97	Dyrness, 1970
Wood chip mulch	Road fills	61	Bethlahmy and Kidd, 1966
Straw mulch	Road fills	72	Ohlander, 1964
Grass and legume seeding	Road cuts	71	Dyrness, 1970
Gravel surface	Surface	70	Burroughs and King, 1985
Dust oil	Surface	85	Burroughs and King, 1985
Bituminous surfacing	Surface	99	Burroughs and King, 1985

[a] Percent decrease in erosion compared to similar, untreated sites.
[b] No difference in erosion reduction between these three treatments.
[c] Intermountain Forest and Range Experiment Station, Forestry Sciences Laboratory, Boise, ID, nd.

Road Construction, Fish Habitat, Stream Crossings, and Fish Passage

Chapter 2 discusses how road construction and road use can cause sediment to be delivered to streams, and it reviews the water quality and fish passage problems associated with sediment and stream crossings. The quality of surface waters to support early life stages of fish can be degraded by nonpoint source pollution from forestry activities as well. Salmonids and other fish that nest on stream bottoms are very susceptible to sediment pollution due to the settling of sediment that can smother nests and deplete the oxygen available to the eggs. The eggs, buried 1 to 3 feet deep in the gravel redd, rely on a steady flow of clean, cold water to bring oxygen and remove waste products. In coastal streams, eggs hatch in a month or so, depending on water temperatures and species of fish. Eggs hatch into alevin and remain in the gravel another 30 days or so, living on the nutrients in their yolk sacs. As they develop into fry, the yolk gets used up, and fry emerge through spaces in the gravel to begin life in the stream. During the 60-day period when the eggs and alevin are in the gravel, any shifts of the stream bottom can kill them.

Recent studies in streams on the Olympic Peninsula in Washington found that if more than 13 percent fine sediment (< 0.85 mm) intruded into the redd, no steelhead or coho salmon eggs survived (McHenry et al., 1994). Chinook salmon are the most susceptible to increased fine sediment, followed by coho salmon, steelhead, and cutthroat trout, respectively (Lotspeich and Everest, 1983). The different tolerances to fine sediment is due to the different head diameters of the fry of the species.

> The predominant source of sediment from logging is from the construction and maintenance of access roads.

The redd is a depression in the gravel streambed where the eggs are laid, and the depression creates a Venturi effect, drawing water down into the gravel. If the water in the stream above is full of fine sediment, the sediment is drawn down into the redd and smother the eggs.

In a healthy stream, young salmon and trout hide in the interstitial spaces between cobbles and boulders to avoid predation. In streams that become extremely cold in winter, young steelhead may actually burrow into the streambed and spend the winter in flowing water down within the gravel. The area of the stream where flowing water extends down into the gravel is also extremely important for aquatic invertebrates, which supply most of the food for young salmon, steelhead, and cutthroat trout. If fine sediment is clogging interstitial spaces between streambed gravel, juvenile salmonids lose their source of cover and food.

During the year coho salmon spend in freshwater, they prefer pools. High sediment concentrations in the water can cause pools to fill with sediment and reduce or destroy essential coho rearing habitat. Case studies in southwest Oregon showed that streams damaged by logging can also have significant problems with mortality of salmon eggs and alevin (Nawa and Frissell, 1993). When streams are affected by high sediment deposition, these formerly productive low-gradient reaches become wide and shallow and recovery of fish habitat can take decades (Frissell, 1992).

A fishway is any structure or modification to a natural or artificial structure for the purpose of fish passage. Five common conditions at stream crossing culverts create migration barriers (WADOE, 1999):

* Excess drop at culvert outlet

- High velocity within culvert barrel
- Inadequate depth within culvert barrel
- Turbulence within the culvert
- Debris accumulation at culvert inlet

Figure 3-12 illustrates four of these conditions. Barriers to fish passage can be complete, partial, or temporal. Complete barriers block the use of the upper watershed, often the most productive spawning habitat in the watershed for migratory species of fish. Partial barriers block smaller or weaker fish of a population. Culverts are therefore designed to accommodate smaller or weaker individuals of target species, including juvenile fish. Temporal barriers block migration during some part of the year. Fish passage can be provided in streams that have wide ranges of flow by providing multiple culverts (Figure 3-13). They can delay some fish from arriving at upstream locations, which for some fish (anadromous salmonids that survive a limited amount of time in fresh water) can cause limited distribution or mortality (WADOE, 1999). The FishXing Web site (http://www.stream.fs.fed.us/fishxing/index.html) provides software and learning systems for fish passage through culverts.

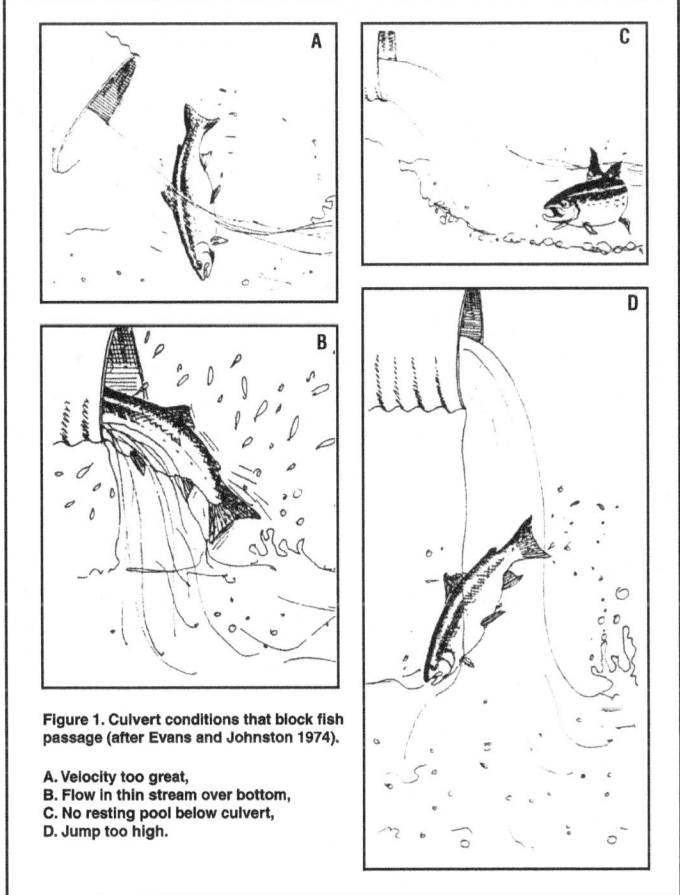

Figure 1. Culvert conditions that block fish passage (after Evans and Johnston 1974).

A. Velocity too great,
B. Flow in thin stream over bottom,
C. No resting pool below culvert,
D. Jump too high.

Figure 3-12. Culvert conditions that block fish passage (Yee and Roelofs, 1980).

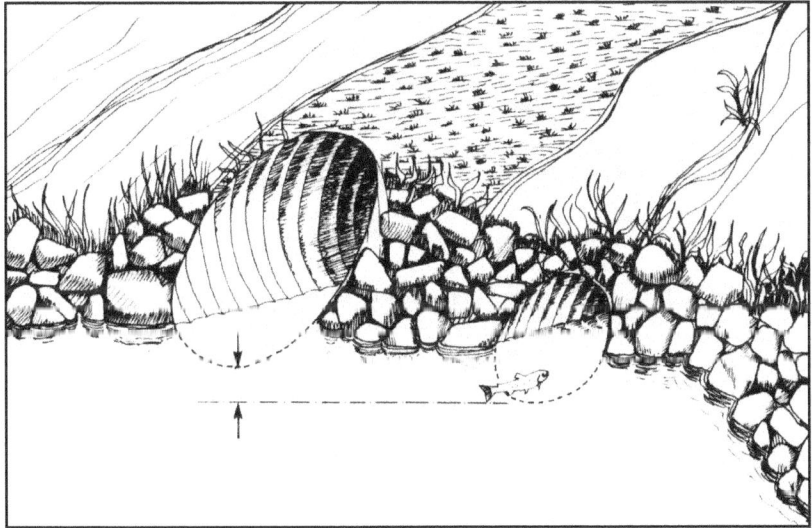

Figure 3-13. Multiple culverts for fish passage in streams that have a wide range of flows (Hyson et al., 1982).

> ### Stream Crossing Considerations
>
> - Whether fish use the channel at the crossing site
>
> - Whether the crossing will be temporary or permanent
>
> - The type of vehicles that will use the crossing
>
> - The slope, configuration, and stability of the natural hillslopes on either side of the channel
>
> - The slope of the channel bed
>
> - The orientation of the stream to the proposed road
>
> - The expected 50- and 100-year flood discharge
>
> - The amount and type of sediment and woody debris that is in transport within the channel
>
> - The installation and subsequent maintenance costs for the crossing
>
> - The expected frequency of use
>
> - Permits and other legal requirements
>
> (Source: Weaver and Hagans, 1984)

Barriers at culverts can result from improper initial design or installation, or they can be the result of channel degradation that leaves culvert bottoms elevated above the downstream channel. Changes in hydrology due to an extensive road network can be a primary reason for channel degradation, and older culverts that might have been adequate when installed can become inadequate for fish passage when channel degradation or land use changes cause changes in stream channel hydrology (Baker and Votapka, 1990; WADOE, 1999). When such changes occur in a watershed, inspect culverts and, if necessary, replaced them with ones that meet actual specifications.

Other problems at culverts include their not providing the roughness and variability of the adjacent stream channel bottom, which can create short distances of increased water velocity and turbulence (WADOE, 1999). These problems create barriers to the upstream migration of juvenile fish. Fish will not travel upstream under high water velocity conditions (Barber and Downs, 1996).

Water velocity in culverts is a complex issue, involving the length of the culvert in relation to fish capabilities, depth of water, icing and debris flows, and design flows in relation to fish migration upstream or downstream. The size and species of fish passing through a culvert and the magnitude, duration, frequency, and seasonal relationship of the flow to the timing of fish movement have to be considered in setting guidelines for culvert design to meet fish passage requirements (Ashton and Carlson, 1984; Baker and Votapka, 1990).

The addition of baffles to a culvert to affect water velocity and turbulence is not generally recommended because of the regular cleaning that becomes necessary. In addition, it has been found that turbulence at the edge of a baffled culvert actually creates a blockage to fish passage, and in higher-velocity culverts passage success can be higher in smooth pipe (Bates, 1994; Powers, 1996).

Countersunk culverts are recommended where fish passage is desired. Installation of multiple, parallel culverts in place of a larger single culvert is discouraged except in special cases, such as to permit fish passage where flows vary widely (see Figure 3-9). Countersunk culverts allow for natural downstream transport of sediment and a natural stream bottom within the culvert (White, 1996).

Wetland Road Considerations

Sedimentation is also a concern when considering road construction through wetlands. Because of the fragility of these ecosystems, where an alternative route exists, avoid putting a forest access road through a wetland. If it's necessary to traverse a wetland,

---

implement the BMPs suggested by the state. In addition, if road construction or maintenance involves a discharge of dredged or fill material into wetlands or other waters of the United States, section 404(f) requires the application of specific BMPs designed to protect the aquatic environment. (More information on wetlands and forestry, including a list of the aforementioned BMPs, is provided in Chapter 3, section J.)

Benefits of Road Construction Practices

Many states have found roads to consistently be sources of sediment discharge to streams. The Vermont Agency of Natural Resources assessed BMP implementation and effectiveness and found that roads were consistently the most problematic with respect to proper BMP implementation. Drainage ditches, culverts, and stream crossings were most frequently the points of origin of stream sedimentation. The Virginia Department of Forestry also found that water control structures on roads are often inadequately used and applied. The Department found that water bars, rolling dips, and broad-based dips were usually installed improperly. Water bars, for instance, were built using fill only, rather than by cutting into the road bed and then using fill material to shape the bar. These structures were often placed too infrequently and too far apart as the road grade increased, and in some cases they were installed backwards, being angled uphill with the outlet pointing upslope.

The Montana Department of Natural Resources and Conservation, Forestry Division, also monitored BMP implementation and effectiveness and similarly found that the most frequent departures from BMP implementation standards and sources of effects were associated with providing adequate road surface drainage, routing road drainage through adequate filtration zones before the runoff entered a stream, maintaining erosion control structures, and providing energy dissipaters at drainage structure outlets. The division also found that high-risk BMPs were more frequently not applied properly, and water quality effects from them were common.

The Virginia Department of Forestry assessed BMP implementation and effectiveness in 1994 and concluded from the study that although improvement was needed in meeting minimum standards of BMP implementation, properly implemented stream crossings (as well as SMAs and preharvest plans) are crucial to protecting water quality. Where not implemented properly, stream crossings are less effective than they could be. Improper sizing, placement, and installation of culverts are the causes of most failures. Culverts often were found to be too short for the intended roadbed width, and consequently they became clogged or buried. Some culverts were placed improperly, and without correction could have been rendered ineffective or swept away by storm water cutting through fill material.

In general, poor BMP effectiveness can be due to many factors, including the following:

- A lack of time or willingness to plan timber harvests carefully before cutting begins.
- A lack of skill in or knowledge of designing effective BMPs.
- A lack of equipment needed to implement effective BMPs.
- The belief that BMPs are not an integral part of the timber harvesting process and can be engineered and fitted to a logging site after timber harvesting has been completed.
- A lack of timely implementation and maintenance of BMPs.

Road Construction and Stream Crossing BMP Costs

Costs of forestry BMPs for water quality protection are difficult to specify because the need for and design of BMPs varies from site to site with changes in topography, soil, and proximity to water, among other factors. However, with respect to road construction BMPs, some generalizations can be made. In a study of the costs of various forestry practices in the southeastern United States, practices associated with road construction were generally found to be the most expensive, regardless of terrain, and the costs for broad-based dips and water bars increased as slope increased (Lickwar, 1989) (Table 3-16). The proximity of roads to watercourses also increases the cost of road construction because of the increased need to prevent sediment runoff from reaching the surface waters.

Unit cost comparisons for road surfacing practices (Swift, 1984a) revealed that grass is the least expensive alternative at $272 per kilometer of road (1998 dollars) (Table 3-17). Initial material costs alone, however, are misleading because a durable road surface can endure several years of use, whereas a grassed or thinly graveled surface will generally need regular maintenance and resurfacing. Grass and thin gravel coverings are also likely to result in more erosion and sedimentation. Table 3-18 compares the cost of using a single BMP (dry seeding alone) versus using multiple BMPs (seeding in conjunction with plastic netting) to control erosion (Megahan, 1987).

Table 3-16.   Cost Estimates (and Cost as a Percent of Gross Revenues) for Road Construction (Lickwar, 1989)

Practice Component	Location					
	Steep Sites[a]		Moderate Sites[b]		Flat Sites[c]	
Stream crossings	$45	(0.01%)	$185	(0.03%)	$4,303	(0.33%)
Broad-based dips	$16,550	(2.88%)	$10,101	(1.49%)	$4,649	(0.36%)
Water bars	$12,225	(2.13%)	$6,371	(0.94%)	$2,999	(0.24%)
Added road costs	$5,725	(1.00%)	Not provided		Not provided	

Note: All costs updated to 1998 dollars.
[a] Based on a 1,148-acre forest and gross harvest revenues of $399,685. Slopes average over 9 percent.
[b] Based on a 1,104-acre forest and gross harvest revenues of $473,182. Slopes ranged from 4 percent to 8 percent.
[c] Based on a 1,832-acre forest and gross harvest revenues of $899,491. Slopes ranged from 0 percent to 3 percent.

Table 3-17.   Cost of Gravel and Grass Road Surfaces (North Carolina, West Virginia) (Swift, 1984a)

Surface	Quantity/km	Unit Cost	Total Cost/km
Grass	28 kg Ky-31	$1.32/kg	$36.90
	14 kg rye	$1.03/kg	$14.50
	405 kg 10-10-10	$0.189/kg	$76.89
	900 kg lime	$0.052/kg	$46.59
	Labor and equipment	$97.49/km	$97.49
Crushed rock (5 cm)[a]	425 ton	$7.34/ton	$3,120
Crushed rock (15 cm)[a]	1,275 ton	$7.34/ton	$9,361
Large stone (20 cm)[a]	1,690 ton	$8.22/ton	$13,893

Note: All costs updated to 1998 dollars.
[a] Values in parentheses are thickness or depth of surfacing material.

Table 3-18.   Costs of Erosion Control Measures in Idaho (Megahan, 1987)

Measure	Cost ($/acre)
Dry seeding	$178
Plastic netting placed over seeded area	$8,124

Best Management Practices

Road Surface Construction Practices

◆ Follow the design developed during preharvest planning to minimize erosion by properly timing and limiting ground disturbance operations.

Verify with site visits that information used during preharvest planning to develop road layout and surfacing designs is accurate. Make any changes to road and road surface construction designs that are necessary based on new information obtained during these site visits.

◆ *During road construction, operate equipment to minimize unintentional movement of excavated material downslope.*

◆ *Properly dispose of organic debris generated during road construction.*

• Stack usable materials such as timber, pulpwood, and firewood in suitable locations and use them to the extent possible. Organic debris can be used as mulch for erosion control, piled and burned, chipped, scattered, place in windrows, or removed to designated sites. Slash can be useful if placed as windrows along the base of the fill slope. A windrow is created by piling logging debris and unmerchantable woody vegetation in rows on the contour of the land. Arranged in this manner, the slash material provides a barrier to overland flow, prevents the concentration of runoff, and reduces erosion.

• Don't use organic debris as fill material for road construction since the organic material eventually decomposes and causes fill failure.

• Perform any work in the stream channel by hand to the extent practicable. Machinery can be used in the SMA as long as the desired SMA objective is not compromised.

◆ *Prevent slash from entering streams and promptly remove slash that accidentally enters streams to prevent problems related to slash accumulation.*

To the extent possible, prevent slash from entering streams. If allowed to stay in streams, it can cause flow or fish passage problems, or dissolved oxygen depression as it decomposes. Leave natural debris in stream channels, and remove only that slash that is contributed during road construction or harvesting. Large woody debris is an important source of energy for aquatic organisms, especially in smaller headwater streams, and it creates habitat diversity important to aquatic invertebrates and young fish. It is important, therefore, to inspect streams before any work is done near them and to attempt to leave them in a condition similar to that prior to the work.

◆ *Compact the road base at the proper moisture content, surfacing, and grading to give the designed road surface drainage shaping.*

The predominant source of sediment associated with forest harvesting is the construction and maintenance of access roads, which contribute as much as 90 percent of the total eroded sediments (Appelbloom et al., 1998). The annual production of sediment from roads can be as high as 100 tons per hectare (40.5 tons per acre) of road surface or more (Grayson et al., 1993; Kockenderfer and Helvey, 1984). Management practices, including gravel surfacing, proper road maintenance, and proper drainage control, can reduce sediment loss. Gravel surfacing has to be of a sufficient depth (e.g., 15–20 cm). Improperly maintained roads can produce up to 50 percent more sediment than properly maintained roads. Since roads can produce large quantities of sediment even when they are well maintained, careful consideration of their placement and management is extremely important to minimizing their effects on water quality.

◆   *When soil moisture is high, promptly suspend earthwork operations and weatherproof the partially completed work.*

Regulating traffic on logging roads during unfavorable weather is an important phase of erosion control. Construction and logging under these conditions destroy drainage structures, plug up culverts, and cause excessive rutting, thereby increasing the amount and the cost of maintenance.

◆   *Consider geotextiles for use on any section of road requiring aggregate material layers for surfacing.*

Geotextile is a synthetic permeable textile material used with soil, rock, or any other geotechnical engineering-related materials (Wiest, 1998). Also known as geosynthetics, geotextiles are associated with high-standard all-season roads, but can also be used in low-standard logging roads. Geotextiles have three primary functions: drainage (filtration), soil separation (confinement), and soil reinforcement (load distribution). These functions are performed separately or simultaneously, but not all functions are provided by each type of geotextile, so use care when making a purchase. Geotextiles reduce the amount of aggregate needed, thus reducing the cost of the road (Wiest, 1998).

The location of a geotextile along a forest road does not affect installation procedures. When installing geotextiles, proper procedure includes the following steps:

*   Clear the subgrade of sharp objects, stumps, and debris.

*   Grade the surface to provide proper drainage and cross-slope shaping.

*   Unroll the geotextile on the subgrade. The amount of overlap depends on the load-bearing capacity of the subgrade, and varies from 1.5 to 3 feet. Sewing may be necessary if the geotextile is to provide reinforcement.

*   Place and compact the aggregate fill. Depth of the aggregate is determined by subgrade strength and the anticipated wheel loading (usually between 9 and 24 inches). It might be necessary to back-dump the aggregate onto the geotextile and spread with a dozer or grader. The rock is feathered out, since pushing it onto the site produces an uneven distribution of the aggregate. Spread the aggregate in the same direction as the geotextile overlap to avoid separation.

*   Compact the aggregate by conventional methods.

Streambanks and other slopes with light wave action can be stabilized by placing the revetment material directly on top of the geotextile. Installing the geotextile underneath the revetment material prevents the occurrence of scour which normally takes place along streambanks behind BMPs such as rip-rap. To ensure that the geotextile stays in place, toe it in at the top and bottom.

Geotextiles extend the service life of roads, increase their load-carrying capacity, and reduce the incidence of ruts. These benefits are realized due to the textiles separating aggregate structural layers from subgrade soils while allowing the passage of water.

◆ *Protect access points to the site that lead from a paved public right-of-way with stone, wood chips, corduroy logs, wooden mats, or other material to prevent soil or mud from being tracked onto the paved road.*

This practice prevents tracking of sediment onto roadways, thereby preventing the subsequent washoff of that sediment during storm events. When necessary, clean truck wheels to remove sediment before entering a public right-of-way.

◆ *Use pioneer roads to reduce the amount of area disturbed and ensure the stability of the area involved.*

Pioneer roads are temporary access ways used to facilitate construction equipment access when building permanent roads. Confine pioneer roads to the construction limits of the surveyed permanent roadway, and it is important that pioneer roads be fitted with temporary drainage structures to prevent erosion, sedimentation, and road deterioration.

◆ *If the use of borrow or gravel pits is needed during forest road construction, locate rock quarries, gravel pits, and borrow pits outside SMAs and above the 50-year flood level of any waters to minimize the adverse effects caused by the resulting sedimentation. Avoid excavating below the water table.*

Gravel mining directly from streams causes a multitude of effects, including destruction of fish spawning sites, turbidity, and sedimentation. During the construction and use of rock quarries, gravel pits, or borrow pits, either divert runoff water onto the forest floor or pass it through one or more settling basins. Revegetate and reclaim rock quarries, gravel pits, spoil disposal areas, and borrow pits upon abandonment.

Road Surface Drainage Practices

◆ *Install surface drainage controls at intervals that remove storm water from the roadbed before the flow gains enough volume and velocity to erode the surface. Avoid discharge onto fill slopes unless the fill slope has been adequately protected. Route discharge from drainage structures onto the forest floor so that water disperses and infiltrates. Methods of road surface drainage include the following:*

• *Broad-based dips.* A broad-based dip is a gentle roll in the centerline profile of a road that is designed to be a relatively permanent and self-maintaining water diversion structure that can be traversed by any vehicle (Figure 3-14). Outslope dips 3 percent to divert storm water off the roadbed and onto the forest floor, where transported soil can be trapped by forest litter. Use broad-based dips on roads having a gradient of 10 percent or less because on steeper grades they can be difficult for loaded trucks to traverse

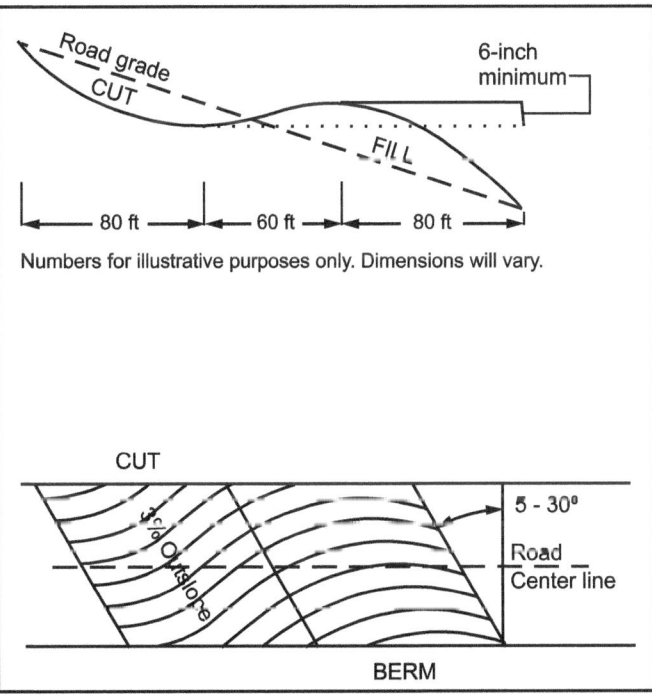

Numbers for illustrative purposes only. Dimensions will vary.

Figure 3-14. **Broad-based dip installation. A broad-based dip is a portion of road sloped to carry water from the inside edge to the outside onto natural ground (Minnesota DNR, 1995; Montana State University, 1990).**

(Kochenderfer, 1995). Dips can be difficult to construct on very rocky sections of roads as well.

- *Road outsloping, Insloping, Crowning, and Grading.* Water accumulation on road surfaces can be minimized by grading and insloping or outsloping roadbeds (Figure 3-15). This minimizes erosion and the potential for road failure. Outsloping involves grading a road so that the entire width of the road slopes down the hill it is cut into, and it is appropriate when fill slopes are stable and drainage won't flow directly into stream channels. Outsloping the roadbed keeps water from flowing next to and undermining the cutbank, and it is intended to spill water off the road in small volumes along its length. Give the width of the road a 2 to 3 percent outslope.

In addition to outsloping the roadbed, construct a short broad-based dip to turn water off the surface. The effectiveness of outsloping is limited by roadbed rutting during wet conditions. Providing a berm on the outside edge of an outsloped road during construction, and until loose fill material is protected by vegetation, can eliminate erosion of the fill. A continuous berm (i.e., a low mound of soil or gravel built along the edge of a road) along a roadside can reduce total sediment loss by an average of 99 percent over a standard graded soil road surface (Applebloom et al., 1998). Berms need to have openings provided to allow water to drain off the road surface at appropriate locations where a suitable infiltration or sediment trap site is reached (Swift and Burns, 1999). Construct berms high enough to contain the storm water, and wide enough and with a coarse material to prevent their erosion. Berms are also installed over culvert crossings to prevent runoff from draining directly into streams. A graveled road surface or a grassed strip on the edge of the driving surface can reduce total loss of sediment from roads by up to 60 percent over a standard graded soil road surface. Also, natural berms can form along the edge of older roadbeds or at

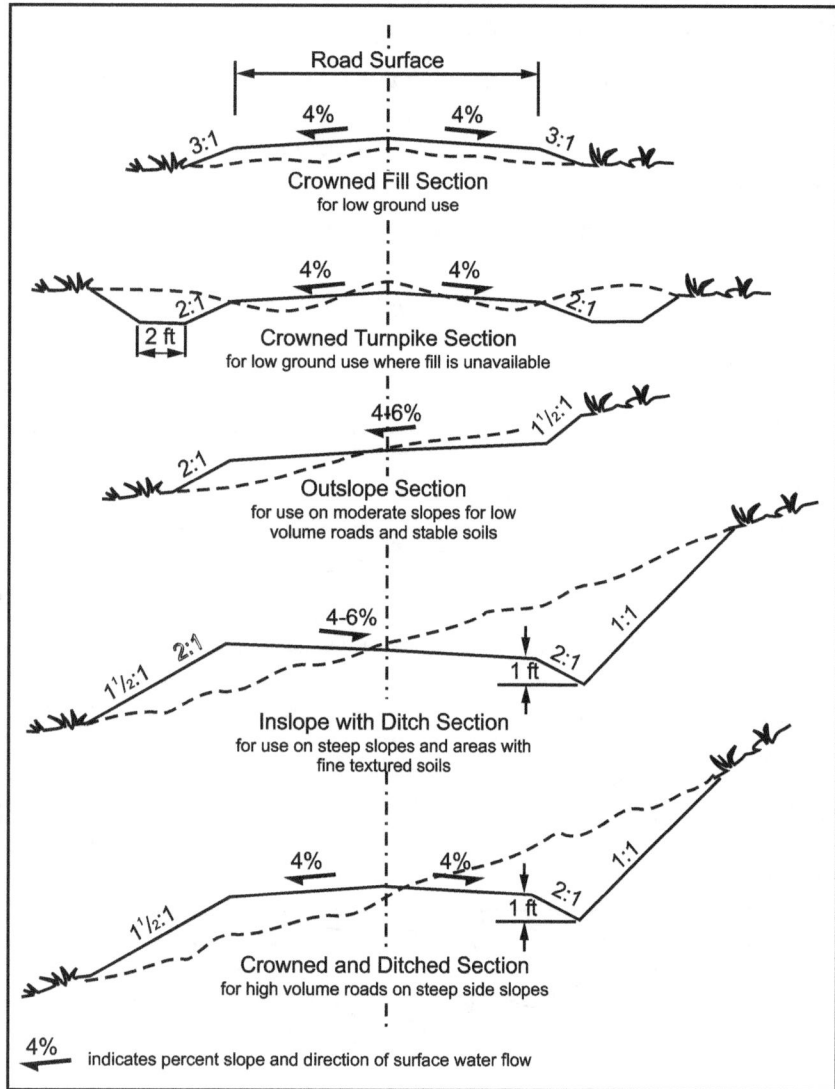

Figure 3-15. **Typical road profiles for drainage and stability. Choice of cross section depends on drainage needs, soil stability, slope, and expected traffic volume. Dashed lines indicate natural land contour and solid lines indicate constructed road (Wiest, 1998).**

drainage locations on constructed berms over time and block drainage. Proper maintenance, therefore, is necessary.

Insloped roads carry road surface water to a ditch along the cutbank. Ditch gradients of between 2 and 8 percent usually perform best. Slopes greater than 8 percent give runoff waters too much momentum and enough erosive force to carry excessive sediment and debris for long distances, and slopes of less than 2 percent tend to cause water to drain too slowly and do not provide the runoff with enough energy to move accumulated debris with it. The ditch grade also depends on the soil type—nearer to 2 percent on less stable soils and nearer to 8 percent on stable soils.

A crowned road surface is a combination of both an outsloped and insloped surface with the high point (crown) at the center of the road (Moll et al., 1997). The crowned road provides drainage to both sides of the roadway, and a drainage ditch is usually placed next to the road on the insloped side. Properly spaced and sized culverts then direct the runoff to an appropriate grassed buffer, detention basin, or other sediment control structure.

- *Relief culverts.* Relief culverts move water from an inside ditch to the outside edge of a road for dispersion. The culverts should protrude from both ends at least 1 foot beyond the fill and be armored at inlets to prevent undercutting and at outlets to prevent erosion of fill or cut slopes (Figure 3-16).

Where the slope on the cutslope above a culvert is steep, as is often the case because of the need to cut into the slope to accommodate the culvert opening, soil erosion above culverts and culvert plugging might be a problem. Installing a riser pipe on the inlet end of a culvert with holes or slits cut at a proper height to allow water to enter (which depends on the amount of soil eroding and flow in the ditch) can prevent plugging while allowing runoff drainage. A ditch dam will reinforce the entrance of water into the culvert through the riser holes (Firth, 1992).

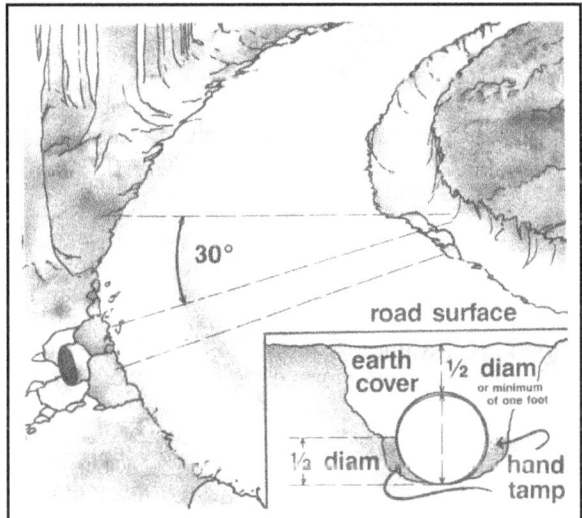

Figure 3-16.   Design and installation of relief culvert (Vermont DFPR, 1987).

- *Open-top or pole culverts.* Open-top or pole culverts are temporary drainage structures that are most useful for intercepting runoff flowing down road surfaces (Figure 3-17). They can also be used as a substitute for pipe culverts on roads of smaller operations, if properly built and maintained, but don't use them for handling intermittent or live streams. Place open-top culverts at angles across a road to provide gradient to the culvert and to ensure that no two wheels of a vehicle hit it at once. For an open-top culvert to function properly, careful installation and regular maintenance are necessary. Open-top culverts are recommended for ongoing operations only and are best removed upon completion of forestry activities (Wiest, 1998). These culverts generally slope below the perpendicular to the road at 10 to 45 degrees. Additional maintenance can be necessary as the angle approaches 10 degrees because at this angle debris tends to accumulate; an angle of 30 to 45 degrees is usually recommended (Wiest, 1998).

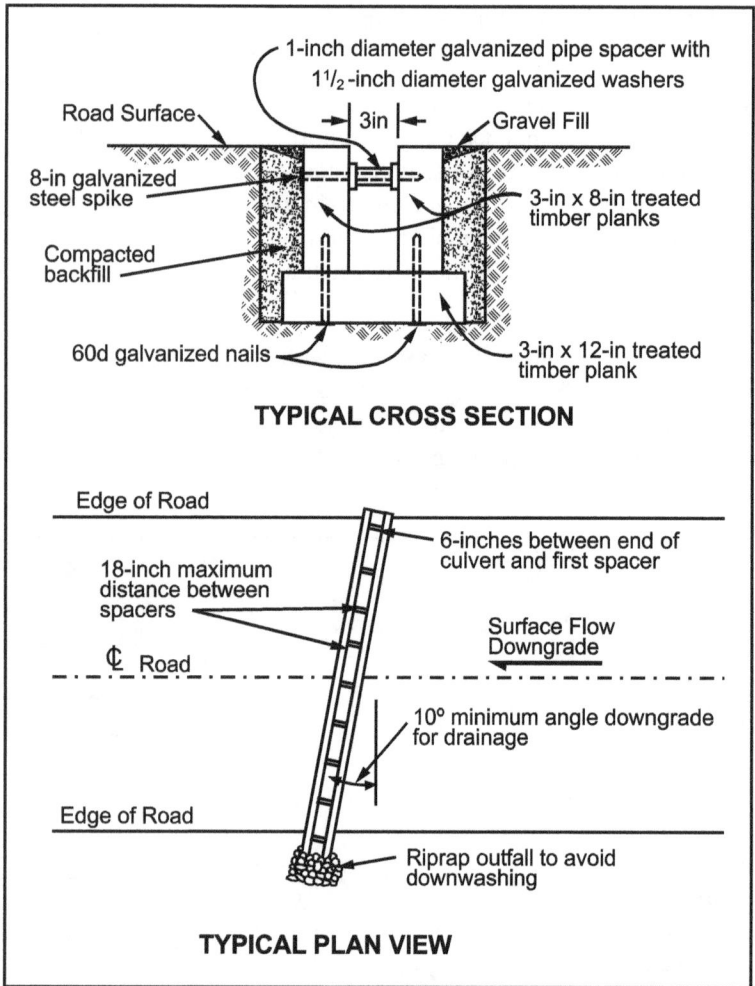

1-inch diameter galvanized pipe spacer with
1½-inch diameter galvanized washers

Road Surface

←3in→

Gravel Fill

8-in galvanized
steel spike

3-in x 8-in treated
timber planks

Compacted
backfill

60d galvanized nails

3-in x 12-in treated
timber plank

TYPICAL CROSS SECTION

Edge of Road

6-inches between end of
culvert and first spacer

18-inch maximum
distance between
spacers

Surface Flow
Downgrade

₡ Road

10° minimum angle downgrade
for drainage

Edge of Road

Riprap outfall to avoid
downwashing

TYPICAL PLAN VIEW

Figure 3-17.   Details of installation of open-top and pole culverts (Wiest, 1998; Vermont DFPR, 1987).

Open-top culverts constructed of 8-inch or 10-inch pipe are useful as a supplemental means of runoff control on steep sections of roads where broad-based dips are difficult to install and difficult for trucks to traverse (Kockenderfer, 1995). They are also useful on excessively rocky sections of roads where broad-based dips are difficult to construct. Rectangular openings spaced evenly along the top of a piece of pipe direct runoff into the pipe, and unbroken spacings between the openings provide structural integrity. The culverts can be installed by hand and can be removed and used elsewhere when a road is decommissioned. Their trenches are shallower than those for pole culverts. Discharges from all types of culverts can be controlled using plastic corrugated culvert piping cut in half or, where something that blends in with the surroundings is desired, with riprap (Kockenderfer, 1995). Diversions or in-ditch dams can be placed in ditches to ensure that flow in ditches is directed into culverts and it does not bypass culverts and continue to gain momentum and erosive force.

- *Ditches and turnouts.* Use ditches only where necessary to discharge water to vegetated areas via turnouts (Figure 3-18). Turnouts should be used wherever there is an adequate, safe outlet site where the water can infiltrate. In most cases, the less water a ditch carries and the more frequently water is discharged, the better. Construct wide, gently sloping ditches, especially in areas with highly erodible soils. Slow the velocity of water by installing check dams, rock dams that intercept water flow, along the ditch or lining the ditch with rocks. Check dams also trap sediment and need to be

Table 3-B.

Spacing of Turnouts

Road Grade *(percent)*	Spacing *(feet)*
2-5	500-300
6-10	300-200
11-15	200-100
16-20	100

Source: Cooperative Extension Service Division of Agricultural Sciences and Natural Resources, Oklahoma State University

Figure 3-18.   Grading and spacing of road turnouts (Georgia Forestry Commission, 1999).

inspected for sediment build-up. Additionally, stabilize ditches with rock and/or vegetation and protect outfalls with rock, brush barriers, live vegetation, or other means. Roadside ditches need to be large enough to carry runoff from moderate storms. A standard ditch used on secondary logging roads is a triangular section 45 cm deep, 90 cm wide on the roadway side, and 30 cm wide on the cutbank side. The minimum ditch gradient is 0.5 percent, and 2 percent is preferred to ensure good drainage. Runoff is diverted frequently to prevent erosion or overflow.

◆ *Install turnouts, wing ditches, and dips to disperse runoff and reduce the amount of road surface drainage that flows directly into watercourses.*

◆ *Install appropriate sediment control structures to trap suspended sediment trans-ported by runoff and prevent its discharge into the aquatic environment.*

Methods to trap sediment include the following:

- *Sediment traps.* Sediment traps are used downstream of erodible soil sites, such as cuts and fills, to keep sediment from flowing downstream and entering water bodies (Figure 3-19) (Ontario MNR, 1990). They are located close to the source of sedi-ment and preferably in a low area. Use them for drainage areas of less than 5 acres. Size sediment traps so that the expected sediment runoff fills them at about the time that the disturbed area reestablishes vegetation. If sediment accumulates beyond this time, periodic cleaning becomes necessary. Sediment traps are most effective at removing large sediment particles.

- *Brush barriers.* Brush barriers are slash materials piled at the toe slope of a road or at the outlets of culverts, turnouts, dips, and water bars. Install brush barriers at the toes of fills if the fills are located within 150 feet of a defined stream channel. Brush barriers must have good contact with the ground and be constructed approximately on the contour if they are to be effective in minimizing sediment runoff. Figure 3-20 shows the use of a brush barrier at the toe of fill. Proper installation is important because if the brush barrier is not firmly anchored and embedded in the slope, brush material can be ineffective for sediment removal and can detach to block ditches or culverts. In addition to use as brush barriers, slash can be spread over exposed mineral soils to reduce the effect of precipitation events and surface flow.

- *Silt fences.* Silt fences are temporary barriers used to intercept sediment-laden runoff from small areas. They act as a strainer: silt and sand are trapped on the surface of the fence while water passes through. They usually consist of woven geotextile filter fabric or

Figure 3-19. Sediment trap constructed to collect runoff from ditch along cutslope (Ontario MNR, 1990).

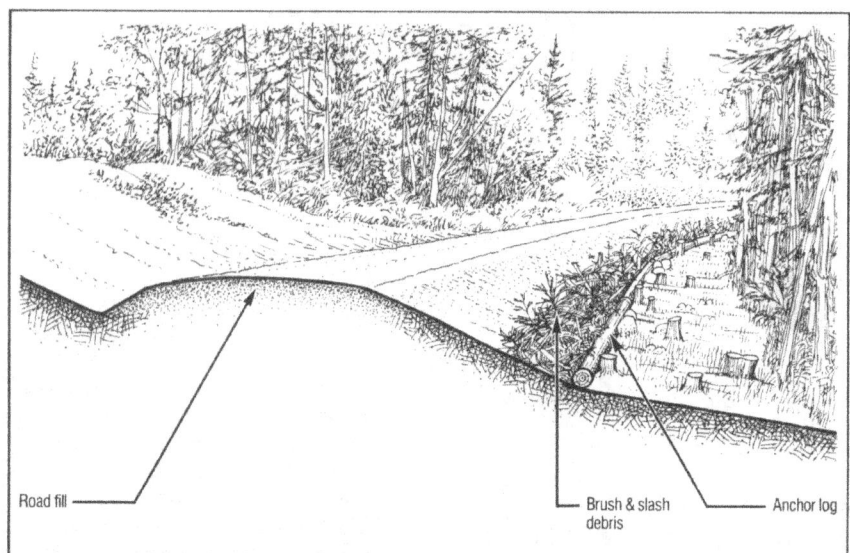

Figure 3-20. **Brush barrier placed at toe of fill to intercept runoff and sediment (Ontario MNR, 1990).**

straw bales. Install silt fences before earthmoving operations and place them as much along the contour as possible (Figure 3-21).

- *Filter strips.* Sediment control is achieved by providing a filter or buffer strip between streams and construction activities to use the natural filtering capabilities of the forest floor and litter (Figure 3-22). The Streamside Management Area management measure recommends the presence of a filter or buffer strip around all water bodies. Filter strips are effective at trapping sediment only when the runoff entering them is dispersed. Concentrated flows, such as from culverts, ditches, gullies, etc., entering filter strips will tend to cut a path through the filter strip and render it ineffective.

Foresters with the USDA Forest Service working in the Allegheny National Forest in Pennsylvania inspected numerous roads and streams to determine the minimum length of filter strip between the two that was necessary for preventing sediment from reaching the streams (USDA-FS, 1994, 1995). They found that no matter what the slope, filter strips 100 feet in length were the minimum necessary to prevent sedimentation; in more than a few instances, filter strips as long as 200 feet were necessary. In a test of filtering capacities of roadside erosion control techniques in Tuskegee National Forest in Macon County, Alabama, sediment fences retained 29 percent of runoff sediment and vegetative strips

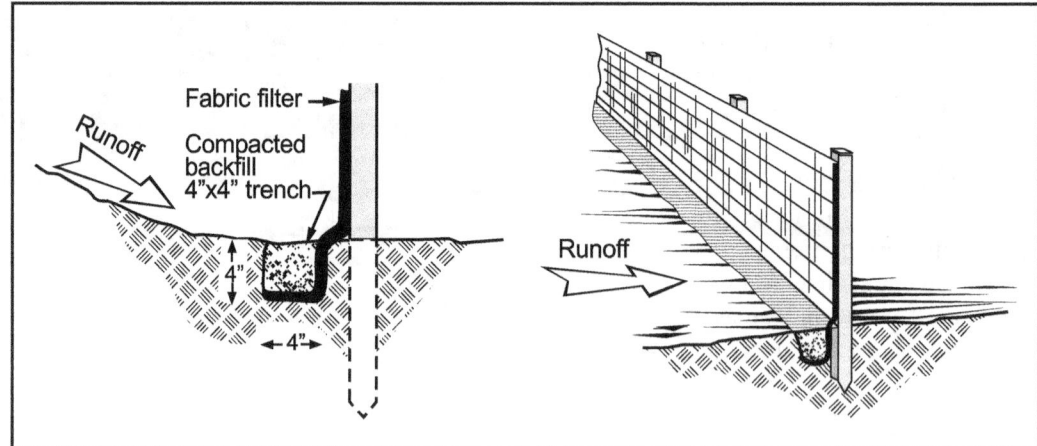

Figure 3-21. **Silt fence installation (Wisconsin DNR, 1989).**

retained 13.5 percent. Sediment below riprap increased by 10 percent, indicating that riprap has no ability to filter sediment from runoff.

These findings illustrate the importance of both using guidelines developed for the area where the harvest is to occur and inspecting points where runoff is concentrated (e.g., culvert outlets, turnouts) to see if sedimentation controls are sufficient to protect streams. Slope, type of vegetation, ground litter, and nature of flow (channelized or overland) combine to determine how effective filter strips are, and how wide they must be. If sedimentation is found to be occurring despite having installed BMPs according to specifications additional sediment control BMPs might be needed.

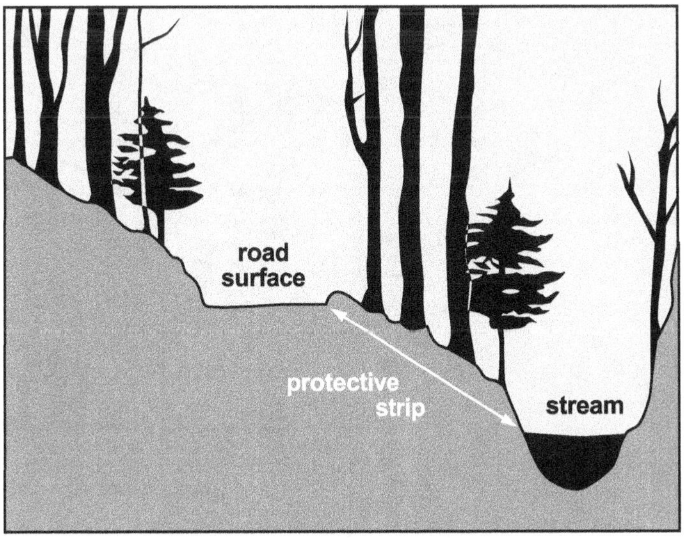

Figure 3-22. **Protective filter strip maintained between road and stream to trap sediment and provide shade and streambank stability (Vermont DFPR, 1987).**

Road Slope Stabilization Practices

◆ *Visit locations where roads are to be constructed on steep slopes or cut into hillsides to verify that these are the most favorable locations for the roads.*

Aerial photos and topographic and soil maps can inaccurately represent actual conditions, especially if these media are more than a few years old. Visiting a location where roads are to be cut into slopes or built on steep slopes or where skid trails, landings, and equipment maintenance areas are to be located is valuable for verifying that the information used during planning is accurate. Such visits can also help in determining whether roads can be located to pose less risk of erosion than the risk associated with the locations originally chosen.

◆ *Use straw bales, straw mulch, grass seeding, hydromulch, and other erosion control and revegetation techniques to stabilize slopes and minimize erosion (Figure 3-23). Straw bales and straw mulch are temporary measures used to protect freshly disturbed soils and are effective when implemented and maintained until adequate vegetation has established to prevent erosion.*

◆ *Compact the fill to minimize erosion and ensure road stability.*

During construction, fills or embankments are built up by gradual layering. Compact the entire surface of each layer with a tractor or other construction equipment. If the road is to be grassed, do not compact the final layer in order to provide an acceptable seedbed.

◆ *Revegetate or stabilize disturbed areas, especially at stream crossings.*

Cutbanks and fill slopes along forest roads are often difficult to revegetate. Properly condition slopes to provide a seedbed, including rolling embankments and scarifying cut slopes. The rough soil surfaces provide niches in which seeds can lodge and germinate. Seed as soon as it is feasible after the soil has been disturbed, preferably before it rains. Early grassing and spreading of brush or erosion-resisting fabrics on exposed soils at stream crossings are imperative. See the Revegetation of Disturbed Areas management measure for a more detailed discussion.

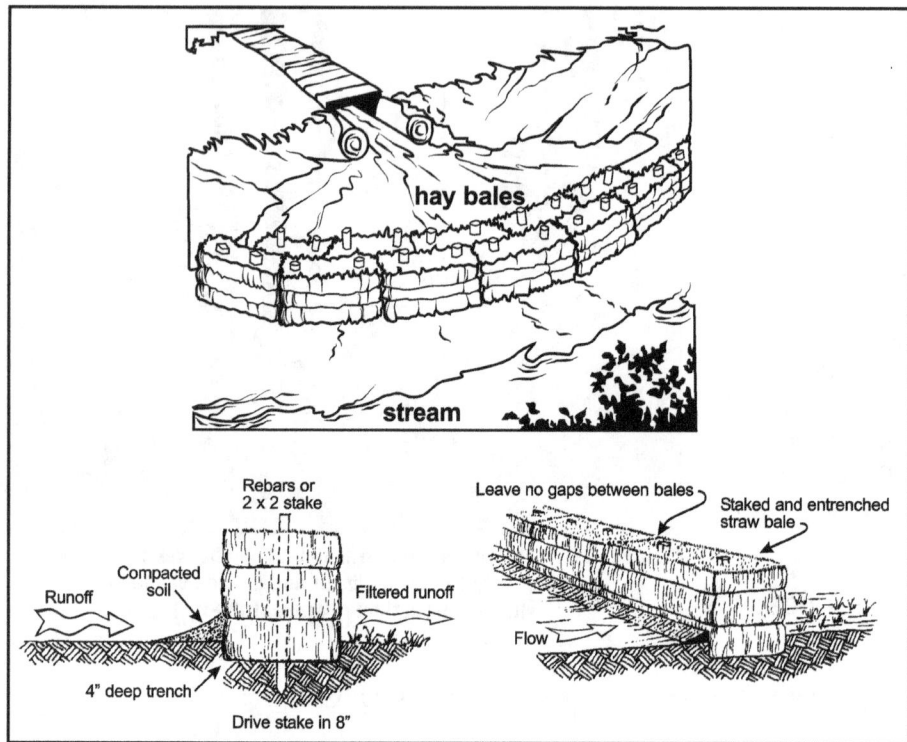

Figure 3-23. Details of hay bale installation, used to prevent sediment from skid trails and roads from entering surface waters (Georgia Forestry Commission, 1999; Vermont DFPR, 1987).

Stream Crossing Practices

◆ *Based on information obtained from site visits, make any alterations to the harvesting plan that are necessary or prudent to protect surface waters from sedimentation or other forms of pollution and to ensure the adequacy of fish passage.*

After preharvest planning has been completed with the aid of aerial photos and/or topographic maps, site visits can be conducted to verify the information used to determine the locations of stream crossings. Photos and maps record the landscape at a moment in time, and changes might have occurred since these media were created. Land use changes in the upper portion of the watershed in which harvesting occurs could have altered streamflow, which in turn might have modified stream corridor characteristics. As a result, alternative stream crossing locations might have to be found. Slopes might be inaccurately represented on topographic maps, and therefore stream crossing approaches or roads near streams might have to be relocated to avoid steep grades, or the width of SMAs might have to be increased. Land use changes in the watershed that increase streamflow or changes in weather patterns (such as numerous recent years of above-average rainfall) that affect streamflow characteristics might call for larger culverts than those originally intended or a switch from fords to culverts or from culverts to temporary bridges to ensure that fish can pass and that stream crossings can adequately handle streamflow. Refer to *Fish Passage Practices* later in this section for further information on constructing stream crossings that ensure adequate fish passage.

◆ *Construct stream crossings to minimize erosion and sedimentation.*

Erosion and sedimentation can be minimized by avoiding any operation of machinery in water bodies. It is especially important to not work in or adjacent to live streams and water channels during periods of high streamflow, intense rainfall, or migratory fish spawning.

Avoid stream crossings whenever practical alternatives are available. When it is necessary to construct stream crossings, install as few of them as possible, select their locations carefully, and select the most appropriate type of stream crossing for the particular site (Blinn et al., 1999). Use existing stream crossings whenever this would affect water quality less than constructing a new one. Make crossings at the narrowest practical portion of a stream and, if possible, cross at a right angle to the stream. Crossing at right angles reduces the potential for sediment to be carried down the road and deposited into the stream during a rain event. If the right angle crossing is too long it is likely to be ineffective. Crossing at right angles is not always practical, particularly in gentle topography. Gentle topography does not accelerate runoff into streams as steep angles do. If there is a gentle grade to a stream, the installation of water turnouts and a broad-based dip on each side of the crossing might suffice. This diverts the majority of the water that is runoff down the road. Avoid sags in grades on stream crossings, as they can cause road runoff to enter the stream (Swift and Burns, 1999). Road grade, whether up or down, should be maintained over the length of the crossing and the runoff diverted from the road at the first feasible location after the crossing.

Diverting a stream from its natural course is a potential problem when any stream crossing is constructed. When the capacity of a culvert under a stream crossing is too small or a culvert becomes plugged, flow is diverted around the culvert (Furniss et al., 1997). The stream might maintain its natural course (flow across the road parallel to the culvert), or, if the road has an inclining grade across the stream crossing in the direction of streamflow or it slopes downward away from a stream crossing in at least one direction, flow is diverted along the road for a distance until it reaches a low point, flows out of the road, and finds a new course to rejoin the original stream course. If left unchecked, such unintentional diversion can result in very large amounts of erosion and sedimentation and long-term adverse effects to roads and aquatic habitats. Stream diversion can also be caused by accumulations of snow and ice on the road that direct water out of the channel. Diversion potential is greatest on outsloped roads that redirect stream water down a road instead of across it (Best et al., 1995).

Stream diversion is best avoided by properly sizing culverts based on streamflow, constructing crossings such that their grade rises away from the crossing at each approach, inspecting stream crossings regularly after their construction, and maintaining roads and stream crossings properly (Bohn, 1998). Eliminating the potential for stream diversion by properly planning, installing, and maintaining roads and stream crossings is, in the long term, much less expensive and straightforward than attempting to correct improper design and installation after a stream crossing fails (Furniss et al., 1997).

◆ *Install a stream crossing that is appropriate to the situation and conditions.*

Determining the stream classification and the type of road to be constructed (e.g., temporary, seasonal, or permanent all-weather) is the first step in defining the type of stream crossing to be installed (Weaver, 1994). Design stream crossings to minimize effect on water quality, to handle peak runoff from flood waters, and to allow for adequate fish passage (where fish could be seasonally present). There are three basic subcategories of both permanent and temporary stream crossings: (1) bridges, (2) fords, and (3) culverts.

- *Bridges.* Temporary or portable bridges are being used increasingly because they can be installed and removed with minimal site disturbance or water quality effect and reused (Figure 3-24) (Taylor et al., 1999). Temporary stream crossings can be constructed of polyvinyl chloride and high-density polyethylene pipe bundles, and portable bridges are often constructed of steel (Blinn et al., 1999; Taylor et al., 1999). Approaches on weak soils can be protected with logs, wood mats, wood panels, or expanded metal grating placed over a woven geotextile.

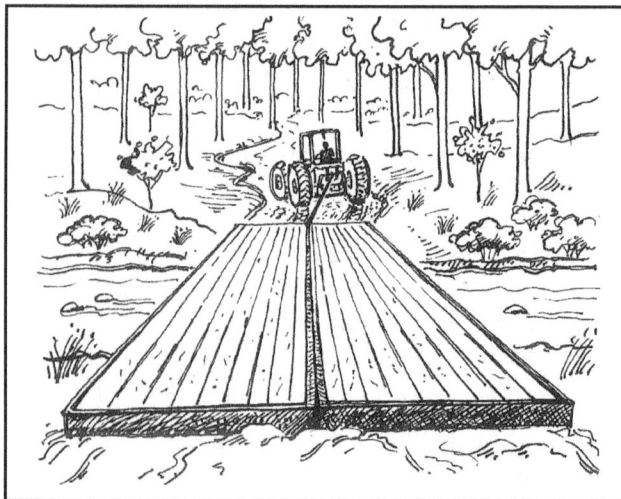

Figure 3-24.   Portable bridge for temporary stream crossing (Indiana DNR, 1998).

- *Fords.* A ford is a low-water crossings that uses existing or constructed stream bottoms to support vehicles when crossing a stream (Figure 3-25). A ford is an appropriate stream crossing structure under the following circumstances (Wiest, 1998):

  – The streambed has a firm rock or coarse gravel bottom, and the approaches are low and stable enough to support traffic.

  – Traffic volume is low.

  – Water depth is less than 3 feet.

  – Ford will not prevent fish migration.

If log, coarse gravel, or gabion is used to create a driving surface at a stream ford, install the crossing flush with the streambed to minimize erosion and to allow fish passage. Stabilize approaches to the ford using nonerodible material that extends at least 50 feet from the ford on both sides of the stream crossing.

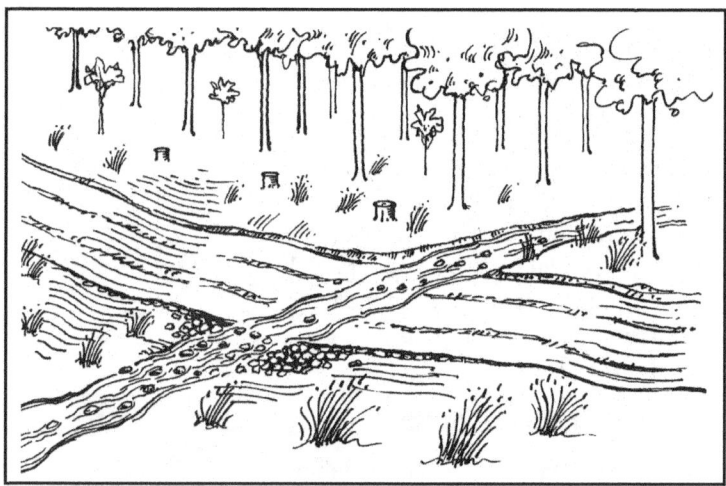

Figure 3-25.   A stream ford. Hard and stable approaches to a ford are necessary (Indiana DNR, 1998).

The following is a common procedure for crossing a small stream where a streambed is not armored with bedrock or an otherwise stable foundation:

  – Place several inches of rock down on the streambed. The rock size depends on actual costs, haul distance, and how much is to be installed. Normally, 2 feet or more of rock is installed.

  – Place geotextiles over the rock. Geotextile costs approximately $550 per 1,000 square yards.

  – Spread out approximately 1 foot of gravel. The amount and size of gravel varies with the conditions of the stream crossing.

Unless they are very large, stream fords are often the least expensive stream crossing to construct (Taylor et al., 1999). However, they can have greater effects on water quality than other crossings because sediment is introduced during construction and vehicle crossings. They also permit sediment-laden runoff to flow downslope directly into a stream unless adequate runoff diversions are installed.

- *Stream Crossing Culverts.* Stream crossing culverts are placed on roads where a semi-permanent or permanent stream crossing is necessary and to minimize interference with streamflow and stream ecology. Culverts often need outlet and inlet protection to keep water from scouring away supporting material and to keep debris from plugging the culvert. Firmly anchor culverts and compact the earth at least halfway up the side of the pipe to prevent water from leaking around it (Figure 3-26). Energy dissipaters, such as riprap and slash, can be useful for this if installed at culvert outlets. If riprap is used for inlet protection, a layer of geotextile should be placed behind the riprap to prevent erosion. Culvert spacing depends on rainfall intensity, drainage area, topography, and amount of forest cover. Most state forestry departments can provide recommendations for culvert pipe diameters.

According to Murphy and Miller (1997), culverts should be able to handle large flows— at least the 50-year flood. The larger the drainage area leading to a culvert and the steeper the topography, the larger the culvert needs to be to adequately handle the storm flow. If culverts are not properly sized for site-specific factors, culvert blowouts and overtopping can occur. Improper culvert sizing and spacing in Breitenbush, Oregon, led to severe road damage after a storm, and the estimated cost for the additional culverts that would have properly drained the watershed was $23,500, or 21 percent of the estimated $110,000 that was necessary to restore the road after the storm (Copstead et al., 1998).

If possible, install arch culverts (Figure 3-4) to avoid disturbance to the stream bottom, or place culverts within the natural streambed (Figure 3-27). Place the inlet on or below the streambed to minimize flooding upstream and to facilitate fish passage. Align large culverts with the natural course and gradient of the stream unless the inlet condition can be improved and the erosion potential reduced with some channel improvement. Use energy dissipators at the downstream end of the culverts to reduce the erosion energy of emerging water.

- Design stream crossings to fail during very large storm events.

Stream crossings cannot be designed for the largest possible storm that could occur, and rarely but eventually many streams will carry flows that exceed even the largest stream crossings along it. If stream crossings are not designed to fail under such circumstances, major erosion can result. One of the most important aspects of designing a stream crossing for failure is to design the path that excessive stream flow will

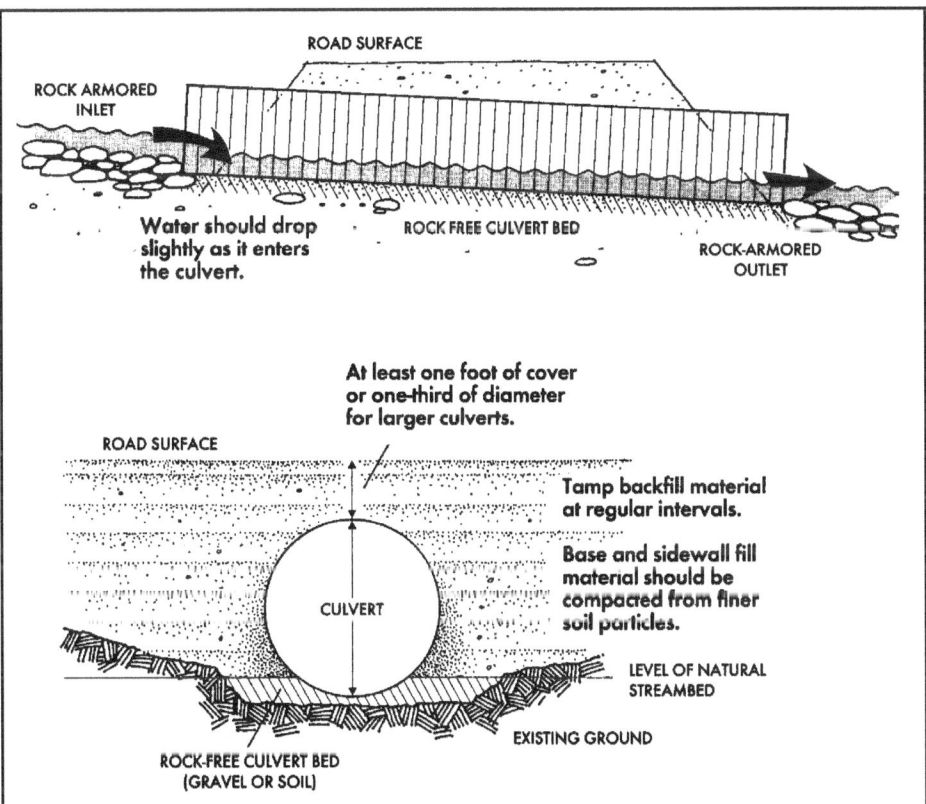

Figure 3-26.   Design and installation of pipe culvert at stream crossing (Montana State University, 1991).

Figure 3-27. Proper installation of culvert in the stream is critical to preventing plugging or undercutting (Montana State University, 1991).

follow (Furniss et al., 1997). Maximize the likelihood that the excessive flow will follow the natural course of the stream. The following are means to achieve this objective (Furniss et al., 1998):

– Locate stream crossings where the road grade rises away from the crossing at each approach.

– Create a rolling grade where a stream is crossed on a climbing road to prevent overflow from flowing down the road.

– Design stream crossings with the least amount fill possible and construct fills with coarse material.

◆ *Construct bridges and install culverts during periods when streamflow is low.*

◆ *Do not perform excavation for a bridge or a large culvert in flowing water. Divert the water around the work site during construction with a cofferdam or stream diversion.*

Isolating the work site from the flow of water is necessary to minimize the release of soil into the watercourse and to ensure a satisfactory installation in a dry environment. Minimize environmental effects by limiting the duration of construction and by establishing limits on the quantity of surface area disturbed and the equipment to be used. Also, operate when disturbance can most easily be controlled, and use erosion and sediment controls such as silt fences and sediment catch basins. Only use diversions where constructing the stream crossing structure without diverting the stream would result in instream disturbance greater than the disturbance from diverting the stream. Figure 3-28 portrays a procedure for installing a large culvert when excavation in the channel of the stream would cause sedimentation and increase turbidity.

◆ *Protect embankments with mulch, riprap, masonry headwalls, or other retaining structures.*

Some form of reinforcement along stream banks at road stream crossings can reduce sediment loss from these sites (Table 3-19). Soft protection, such as mulch or forest debris, or hard protection, such as gravel or riprap, can be used to protect these vulnerable locations.

◆ *Construct ice bridges in streams with low flow rates, thick ice, or dry channels during winter. Ice bridges might not be appropriate on large water bodies or areas prone to high spring flows.*

Ice bridges can provide acceptable temporary access across streams during winter. Ice bridges are made by pushing and packing snow into streams and applying water to freeze the snow (Figure 3-29). Their use is limited to winter under continuous freezing conditions. A permit might be necessary before an ice bridge crossing can be built, and operators can check this with the appropriate state agency prior to ice bridge construction.

The Minnesota Extension Service (1998) suggests the following when building an ice bridge:

• Choose a period when night temperatures are below 0 °F.

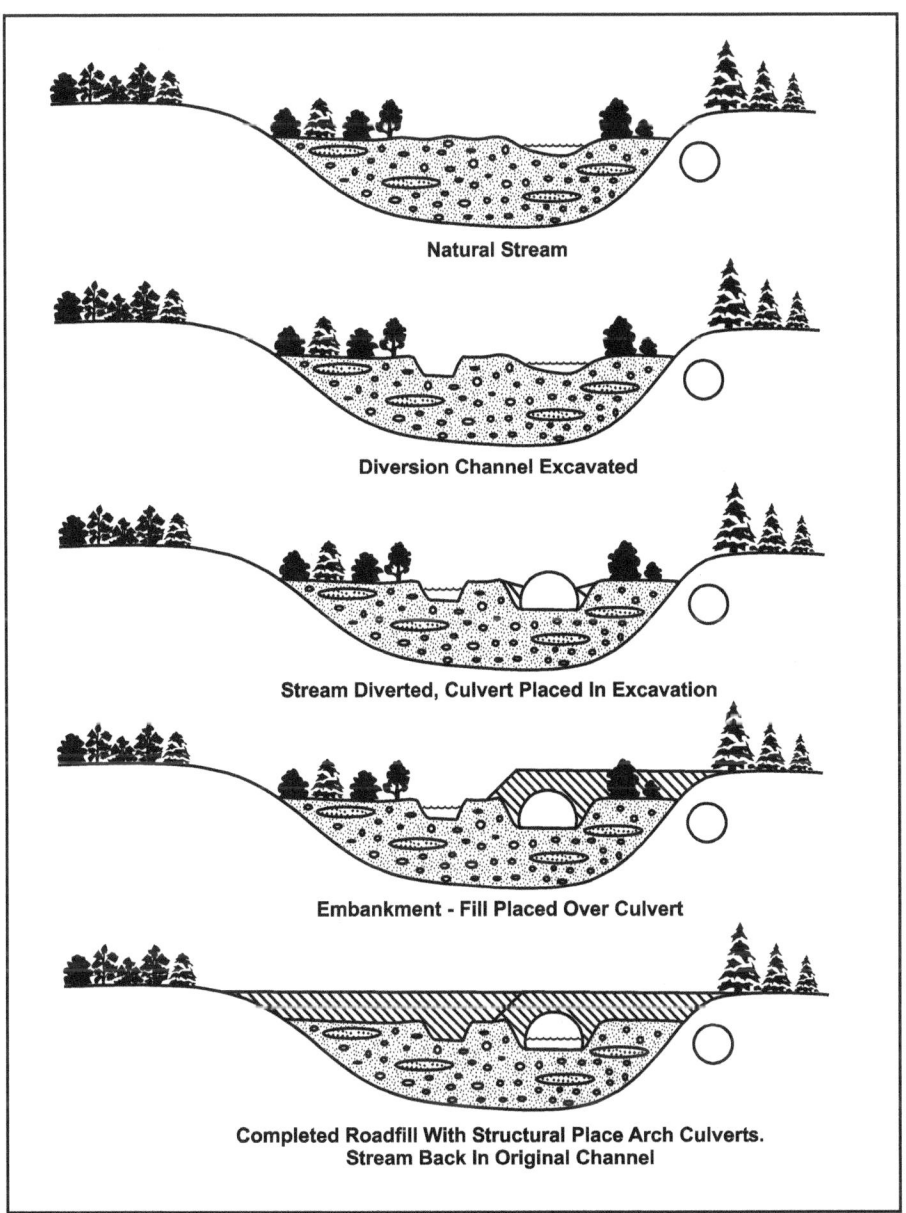

Figure 3-28. Procedure for installing culvert when excavation in channel section of stream could cause sediment movement and increase turbidity (Hynson et al., 1982).

Table 3-19. Sediment Loss Reduction from Reinforcement at Road Stream Crossings (Rothwell, 1983)

Quantity of Sediment Lost	Embankment Reinforcement with Mulch	No Reinforcement
	566 kg/day/ha	2,297 kg/day/ha

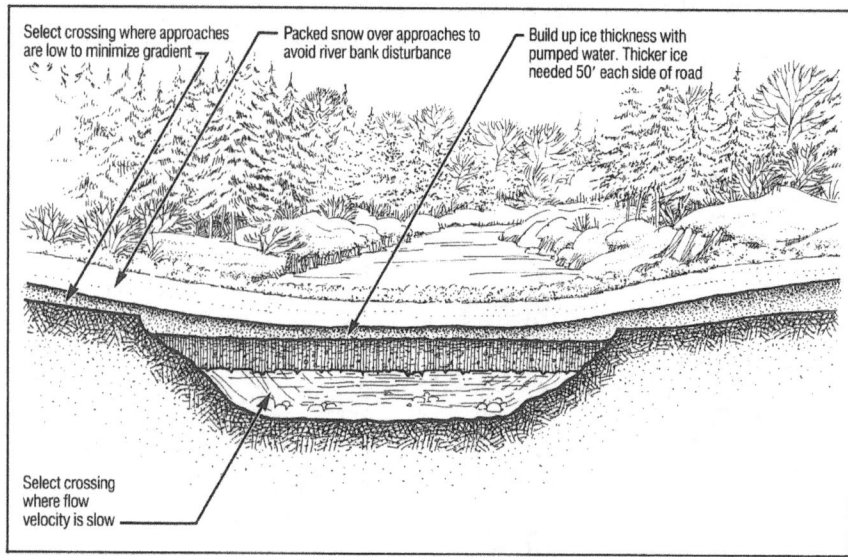

Figure 3-29.   Details of ice bridge construction for temporary stream crossing in winter (Ontario MNR, 1990).

- Make the approaches to the ice bridge nearly level or level.
- Don't add brush or other vegetation to the ice bridge. Doing so weakens the structure and can create a dam when the bridge melts.
- Let the surface freeze; then repeat the construction process until the crossing is of the desired thickness and width.
- Make the bridge thick enough to permit a level approach.
- Also, make the ice thick enough to support the weight and speed of anticipated traffic.
- Inspect the bridge often, because weather and water flow can affect its strength.

Properly constructed winter roads have provisions for adequate drainage during winter weather warmups, and for the spring thaw. If a winter thaw occurs, expect to temporarily shut down road travel. The thaw creates working conditions similar to a wet weather event and causes erosion, severe soil compaction, rutting, and possibly vehicle damage.

Fish Passage Practices

◆ *On streams with spawning areas, avoid construction during egg incubation periods.*

◆ *Design and construct stream crossings for fish passage according to site-specific information on stream characteristics and the fish populations in the stream where the passage is to be installed.*

The types of structures recommended for use on forest roads as fish passage structures are listed below in order of preference (WADOE, 1999). The choice and design of each is determined by a number of factors, including sensitivity of the site to critical fish habitats, engineering specifications, cost, and availability of materials.

1. Bridges—permanent, semipermanent, and temporary
2. Bottomless culverts or log culverts
3. Embedded metal culverts
4. Nonembedded culverts
5. Baffled culverts

Baffled culverts are the most complicated type of fish passage and are the most difficult to design and construct.

To ensure safe fish passage can be provided without resulting in unacceptable effects on existing fisheries habitat values, consider physical, hydrological, and biological factors to determine whether a structure is acceptable for a site. Review the harvest plan and, based on actual site conditions, make any changes necessary to ensure adequate fish passage. Streamflow, bottom substrate, approach slopes, and soil types on either side of the stream are some details from the harvest plan to verified at the site prior to constructing stream crossings and installing culverts. The minimum site data for any proposed bridge or major culvert include

- Cross section showing the high water mark and profile of water crossing.
- Description of water body bed materials.
- Presence or absence of and depth to bedrock.
- Water velocity and direction.
- Bankfull width and depth.
- Bottom channel width.
- Channel topography, including gradient for the site and reach.
- Assessment of natural sediment and debris loading and any other condition that might influence the choice, design, and location of a structure.
- Existing improvements and resource values that might influence the structure.

Minimum biological data for successful stream crossing design include

- Species of fish that you'll want to safely pass
- Size of fish that will pass (life stage)
- Time of year in which fish passage occurs
- High and low design passage flows

The success of any fish passage structure depends very much on channel adjustments that occur after construction of the stream crossing, so it is important to survey far enough upstream and downstream to account for any possible channel conditions that might affect the design and placement of the structure.

3D: ROAD MANAGEMENT

Management Measure for Road Management

(1) Avoid using roads where possible for timber hauling or heavy traffic during wet or thaw periods on roads not designed and constructed for these conditions.

(2) Evaluate the future need for a road and close roads that will not be needed. Leave closed roads and drainage channels in a stable condition to withstand storms.

(3) Remove drainage crossings and culverts if there is a reasonable risk of plugging or failure from lack of maintenance.

(4) Following completion of harvesting, close and stabilize temporary spur roads and seasonal roads to control and direct water away from the roadway. Remove all temporary stream crossings.

(5) Inspect roads to determine the need for structural maintenance. Conduct maintenance practices, when conditions warrant, including cleaning and replacement of deteriorated structures and erosion controls, grading or seeding of road surfaces, and, in extreme cases, slope stabilization or removal of road fills where necessary to maintain structural integrity.

(6) Conduct maintenance activities, such as dust abatement, so that chemical contaminants or pollutants are not introduced into surface waters to the extent practicable.

(7) Properly maintain permanent stream crossings and associated fills and approaches to reduce the likelihood (a) that stream overflow will divert onto roads and (b) that fill erosion will occur if the drainage structures become obstructed.

Management Measure Description

The objective of this management measure is to ensure the management of existing roads to maintain their stability and utility; to minimize erosion, polluted runoff from roads and road structures, and sedimentation in water bodies; and to ensure that roads no longer needed are properly closed and decommissioned so they pose minimal risk to water quality.

Roads that are actively maintained reduce the potential for erosion to occur. Road drainage structures, road fills in stream channels, and road fills on steep slopes are of greatest concern with respect to water quality protection in road management. Roads actively used for timber hauling usually need the most maintenance, and mainline roads typically need more maintenance than spur roads. Regular road use by heavy trucks, especially at stream crossings, creates a chronic source of sediment runoff to streams (Murphy and Miller, 1997). It is important to inspect and repair roads prior to heavy use, especially during wet or thawing ground conditions (Weaver and Hagans, 1984). Use of roads during wet or thaw periods can result in excessive sediment loading to water bodies when road surfaces become deeply rutted and drainage becomes impaired. The first rule of maintaining a stable road surface is to minimize hauling and grading during wet weather conditions, especially if the road is unsurfaced (Weaver and Hagans, 1984).

Sound planning, design, and construction measures often reduce road maintenance needs after construction. Roads constructed with a minimum width in stable terrain, and with frequent grade reversals or dips, need minimum maintenance. Unfortunately, older roads remain one of the greatest sources of sediment from managed forestlands. After harvesting is complete, roads are often forgotten, and erosion problems might go unnoticed until after severe resource damage has occurred.

Routine maintenance of road dips and road surfaces and quick response to drainage problems can significantly reduce road deterioration and prevent the creation of ruts that could channelize runoff (Ontario Ministry of Natural Resources, 1988; Oregon Department of Forestry 1981). Roads and drainage structures on all roads, including decommissioned roads for as long as water quality effects might result from them, should be inspected annually, at a minimum, prior to the beginning of the rainy season (Weaver and Hagans, 1984). Also inspect and perform emergency maintenance during and following peak storms.

In some locations, problems associated with altered surface drainage and diversion of water from natural channels results in serious gully erosion or landslides. In western Oregon, 41 out of the 104 landslides reported on private and state forestlands during the winter of 1989-90 were associated with older (built before 1984) forest roads. These landslides were related to both road drainage and original construction problems. Smaller erosion features, such as gullies and deep ruts, are far more common than landslides and very often are related to poor road drainage.

Sedimentation from roads can be reduced significantly if drainage structures are maintained to function properly. Culverts and ditches that are kept free of debris are less likely to restrict water flow and fish passage. Routinely cleaning these structures can minimize clogging and prevent flooding, gullying, and washout (Kochenderfer, 1970). Fish passage was discussed in the last management measure as an issue of proper sizing and installation of culverts and other stream crossings, and it is equally important to inspect culverts, fords, and bridges on a regular basis to ensure that debris and sediment do not accumulate and prevent fish migration. Undercutting of culvert entrances or exits can create vertical barriers to fish passage, and debris buildup at the entrances of culverts or at trash racks can prevent fish migration. If roads are no longer in use or won't be needed in the foreseeable future, removing drainage crossings and culverts where there is a risk of plugging or failure from lack of maintenance is a precautionary measure. Where a road will be used in the future, it is usually more economical to periodically maintain crossing and drainage structures than not to do so and to have to make extensive repairs after failure.

Road Reconstruction

Road reconstruction provides the opportunity to upgrade and improve substandard and old roads that are no longer used. After an on-site inspection of the entire route and consideration of the economic and environmental costs of the reconstruction, a decision about reopening a road can be made. Reconstruction might be economically feasible for a particular road but could entail unacceptable environmental costs. Roads where stream crossings have been washed out or short, steep sections of road have been entirely lost to progressive erosion or landsliding are examples of roads where the environmental costs of reconstruction might be too high (Weaver, 1994). In such cases, it might be possible to lessen the environmental damage incurred in reconstruction by rerouting the road around problem areas with a section of new road. Factor overall project costs into the economic and environmental costs of any rerouting to determine its feasibility, and do all road reconstruction in a manner consistent with the Management Measure for Road Construction.

Washed-out stream crossings are the most common obstacle to effective road reconstruction. Initial improper sizing of drainage structures or their not being installed or maintained properly results in erosion at stream crossings. When reconstructing stream crossings, it is important to follow the same design and installation procedures as are used for new crossings.

Road Decommissioning

Proper closure, decommissioning, and obliteration are essential to preventing erosion and sedimentation on roads and skid trails that are no longer needed or that have been abandoned (Swift and Burns, 1999). Road closure involves preventing access by placing gates or other obstructions (such as mounds or earth) at road access points while maintaining the road for future use. Roads that will no longer be used or that have remained unused for many years may be decommissioned and obliterated. Decommissioning typically involves stabilizing fills, removing stream crossings and culverts, recontouring slopes, reestablishing original drainage patterns, and revegetating disturbed areas (Harr and Nichols, 1993; Kochenderter, 1970; Rothwell, 1978). Revegetating disturbed areas protects the soil from rainfall and binds the soil, thereby reducing erosion and sedimentation and the potential for mass wasting in the future. Because closed roads and trails are rarely inspected, it is important to leave them in as stable a condition as possible to prevent erosion that could become a large problem before any damage is noticed (Rothwell, 1978).

Road decommissioning can significantly reduce water quality effects from unused roads, and road closure and decommissioning can help realize many objectives and purposes (Harr and Nichols, 1993; Moll, 1996):

- Eliminate or discourage access to roads to reduce maintenance expenditures.

- Eliminate the potential for drainage structure failure and stream diversion.

- Reduce soil loss, embankment washout, mass wasting, failures, slides, slumps, sedimentation, turbidity, and damage to fish habitat.

- Provide cover and organic matter to soil, and improve the quality of wildlife and fish habitat.

- Enhance the visual qualities of road corridors and disturbed areas.

- Attempt to restore the natural pre-road hydrology to the site.

Benefits of Road Management

Proper road maintenance has definite economic benefits. In one comparison of road maintenance costs over time, maintenance costs on a road where BMPs were not installed initially were 44 percent higher than costs on a road where BMPs were installed initially (Dissmeyer and Frandsen, 1988) (Table 3-20).

Table 3-20.   Comparison of Road Repair Costs for a 20-Year Period With and Without BMPs[a] (Dissmeyer and Frandsen, 1988)

Maintenance Costs Without BMPs		Costs of BMP Installation	
Equipment	$365	Labor to construct terraces and water diversions	$780
Materials (gravel)	122		
Work supervision	0	Materials to revegetate	120
Repair cost per 3 years	527	Cost of technical assistance	300
Total cost over 20 years [b]	$2,137	Total cost over 20 years	$1,200

IRR: 11.2%
PNV: $937
B/C ratio: 1.78 to 1.00 for road BMP installation versus reconstruction/repair.

[a] BMPs include construction of terraces and water diversions, and seeding.
[b] Discounted at 4%.

In another economic study, the costs of various revegetation treatments and associated technical services (e.g., planning and reviewing the project in the field) were compared to the benefits over time of the initial planning and BMP installation (Dissmeyer and Foster, 1987) (Table 3-21). Savings resulted from avoiding problem soils, wet areas, and unstable slopes, and the analysis demonstrated that including soil and water resource management (i.e., revegetating and technical services) in road planning and construction is more economical over the long term.

As part of the Fisher Creek Watershed Improvement Project, Rygh (1990) examined the costs of ripping and scarification using different techniques and specifically compared the relative advantages of using track hoes for ripping and scarification versus using large tractor-mounted rippers. Track hoes were found to be preferable to tractor-mounted rippers for a variety of reasons, including the following:

Table 3-21.   Analysis of Costs and Benefits of Watershed Treatments Associated with Roads (SE United States) (Dissmeyer and Foster, 1987)

	Treatment[a]		
	Seed Without Mulch	Seed With Mulch	Hydroseed With Mulch
Costs			
Cost per kilometer ($)	511	816	1,006
Cost per kilometer for soil and water technical services ($)	89	89	89
Total cost of watershed treatment ($)	600	905	1,095
Benefits[b]			
Savings in construction costs ($/km)	446	446	446
Savings in annual maintenance costs ($/km)	267	267	267
Benefit/cost (10-year period)	4.4:1	2.9:1	2.4:1

Note: All costs updated to 1998 dollars.
[a] Treatments included fertilization and liming where needed.
[b] Cost savings were associated with soil and water resource management in the location and construction of forest roads by avoiding problem soils, wet areas, and unstable slopes. Maintenance cost savings were derived from revegetating cut and fill slopes, which reduced erosion, prolonging the time taken to fill ditch lines with sediment and reducing the frequency of ditch line reconstruction.
Source: Adapted by Dissmeyer and Foster from West, S., and B.R. Thomas, 1982. Effects of Skid Roads on Diameter, Height, and Volume Growth in Douglas-Fir. *Soil Sci. Soc. Am. J.*, 45:629–632.

- A reduction in furrows and resulting concentrated runoff caused by tractors
- Improved control over the extent of scarification
- Increased versatility and maneuverability of track hoes
- Cost savings

The study concluded that the cost of ripping with track hoes ranged from $406 to $506 per mile compared to $686 per mile for ripping with D7 or D8 tractors (1998 dollars) (Table 3-22).

Road decommissioning, however, can be expensive. The estimated cost for small roads with gentle terrain and few stream crossings is approximately $22,500; for larger roads with greater slope and larger and more stream crossings, the cost can equal or exceed $282,000 (1998 dollars) (Glasgow, 1993).

Table 3-22.   Comparative Costs of Reclamation of Roads and Removal of Stream Crossing Structures (ID) (Rygh, 1990)

Method	Cost (dollars/mile)
Ripping/scarification	
Ripping with D7 or D8 tractor	$686
Scarifying with D8-mounted brush blade	$1,053
Scarification to 6-inch depth and installation of water bars with track hoe	$2,086
Ripping and slash scattering with track hoe	$549–$823
Ripping, slash scattering, and water bar installation with track hoe	$1,013
Ripping with track hoe	$406–$506

Best Management Practices

Road Maintenance Practices

◆ *Blade and reshape the road to conserve existing surface material; to retain the original, crowned, self-draining cross section; and to prevent or remove berms (except those designed for slope protection) and other irregularities that retard normal surface runoff.*

Ruts and potholes can weaken road subgrade materials by channeling runoff and allowing standing water to persist. Erosion from forest roads is a process associated with their location, construction, and use, and erosion begins with the development of ruts and the erosion of fine material from the road surface (Johnson and Bronsdon, 1995). Severe rutting on a road can cause drivers to seek routes around the ruts and lead to traffic's moving closer to riparian areas and stream channels, essentially widening a road and magnifying the problem (Phillips, 1997). Natural berms can develop on regularly used roads at undesirable locations and can trap runoff on the road instead of allowing it to drain off at design locations. Natural berms can also develop from improper road grading or gradual entrenchment of the road below the surrounding terrain (Swift and Burns, 1999). If serious road degradation due to rutting or other causes has occurred, the road can be regraded, and periodic regrading of roads is usually necessary to fill in wheel ruts and reshape roads. Regrading a road removes ruts, but it exposes more fine sediment that continues to erode for some months after grading until a protective, coarser layer on the road surface is developed. Serious rutting can indicate the need for a more durable surface.

◆ *Maintain road surfaces by mowing, patching, or resurfacing as necessary.*

Annual roadbed mowing and periodic trimming of encroaching vegetation is usually sufficient for grassed roadbeds carrying fewer than 20 to 30 vehicle trips per month.

◆ *Clear road inlet and outlet ditches, catch basins, culverts, and road-crossing structures of obstructions as necessary.*

Avoid undercutting back slopes when cleaning silt and debris from roadside ditches. Minimize machine cleaning of ditches during wet weather. Do not disturb vegetation when removing debris or slide blockage from ditches. The outlet edges of broad-based dips need to be cleaned of trapped sediment to eliminate mud holes and prevent the bypass of storm water. The frequency of cleaning depends on traffic load.

Clear stream-crossing structures and their inlets of debris, slides, rocks, and other materials before and after any heavy runoff period. Surveys by Copstead and Johansen (1998) of the roads in the Detroit Ranger District after storm damage showed that plugged culverts accounted for a greater percentage of damage to the roads than any other cause (Figure 3-30). Culverts were plugged by stream bedload and woody debris. Many times a small branch caught in the culvert inlet caused stream bedload to accumulate, eventually burying the inlet. Undersized culverts accounted for 81 percent of the plugged culverts.

Although regular cleaning of road ditches and culvert inlets and outlets is important, there are circumstances under which leaving accumulated debris in ditches is sometimes called for to help prevent erosion. Some debris might be left in ditches simply to interrupt the free flow of runoff down the ditch, thus reducing the velocity of the runoff and erosion as well.

During road construction, the cut slope is often undercut to provide the design flow capacity in roadside ditches or to provide room for culvert inlets, and undercut slopes are usually unstable. Especially above culvert inlets, soil erosion on the cut slope can lead to high maintenance costs. If, based on experience gained after the road is constructed, the flow in the ditch is less than it was designed for, leaving the accumulated debris in the ditch can help stabilize the cut slope above it. If debris has to be cleared out of a portion of ditch that repeatedly fills with sediment to provide sufficient volume for runoff flow, an option is to build a permanent or temporary passage under the accumulated debris and leave the debris to help stabilize the slope above the ditch. A temporary underpass can be

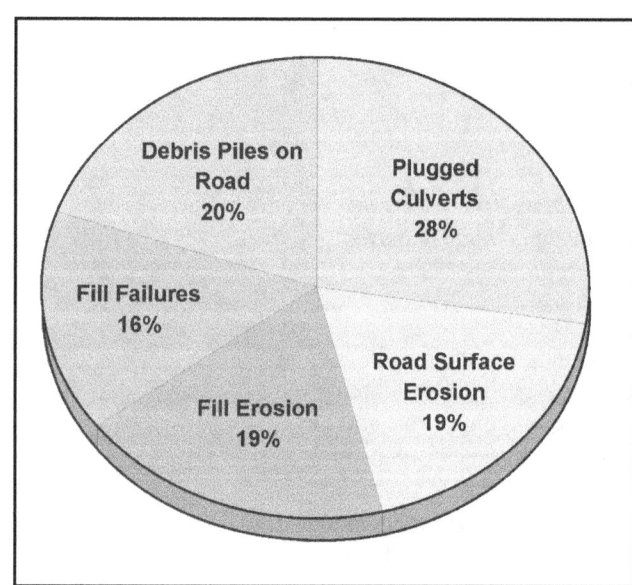

Figure 3-30. Road-related storm damage by type in the Detroit Ranger District (Copstead and Johansen, 1998).

constructed of two logs placed parallel with a gap between them and a third log on top. A permanent underpass can be constructed much like a culvert (Firth, 1992).

◆ *Remove any debris that enters surface waters from a winter road or skid trail located over surface waters before a thaw.*

◆ *Return the spring following a harvest and build erosion barriers on any skid trails that are steep enough to erode.*

◆ *Abate dust problems during dry summer periods.*

Excessive road dust during the summer is a condition that can threaten water quality. Dust can deliver large quantities of fine sediment to nearby stream channels. This fine material can be especially damaging to fish and fish habitat. Seasonal summer roads need almost the same amount of maintenance as permanent roads.

Dust control methods such as applying dust oil and watering during dry summer conditions are almost always necessary during an intensive dry season to prevent excessive loss of surface materials.

Wet and Winter Road Practices

◆ *Before winter, inspect and prepare all permanent, seasonal, and temporary roads for the winter months.*

Winterizing consists of maintenance and erosion control work needed to drain the road surface (Weaver, 1994). Clean trash barriers, culvert inlet basins, and pipe inlets of floatable debris and sediment accumulations. Clean ditches that are partially or entirely plugged with soil and debris, and trim and remove heavy concentrations of vegetation that impede flow. Gate and close seasonal and temporary roads to nonessential traffic.

Surface runoff problems caused by winter use of a bermed, unsurfaced road can cause rutting. The ruts collect runoff and cause additional erosion of the road. Lack of waterbars or rolling dips, together with the graded berm along the outside edge of the road, keep surface runoff on the roadbed. Annual grading can produce an outside berm of soil and rock that can be graded back onto the road surface.

Winter is a popular time to harvest wetlands or areas that are not accessible during wet periods, and road structures that will have to be maintained during the winter can be marked prior to snowfall. Snow accumulation could otherwise hide the BMPs.

◆ *On woodland roads "daylight" or remove trees to a width that permits full sunlight to reach the ground.*

The objective of road "daylighting" is to have sunlight dry the road so that it is less susceptible to erosion and damage from vehicle traffic. Daylighting also promotes the establishment of protective vegetative cover on road fillslopes and cutslopes and vegetation for wildlife. Vegetation clearing to promote daylighting needs to be managed so that slope integrity is not compromised. Daylighting should also be coordinated with wildlife specialists so that openings that might be detrimental to certain wildlife species, such as neotropical migratory birds, are not created.

Stream Crossing and Drainage Structure Practices

◆ *When temporary stream crossings are no longer needed, and as soon as possible upon completion of operations, remove culverts and log crossings to maintain adequate streamflow. Restore channels to pre-project size and shape by removing all fill materials used in the temporary crossing.*

Failure or plugging of abandoned temporary crossing structures can result in greatly increased sedimentation and turbidity in the stream, as well as channel blowout.

◆ *Replace open-top culverts with cross drains (water bars, dips, or ditches) to control and divert runoff from road surfaces.*

Open-top culverts are for temporary drainage of ongoing operations. It is important to replace them with more permanent drainage structures to ensure adequate drainage and reduce erosion potential prior to establishment of vegetation on the roadbed. It is recommended that open-top culverts be used for ongoing operations only and that they be removed upon completion of activities (Wiest, 1998).

◆ *During and after logging activities, ensure that all culverts and ditches are open and functional.*

Culvert plugging is common in woodland streams (Flanagan and Furniss, 1997). The risk of culvert plugging is greatest where small culverts have been installed on wide streams. Channel width controls the size of debris that can be transported in a stream, and culverts with a diameter that is less than the width of the stream are prone to block and accumulate woody debris. Another configuration that leads to debris trapping is increasing channel width toward a culvert inlet. Woody debris, transported in a lengthwise position down a stream, can rotate to a position perpendicular to the channel where the channel widens and block the culvert inlet. Hand, shovel, and chainsaw work can remedy almost all culvert maintenance needs (Weaver and Hagans, 1984). Heavy machinery and equipment is usually unnecessary to keep culverts clean.

Where culvert and ditch plugging is a problem, assess the cause of the problem and develop a strategy to correct it (see Roads Analysis in the Management Measure for Preharvest Planning, subsection 3A). Corrective measures might include installation of a new culvert, trimming dead wood from overhanging vegetation, or performing regularly scheduled maintenance.

Road Decommissioning, Obliteration, and Closure Practices

◆ *Decommission or obliterate roads that are no longer needed (see Road Decommissioning in this section).*

When a road is not needed for harvesting, forest management activities, or recreation, it can be decommissioned. Effective decommissioning reduces actual and potential erosion from the road and saves maintenance costs. Typically, a road is decommissioned by removing temporary stream crossings, installing water bars to minimize erosive surface runoff flows, and planting stream crossings and the road surface with vegetation to retail soil. If decommissioning is properly done, an area previously occupied by a forest road blends into the surrounding landscape naturally, erodes no more than an undisturbed site, and provides wildlife habitat. Decommissioned roads are generally left in a state such that they can be opened and used again in the future should the need arise.

More than 120 miles of roads have been decommissioned in the Targhee National Forest in Idaho (USDA-FS, 1997). Roads in riparian areas were particularly targeted for decommissioning. Decommissioning the roads involved seeding with grasses and adding water bars to prevent erosion. In the Lake Tahoe Basin, existing road surfaces are ripped to a depth of 12 to 18 inches, the surface is seeded, and pine needle mulch is spread on top to prevent erosion and encourage good establishment of vegetation. The road prism and drainage features are left in place to prevent erosion and soil runoff while the vegetation establishes itself. Roads decommissioned by the U.S. Forest Service in Region 8 are similarly seeded to create linear wildlife open areas that provide forage and edge vegetation. The U.S. Forest Service in Region 4, where the Targhee National Forest is located, found that public acceptance of the road decommissioning was enhanced by adding turnarounds and parking areas at the closure gates.

Road obliteration goes further than road decommissioning by returning a forest road to its natural drainage characteristics and topography to the extent possible. It is a suitable goal for roads that will not be used in the future. Road obliteration aims to eliminate alterations in drainage patterns created by a road system and the potential for drainage structure failure and stream diversion, and to reestablish drainage connectivity that might have been interrupted by the presence of the road (Moll, 1996).

Stabilizing areas disturbed by road construction and use is another major goal of road obliteration. Disturbed slopes, road cuts and fills, and areas to which drainage will be directed after the obliteration is terminated are areas that need to be stabilized. In some cases, artificial means to stabilize slopes might be necessary until vegetation has become established.

Road obliteration can lead to improvements in fisheries habitat where sediment runoff from old forest roads enters streams. The practice was used in a watershed in northwest Washington as part of watershed rehabilitation to improve fisheries habitats and water quality and to reduce flood hazards. On unused, 30- to 40-year-old, largely impassable roads and landings, fills were stabilized, stream crossings were removed, slopes were recontoured, and drainage patterns were reestablished at an average cost of $3,950 per kilometer (with a range of $1,500 to $7,500 per kilometer) (1998 dollars). Costs were lowest where little earthmoving was involved, more where a lot of brush had to be cleared away and sidecast material had to be pulled upslope, and highest where fills were removed at stream crossings and landings. Afterward, however, the obliterated roads and landings sustained much less damage from storms than unused roads that were not obliterated (Harr and Nichols, 1993).

Road obliteration in the Redwood National Park demonstrated that the following measures are effective for restoring hydrology and habitat (Belous, 1984, cited in NCASI, 2000): stream crossing removal, road outsloping, straw mulch placement, tree planting on road alignments and stream crossings, and waterbars. Soil decompaction and terrain recontouring wee found to be important first steps in successful road obliteration. Topsoil replacement significantly aided vegetation establishment.

◆ *Wherever possible, completely close roads to travel and restrict access by unauthorized persons by using gates or other barriers (Figure 3-31).*

Figure 3-31. Install visible traffic barriers where appropriate to prevent off-road vehicle and other undesired disturbance to recently stabilized roads (Indiana DNR, 1998).

Closing a road that is not needed in the immediate future for harvesting or other forestry purposes can minimize use that could create erosive conditions and the need for continual maintenance. Closed roads should be decommissioned or maintained regularly. Access to roads at entry points can be restricted using rocks, logs, slash piles, or other on-site materials; planted trees; fences, gates; guardrails; or concrete barriers. Complete obliteration of a road access point can be accomplished by recontouring and removing all drainage structures, bridges, and other road features. Traffic entry should be regulated where restricting access with such barriers is not feasible.

◆ *Convert closed forest access roads into recreation trails.*

An unused forest access road can be converted to recreational use for off-road vehicles, horseback riding, mountain biking, and hiking. All of these activities, however, create the potential for road or trail damage, and regular maintenance of stream crossings, waterbars, and other drainage structures is necessary to ensure that sediment runoff from the road does not threaten water quality. The frequency and type of maintenance depends on the type and intensity of recreational use allowed on the road. Trails need the same kinds of runoff control measures as roads, and regular trail maintenance is as important as regular road maintenance (Figure 3-32).

Figure 3-32. Construct trails using the same drainage structures as closed forest roads (Indiana DNR, 1998).

◆ *Install or regrade water bars on roads that will be closed to vehicle traffic and that lack an adequate system of broad-based dips (Figure 3-33).*

Water bars help to minimize the volume of water flowing over exposed areas and remove water to areas where it will not cause erosion. Water bar spacing depends on soil type and slope. Table 3-23 presents the Oregon Department of Forestry's suggested guidelines for water bar spacing. In other states with different climates, topographies, and soil types, recommended spacing might differ from these guidelines; contact the state forestry department for assistance. Divert water flow off the water bar onto rocks, slash, vegetation, duff, or other less erodible material and avoid diverting it directly to streams or bare areas. Outslope closed road surfaces to disperse runoff and prevent closed roads from routing water to streams.

◆ *Revegetate disturbed surfaces to provide erosion control and stabilize the road surface and banks.*

Refer to the Management Measure for Revegetation of Disturbed Areas for a more detailed discussion of this practice.

◆ *Periodically inspect closed roads to ensure that vegetational stabilization measures are operating as planned and that drainage structures are operational. Conduct reseeding and drainage structure maintenance as needed.*

Figure 3-33. Broad-based dips reduce the potential for erosion (Indiana DNR, 1998).

Table 3-23.   Example of Recommended Water Bar Spacing by Soil Type and Slope (Oregon Department of Forestry, 1979a)

Road Grade (percent)	Soil Type		
	Granitic or Sandy	Shale or Gravel	Clay
2	900	1,000	1,000
4	600	1,000	800
6	500	1,000	600
8	400	900	500
10	300	800	400
12	200	700	400
15	150	500	300
20	150	300	200
25+	100	200	150

Note: Distances (in feet) are approximate and are varied to take advantage of natural features.
Recommendations of spacing will vary with soil type, climate, and topography. Consult your state forester.

3E: Timber Harvesting

Timber Harvesting Management Measure

The timber harvesting management measure consists of implementing the following:

(1) Follow layouts for timber harvesting operations determined under the Preharvest Planning Management Measure, subject to adjustments made based on preharvest on-site inspections.

(2) Install landing drainage structures to avoid sedimentation to the extent practicable. Disperse landing drainage over sideslopes.

(3) Construct landings away from steep slopes and reduce the likelihood of fill slope failures. Protect landing surfaces used during wet periods. Locate landings outside streamside management areas.

(4) Protect stream channels and significant ephemeral drainages from logging debris and slash material.

(5) Use appropriate areas for petroleum storage, draining, and dispensing, and vehicle maintenance. Establish procedures to contain and treat spills that could occur during these activities. Recycle or properly dispose of all waste materials.

For cable yarding:

(1) Limit yarding corridor gouge or soil plowing by properly locating cable yarding landings.

(2) Locate corridors for streamside management areas according to the guidelines of the Management Measure for Streamside Management Areas.

For groundskidding:

(1) To the extent practicable, do not operate groundskidding equipment within streamside management areas except at stream crossings. In streamside management areas, fell and endline trees in a manner that avoids sedimentation.

(2) Use improved stream crossings for skid trails that cross flowing drainages. Construct skid trails to disperse runoff and with adequate drainage structures.

(3) On steep slopes, use cable systems rather than groundskidding where groundskidding could cause excessive sedimentation.

Management Measure Description

The goal of this management measure is to minimize the likelihood of water quality effects resulting from timber harvesting. This goal can be accomplished by taking precautions to control erosion and sedimentation during harvesting operations and by storing, handling, and disposing of petroleum products and vehicle maintenance products in an environmentally safe manner.

Reducing effects on soils and water quality from harvesting begins in the preharvest planning stage, when a system of roads, landings, and skid trails is planned. Preharvest planning, as described in the Preharvest Planning Management Measure, is performed to minimize the amount of disturbed area, which makes it easier to rehabilitate the site after the operation is complete; locate roads on stable soils to minimize erosion and at a safe distance from streams; build stream crossings at the locations where they cause the least amount of instream disturbance and hydrological change; and limit disturbance to sensitive areas. Thoroughly review the Preharvest Planning Management Measure before incorporating the practices in this management measure into a harvesting plan. The practices in that management measure can serve as a guide for reducing soil disturbance and water quality effects during harvesting. Having a harvesting plan reviewed by a professional forester before starting any aspect of harvesting or road building is strongly recommended. The forester might be able to offer ideas specific to the planned harvest on how environmental damage and operational costs can be reduced.

Do an additional review of the harvesting plan in conjunction with a site visit to verify that the information used during planning is still valid. Aerial photos and topographic and soil maps can inaccurately represent actual conditions, especially if these media are more than a few years old. Before construction begins, verify that the soils and slopes where landings and skid trails are to be located are suitable to the use and that equipment maintenance or chemical handling areas are appropriately located. As the harvest progresses, make any alterations to the harvesting plan necessary to protect soils and water quality.

Conducting a harvest with attention paid to the potential for soil disturbance from the operation can result in significantly less water quality impairment than conducting a harvest with little or no attention paid to the potential for environmental damage. For instance, skid trails that are parallel to the slope of the land have far more potential to yield sediment-laden runoff than skid roads that run along the contour. Similarly, practices that minimize soil compaction on and prevent or disperse runoff from landings and loading decks can be implemented to reduce the potential for sediment-laden runoff and to minimize sediment delivery to surface waters. Incorporating these and other erosion reduction practices into a harvesting plan, conducting an on-site inspection during the planning stage before harvesting or road construction begins to ensure that the practices chosen are appropriate to the site, and properly implementing and maintaining the practices can significantly decrease water quality effects.

Spill prevention and containment procedures are necessary to prevent petroleum products from entering surface waters. Chemicals and petroleum products spilled in harvest areas can be transported great distances if they enter areas of concentrated runoff, and therefore can adversely affect water quality far from where they are spilled. Designating appropriate areas for the storage and handling of petroleum products and protecting these areas from precipitation can minimize the water quality effects that could result from spills or leakage.

Many studies have evaluated and compared the effects of different timber harvest techniques on soil loss (erosion), soil compaction, and overall ground disturbance associated with various harvesting techniques. The data presented in Tables 3-24 through 3-28 were compiled from many studies conducted throughout the United States and Canada. Some of the data presented in the table should be considered as older data that were based on operations conducted prior to current understanding and concern for water quality protection. The studies examined different harvesting systems (e.g., clear-cuts, selective harvesting) using a variety of techniques (e.g., cable yarding, skidding). Local factors such as climate, soil type, and topography affected the results of each study. The major

Table 3-24. Soil Disturbance from Roads for Alternative Methods of Timber Harvesting (Megahan, 1980)

| Logging System (State) | Percent of Logged Area Bared | | | Reference |
	Roads	Skid Roads and Landings	Total	
Tractor:				
Tractor — clear-cut (BC)	30.0	—	30.0	Smith, 1979
Tractor — selection (CA)	2.7	5.7	8.4	Rice, 1961
Tractor — selection (ID)	2.2	6.8	9.0	Haupt and Kidd, 1965
Tractor — group selection (ID)	1.0	6.7	7.7	Haupt and Kidd, 1965
Tractor and helicopter — fire salvage (WA)	4.5	0.4	4.9	Klock, 1975
Tractor and cable — fire salvage (WA)	16.9	—	16.9	Klock, 1975
Ground Cable:				
Jammer — group selection (ID)	25–30	—	25–30	Megahan and Kidd, 1972
Jammer — clear-cut (BC)	8.0	—	8.0	Smith, 1979
High-lead — clear-cut (BC)	14.0	—	14.0	Smith, 1979
High-lead — clear-cut (OR)	6.2	3.6	9.8	Silen and Gratkowski, 1953
High-lead — clear-cut (OR)	3.0	1.0	4.0	Brown and Krygier, 1971
High-lead — clear-cut (OR)	6.0	1.0	7.0	Brown and Krygier, 1971
High-lead — clear-cut (OR)	6.0	—	6.0	Fredriksen, 1970
Skyline:				
Skyline — clear-cut (OR)	2.0	—	2.0	Binkley, 1965
Skyline — clear-cut (BC)	1.0	—	1.0	Smith, 1979
Aerial:				
Helicopter — clear-cut	1.2	—	1.2	Binkley[a]

[a] Estimated by Virgil W. Binkley, Pacific Northwest Region, USDA Forest Service, Portland, OR, nd.

conclusions of these studies regarding the relative effects of different timber harvesting techniques on soil erosion, summarized below, are shared among the studies and enable cross-geographic comparison:

- Aerial and skyline cable techniques are far less damaging than other yarding techniques.

- Tractor, jammer, and high-lead cable methods result in significantly more soil disturbance and compaction than skyline and aerial techniques.

- Skyline yarding serves far more area per mile of road than skidding.

Although skidding can be damaging, areas disturbed by skidding operations can be rehabilitated without a net economic loss to the landowner. An analysis of the costs and benefits of rehabilitating skid trails in the southeastern United States by planting different species of trees indicated that the benefit/cost ratios of using shortleaf pine, hardwood pine, and hardwoods were 5.1:1, 2.8:1, and 1.3:1, respectively. Shortleaf pine yielded the highest benefit for costs incurred (Dissmeyer and Foster, 1986).

Table 3-25.   Soil Disturbance from Logging by Alternative Harvesting Methods (Megahan, 1980)

Method of Harvest	Location	Disturbance (%)	Reference
Tractor:			
Tractor — clear-cut	E. WA	29.4	Wooldridge, 1960
Tractor — clear-cut	W. WA	26.1	Steinbrenner and Gessel, 1955
Tractor — fire salvage	E. WA	36.2	Klock,[a] 1975
Tractor on snow — fire salvage	E. WA	9.9	Klock,[a] 1975
Tractor — clear-cut	BC	7.0	Smith, 1979
Tractor — selection	E. WA, OR	15.5	Garrison and Rummel, 1951
Ground Cable:			
Cable - selection	E. WA, OR	20.9	Garrison and Rummel, 1951
High-lead — fire salvage	E. WA	32.0	Klock,[a] 1975
High-lead — clear-cut	W. OR	14.1	Dyrness, 1965
High-lead — clear-cut	W. OR	12.1	Ruth, 1967
High-lead — clear-cut	BC	6.0	Smith, 1979
Jammer — clear-cut	BC	5.0	Smith, 1979
Grapple — clear-cut	BC	1.0	Smith, 1979
Skyline:			
Skyline — clear-cut	W. OR	12.1	Dyrness, 1965
Skyline — clear-cut	E. WA	11.1	Wooldridge, 1960
Skyline — clear-cut	BC	7.0	Smith, 1979
Skyline — clear-cut	W. OR	6.4	Ruth, 1967
Skyline — fire salvage	E. WA	2.8	Klock,[a] 1975
Balloon — clear-cut	W. OR	6.0	Dyrness[b]
Aerial:			
Helicopter — fire salvage	E. WA	0.7	Klock,[a] 1975
Helicopter — clear-cut	ID	5.0	Clayton (in press)

[a] Disturbance shown is classified as severe.
[b] C.T. Dyrness, unpublished data on file, Pacific Northwest Forest and Range Experiment Station, Corvallis, OR, nd.

Benefits of Timber Harvesting Practices

After a 1994 study of BMP implementation and effectiveness, the Virginia Department of Forestry concluded that harvesters often failed to seed bare soil with adequate ground cover. The department determined that ground cover of 70 percent or more is effective, while many sites studied had ground cover on only 0 to 35 percent of bare soil. The Vermont Agency of Natural Resources (1998) also studied the effectiveness of erosion control BMPs and concluded that the construction and proper placement of such BMPs before harvesting is essential for protecting water quality. The Agency also found that regularly maintaining BMPs increased the longevity of their effectiveness.

In general, poor BMP effectiveness can be due to many factors, including

- A lack of time or willingness to plan timber harvests carefully before cutting begins.
- A lack of skill in or knowledge of designing effective BMPs.

Table 3-26.   Relative Effects of Four Yarding Methods on Soil Disturbance and Compaction in Pacific Northwest Clear-cuts (OR, WA, ID) (Sidle, 1980)

Yarding Method	Bare Soil (%)	Compacted Soil (%)	Water Quality Effects
Tractor	35	26	Greater
High-lead	15	9	
Skyline	12	3	
Balloon	6	2	Lesser

Table 3-27.   Percent of Land Area Affected by Logging Operations (Southwest MS) (after Miller and Sirois, 1986)

Operational Area	Cable Skyline (% Land Affected)	Groundskidding (% Land Affected)	Water Quality Effects
Cable corridors or skid trails	9.2	21.4	Greater
Landings	4.1	6.4	
Spur roads	2.6	3.5	Lesser
Water Quality Effects	Lesser	Greater	

Table 3-28.   Skidding/Yarding Method Comparison (after Patric, 1980)

Harvesting System	Acres Served per Mile of Road	Water Quality Effects
Wheeled skidder	20	Greater
Jammer	31	
High-lead	40	
Skyline	80	Lesser

- A lack of equipment needed to implement effective BMPs.
- The belief that BMPs are not an integral part of the timber harvesting process and can be engineered and fitted to a logging site after timber harvesting has been completed.
- A lack of timely BMP maintenance.

Best Management Practices

Harvesting Practices

◆ *Based on information obtained from site visits, make any alterations to the harvesting plan that are necessary or prudent to protect soils from erosion and surface waters from sedimentation or other forms of pollution.*

◆ *Fell trees away from watercourses whenever possible, keeping logging debris from the channel, except where debris placement is specifically prescribed for fish or wildlife habitat.*

◆ *Immediately remove any tree accidentally felled in a waterway.*

◆ *Remove unwanted slash from water bodies and place it above the normal high water line or flood level to prevent downstream transport.*

As discussed in Chapter 2 and in Chapter 3, section B, *Streamside Management Areas*, streams have natural amounts of organic debris (e.g., fallen leaves, twigs, limbs, and trees), and the amount varies with season, tree falls, storms, and so forth. Aquatic organisms are adapted to the presence of and variability in the quantity of organic debris in streams. Large woody debris, or LWD, affects channel morphology, provides structure and complexity to aquatic and terrestrial organism habitats, and is a source of nutrients for aquatic organisms. When the quantity of LWD and organic debris in general that reaches a stream is changed, either to too much or too little, it can be detrimental to the aquatic system's ecology and ability to support life. Removing excessive slash from a stream helps maintain water flow and avoids the addition of excessive nutrients. In instances where the addition of organic debris—especially LWD—to a stream is desirable, an appropriate amount may be left in stream channels or on stream banks. Slash left in streams adds nutrients, regulates stream temperature, and traps fine sediments where these effects are desirable (Jackson, 2000). Consult with a fisheries biologist or the state forestry or ecology department for specific guidance for your area.

Leave pieces of large woody debris in place during stream cleaning to preserve channel integrity and maintain stream productivity. Indiscriminate removal of large woody debris can adversely affect channel stability. Figure 3-34 presents one way to determine debris stability. State forestry or ecology departments can help with such determinations for particular regions and stream types.

Where desirable, leave slash on the harvest site and distribute it to provide good ground cover and minimize erosion after the timber harvest.

Leaving slash on disturbed soils can help reduce erosion until new vegetative growth is established. The quantity of slash to leave depends on the erodibility of the soil, though leaving an amount that provides 40 to 60 percent ground cover for soils that have low to high erodibility, respectively, is recommended. Leaving slash on the ground significantly reduces erosion potential. It also keeps the nutrients contained in the slash material on the site for incorporation into the soil and new vegetative growth.

Practices for Landings

◆ *Make landings no larger than necessary to safely and efficiently store logs and load trucks.*

◆ *Install drainage and erosion control structures as necessary.*

A slight slope on landings facilitates drainage. Also, adequate drainage on approach roads prevents road drainage water from entering the landing area.

◆ *Do not exceed a 5 percent slope on landing surfaces and shape them to promote efficient drainage.*

◆ *Do not exceed 40 percent slope on landing fills and do not incorporate woody or organic debris into fills.*

◆ *If landings are to be used during wet periods, protect the surfaces with a suitable material such as a wooden mat or gravel.*

◆ *Install drainage structures–such as water bars,
culverts, and ditches–on
landings to avoid sedi-
mentation. Disperse
landing drainage over
side slopes. Provide
filtration or settling if
water is concentrated in a
ditch.*

◆ *Upon completion of a
harvest, clean up, re-
grade, and revegetate
landings.*

• Upon abandonment,
minimize erosion on
landings by adequately
ditching or mulching
with forest litter.

• Establish a herbaceous
cover on areas that will
be used again in re-
peated cutting cycles,
and restock landings
that will not be reused.

• If necessary, install
water bars for drainage
control.

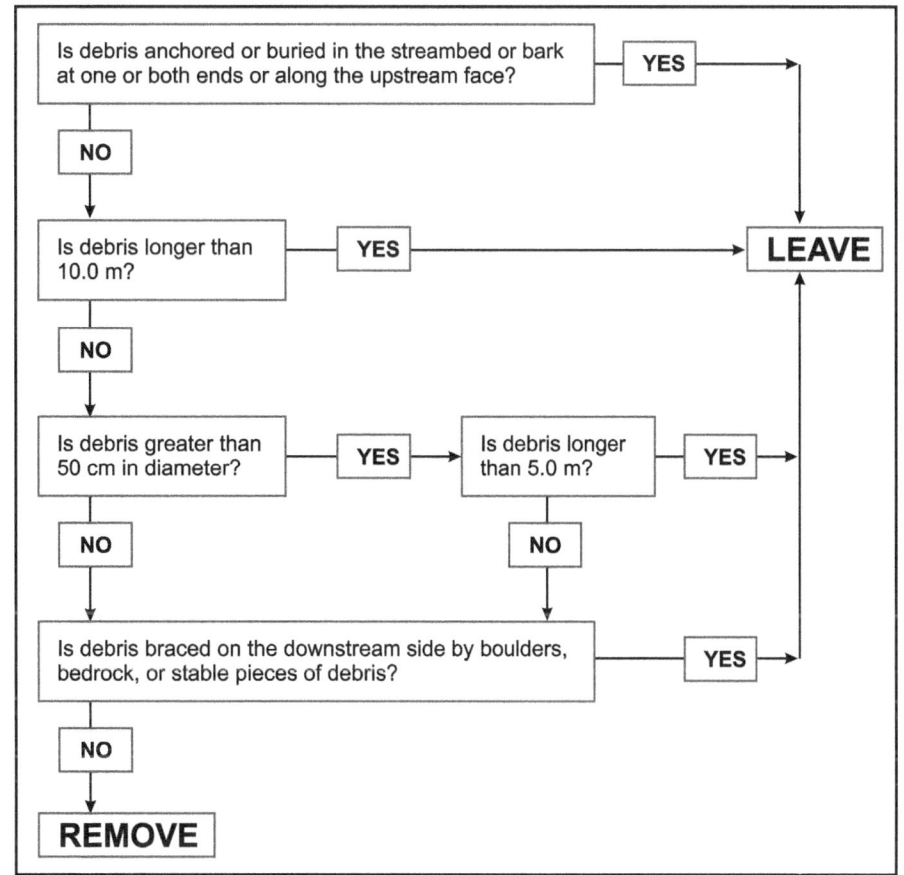

**Figure 3-34. General large woody debris stability guide based on Salmon Creek, Washing-
ton (after Bilby, 1984).**

• Landings should be
ripped to break up compacted soil layers and allow water infiltration. This will also
aid in the establishment of new vegetation.

• Runoff on and from landings should be dispersed with waterbars or dips.

◆ *Locate landings for cable yarding where slope profiles provide favorable deflection
conditions so that yarding equipment does not cause yarding corridor gouge or soil
plowing, which can concentrate drainage or cause slope instability.*

◆ *Locate cable yarding corridors for streamside management areas according to the
Streamside Management Areas management measure. Avoid disturbing major channel banks in SMAs with yarded logs.*

Ground Skidding Practices

◆ *Skid uphill to log landings whenever possible. Skid with ends of logs raised to reduce
rutting and gouging.*

This practice disperses water on skid trails away from the landing. Skidding uphill lets
water from trails flow onto progressively less-disturbed areas as it moves downslope,
reducing erosion hazard. Skidding downhill concentrates surface runoff on lower slopes

along skid trails, resulting in significant erosion and sedimentation hazard. If skidding downhill, provide adequate drainage on approach trails so that drainage does not enter the landing.

◆ *Skid along the contour (perpendicular to the slope), and avoid skidding on slopes greater than 40 percent.*

Following the contour reduces soil erosion and encourages revegetation. If skidding has to be done parallel to the slope, skid uphill, taking care to break the grade periodically.

Avoid skid trail layouts that concentrate runoff into draws, ephemeral drainages, or watercourses and avoid skidding up or down ephemeral drainages. Use endlining to winch logs out of SMAs or directionally fell trees so tops extend out of SMAs and trees can be skidded without operating equipment in SMAs. In SMAs, endline trees carefully to avoid soil plowing or gouge.

Suspend ground skidding during wet periods, when excessive rutting and churning of the soil begins, or when runoff from skid trails is turbid and no longer infiltrates within a short distance from the skid trail. Further limitation of ground skidding of logs, or use of cable yarding, might be needed on slopes where there are sensitive soils and/or during wet periods.

Retire skid trails by installing water bars or other erosion control and drainage devices, removing culverts, and revegetating.

• After logging, obliterate and stabilize all skid trails by mulching and reseeding.

• Build cross drains on abandoned skid trails to protect stream channels or side slopes in addition to mulching and seeding.

• Restore stream channels by removing temporary skid trail crossings.

• Distribute logging slash throughout skid trails to supplement water bars and seeding to reduce erosion on skid trails.

Cable Yarding Practices

◆ *Use cabling systems or other systems when ground skidding would expose excess mineral soil and induce erosion and sedimentation.*

• Use high-lead cable or skyline cable systems on slopes greater than 40 percent.

• To avoid soil disturbance from sidewash, use high-lead cable yarding on average-profile slopes of less than 15 percent.

◆ *Avoid cable yarding in or across watercourses.*

When cable yarding across streams cannot be avoided, use full suspension to minimize damage to channel banks and vegetation in the SMA. Cut or clear cableways across SMAs where SMAs must be crossed. This will reduce the damage to trees remaining and prevent trees next to the stream channel from being uprooted.

◆ *Yard logs uphill rather than downhill.*

When yarding uphill, log decks are placed on ridges or hilltops rather than in low-lying areas. This approach results in less soil disturbance for two reasons: (1) lifting the logs reduces their weight on the ground and thus the amount of friction and ground scouring, and (2) yard trails radiate outward from the elevated position of the log deck, dispersing runoff in numerous directions from the deck.

Downhill yarding does the opposite. The full weight of the logs is transferred to the ground, and runoff from all of the yard trails is directed downslope to the log deck, concentrating the erosive effect of rain. If yarding uphill is not possible, soil disturbance can be minimized during downhill yarding by suspending logs from a pulley system so that the logs are lifted partially or completely off the ground.

The amount of soil disturbance caused by yarding depends on the slope of the area, the volume yarded, the size of the logs, and the logging system. Megahan (1980) ranked yarding techniques (from greatest effect to lowest effect) based on percent area disturbed as follows: tractor (21 percent average), ground cable (21 percent, one study), high-lead (16 percent average), skyline (8 percent average), jammer in clear-cut (5 percent, one study), and aerial techniques (4 percent average). Aerial and skyline cable techniques are far less damaging than other yarding techniques.

The amount of road needed for different yarding techniques varies considerably (Sidle, 1980). Skyline techniques use the least amount of road area, with only 2 to 3.5 percent of the land area in roads. Tractor and single-drum jammer techniques use the greatest amount of road area (10 to 15 percent and 18 to 24 percent of total area, respectively). High-lead cable techniques fall in the middle, with 6 to 10 percent of the land used for roads. Compared to the skyline and aerial techniques, tractor, jammer, and high-lead cable methods result in significantly higher amounts of disturbed soil (Megahan, 1980). Figure 3-35 shows a typical cable yarding operation (OSHA, 1999).

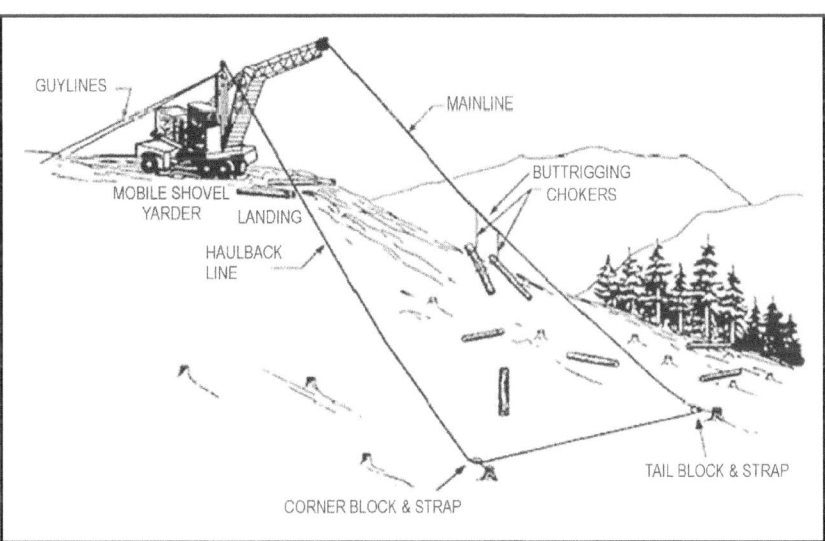

Figure 3-35. Typical cable yarding operation (OSHA, 1999).

Other Yarding Methods

◆ *Horse logging*

Horse logging can be a viable alternative to mechanized logging for small harvests or for sensitive environmental areas of a larger harvest. Horses give a lot of control for logging in partial cuts because logs are cut to log length, not left at tree length, and this improves maneuverability around trees that are left in place. This maneuverability combined with the narrower path needed by horses compared to a skidder means that fewer trees have to be removed solely for access. Soil is compacted and disturbed less with horse logging than with a skidder because a horse weighs about 1,600 pounds compared to a rubber-tired skidder that weighs about 10,000 pounds.

◆ *Helicopter yarding*

Helicopter yarding is a practical and environmentally friendly alternative yarding approach for use on public and private timberlands where other yarding systems would be physically, economically, or environmentally infeasible. According to the Helicopter Logging Association (1998), the benefits of helicopter timber harvesting include:

- Minimum damage is caused to the following:
  - The soil layer. Very little vehicular traffic is associated with the method.
  - Water resources. There is a negligible increase in stream turbidity compared to conventional yarding methods.
  - Riparian areas.
  - Wildlife habitat.

- Damage to retained trees is reduced. Fewer trees are felled per acre and ground-based skidders are absent.

- Road density is lower. A combined helicopter and tractor logging approach can reduce road density by approximately half compared to conventional tractor methods. Environmental damage is thus reduced, and forest access points are fewer.

◆ *Shovel harvesting.*

Shovel harvesting is more widely used in the coastal areas of the Pacific Northwest and the wetland areas of the Southeast than in other parts of the United States (Aust, Virginia Tech, personal communication, 2000). The process of shovel harvesting involves a shovel logger moving in lines parallel to a road, picking up logs that have been felled by a logger and lifting debris out of gullies as it moves forward. The shoveler starts at the nearest access point and moves logs until they are within reach of a road, where they can be retrieved (Figure 3-36) (Humboldt State University, 1999).

Shovel logging is considered an environmentally friendly means to harvest timber. Operations require fewer people and fewer access roads, produce no skid trails, reduce ground disturbance in environmentally sensitive areas such as wetlands, and disturb SMAs less than any conventional logging method. Table 3-29 compares the costs of various yarding methods.

◆ *Balloon harvesting.*

Balloon harvesting involves using hot air or helium balloons to remove logs from a harvest site for loading on trucks (Figure 3-37). Because the logs are lifted off the ground and taken to a log landing, they are not dragged up or down a slope and disturbance to the

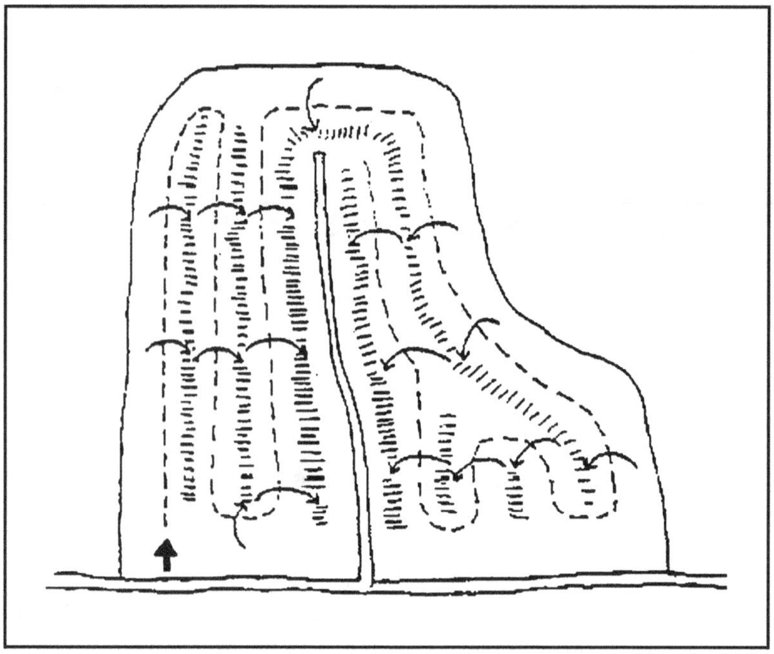

Figure 3-36. Common pattern of shovel logging operations (Humboldt State University, 1999)

Table 3-29.   Costs Associated with Various Methods of Yarding

Yarding Method	Cost Range
Cable Yarding	$90 to $135/ac, depending on yarding distance, crew size, and size of landing. • Clear-cutting costs $50 to $60/mbf • Thinning costs $200/mbf
Helicopter Yarding	$3,000 to $3,500/hr; or $180 to $300/mbf $175 to $285/mbf
Shovel Harvesting	$25.00 to $83.84/hr

ground is reduced. In areas where road construction is expensive, balloon harvesting can save money and protect the environment because of the smaller number of roads and skid trails needed. The environmental benefits realized from balloon harvesting are similar to those associated with helicopter yarding. Additionally, balloon harvesting permits access to wet sites such as wetlands and steep slopes where ground skidding would not be feasible because of the potential for environmental damage or the cost of road construction (Aust, Virginia Tech, personal communication, 2000).

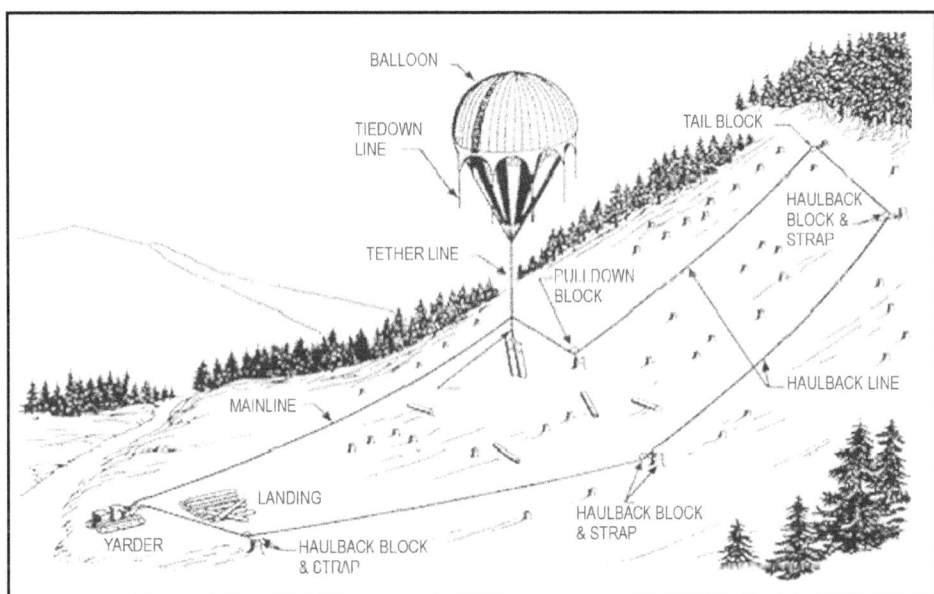

Figure 3-37.   Balloon harvesting practices on a steep slope (OSHA, 1999).

Winter Harvesting

Winter harvesting is a component of several state timber removal programs. In winter frozen ground provides conditions that do not exist during other times of the year for timber harvest activities and an opportunity for low-impact logging (Logan and Clinch, 1991). Areas where winter road construction and harvesting are particularly advantageous include wetlands (see Chapter 3, section J, Management Measure for Wetlands Forest Management of this document for a discussion of BMPs specifically for wetland harvesting), sensitive riparian areas, and sites where erosion and soil compaction would be expected to be a serious problem during nonfrozen conditions.

BMP guidelines for warmer months apply during winter harvesting as well. Additional practices that can be implemented to ensure the protection of water quality include the following (Logan and Clinch, 1991; North Dakota Forestry Service, 1999):

◆ *Consult with operators experienced in winter logging techniques.*

◆ *Compact skid trail snow before skidding logs.*

Compacting the snow prevents damage to soils that are still wet or not completely frozen.

◆ *Avoid steeper areas where frozen skid trails may be subject to erosion the following spring.*

◆ *Before felling in wet, unfrozen soil areas, use tractors or skidders to compact the snow on skid trails. Avoid steep areas where frozen skid trails might be subject to erosion the following spring.*

Petroleum Management Practices

◆ *Service equipment where spilled fuel or oil will not reach watercourses, and drain all petroleum products and radiator water into containers.*

◆ *Dispose of wastes and containers in accordance with proper waste disposal procedures.*

Do not leave waste oil, filters, grease cartridges, and other petroleum-contaminated materials as refuse in the forest.

◆ *Take precautions to prevent leakage and spills.*

Ensure that fuel trucks and pickup-mounted fuel tanks do not have leaks. Use and maintain seepage pits or other confinement measures to prevent diesel oil, fuel oil, or other liquids from running into streams or important aquifers, and use drip collectors on oil-transporting vehicles.

◆ *Develop a spill contingency plan that provides for immediate spill containment and cleanup, and notification of proper authorities.*

Have materials for absorbing spills easily accessible, and collect wastes for proper disposal.

3F: Site Preparation and Forest Regeneration

Management Measure for Site Preparation and Forest Regeneration

Confine on-site potential NPS pollution and erosion resulting from site preparation and the regeneration of forest stands. The components of the management measure for site preparation and regeneration are:

(1) Select a method of site preparation and regeneration suitable for the site conditions.

(2) Conduct mechanical tree planting and ground-disturbing site preparation activities on the contour of sloping terrain.

(3) Do not conduct mechanical site preparation and mechanical tree planting in streamside management areas.

(4) Protect surface waters from logging debris and slash material.

(5) Suspend operations during wet periods if equipment used begins to cause excessive soil disturbance that will increase erosion.

(6) Locate windrows at a safe distance from drainages and SMAs to control movement of the material during high-runoff conditions.

(7) Conduct bedding operations in high-water-table areas during dry periods of the year. Conduct bedding in sloping areas on the contour.

(8) Protect small ephemeral drainages when conducting mechanical tree planting.

Management Measure Description

Regeneration of harvested forestlands is important not only in terms of restocking a valuable resource, but also in terms of minimizing erosion and runoff from disturbed soils that could degrade water quality. Vegetative cover on disturbed soils reduces raindrop impact and slows storm runoff, and the roots of vegetation stabilize soils by holding them in place and aiding their aggregation. Both of these factors decrease erosion.

Harvesters and landowners can follow certain practices to protect the soil and aid tree regeneration. For instance, leaving the forest floor litter layer intact during site preparation operations minimizes soil disturbance and detachment, maintains infiltration, and slows runoff. These factors in turn reduce erosion and sedimentation after site preparation is completed. It is especially important to leave the forest floor litter layer intact in areas that have steep slopes, or erodible soils, or where the prepared site is located near a water body, all of which increase the risk of erosion, landslides, and degraded water quality. Site preparation methods such as herbicide application and prescribed burning cause less disturbance to the soil surface than mechanical practices and can be considered where mechanical site preparation could pose a threat to water quality. Drum chopping, a form of mechanical site preparation, normally results in less soil exposure than other mechanical methods. The intensity of a prescribed burn in part determines whether use of the method will pose a threat to water quality.

Natural regeneration, hand planting, and direct seeding are other methods that can be used to minimize soil disturbance, especially on steep slopes with erodible soils. Mechanical planting with machines that scrape or plow the soil surface can produce erosion rills, increasing surface runoff and erosion and decreasing site productivity.

Data in Figures 3-38 to 3-42 compare sediment loss or erosion rates for numerous site preparation methods. Many of the data are site-specific, so site characteristics and experimental conditions are mentioned (when available) in the text below and regional locations are noted on the figures.

Ballard (2000) reviewed the effects of forest management on forest soils. Mechanical site preparation, he noted, both has benefits and causes problems. Nutrient depletion is one adverse effect. A study in northern British Columbia concluded that 500 kg N/ha were removed on a large area that had been bladed, raked, and piled for burning. Conducting research on intensively-managed loblolly pine plantations in the Piedmont region of North Carolina, Piatek and Allen (2000) found the following nutrient removal rates from sites that received different methods of site preparation: Shear-pile-disk, 591 kg N/ha and 34 kg P/ha; stem-only harvest, 57 kg N/ha and 5 kg P/ha; chop and burn, 46 kg N/ha and 0 kg P/ha. Piatek and Allen (2000) also found that the nutrients removed during site preparation had no observable effect on foliage production when measured 15 years after planting on the site.

Beasley (1979) studied the relative soil disturbance effects of site preparation following clear-cutting on three small watersheds in the hilly northern coastal plain of Mississippi and Arkansas (Figure 3-38). Slopes in the three watersheds were mostly 30 percent or more. One site was single drum-chopped and burned; another was sheared and windrowed (windrows were burned); and a third was sheared, windrowed, and bedded to contour. The control watershed was instrumented and left uncut. Soil exposure was 37 percent on the chopped site, 53 percent on the sheared and windrowed site, and 69 percent on the bedded site. A temporary cover crop of clover was sown after site preparation to protect the soil from rainfall impact and erosion. Increases in soil erosion and sediment production were similar for all three treatments in the first year after site preparation. Decreases in these processes were noted during the second year on all sites. During the second year,

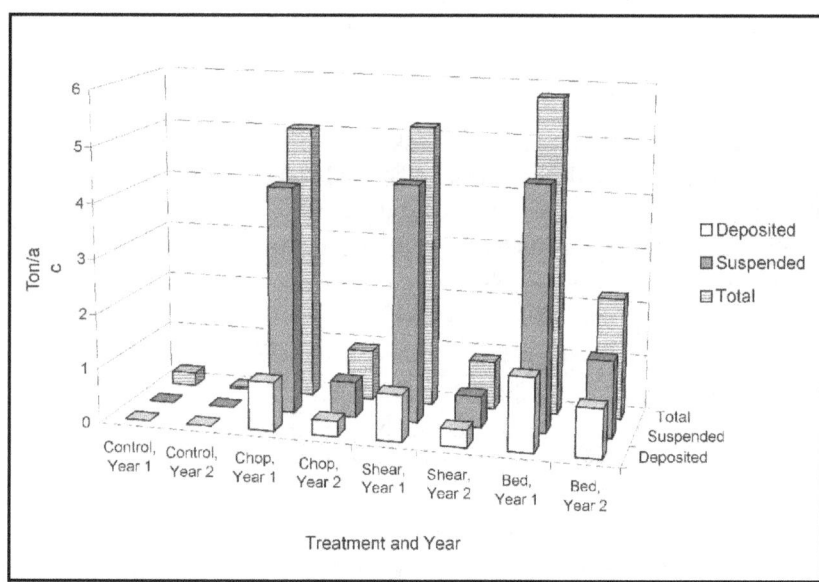

Figure 3-38. Deposited, suspended, and total sediment losses in experimental watersheds during water years 1976 and 1977 for various site preparation techniques (Mississippi, Arkansas) (after Beasley, 1979).

the clover and other vegetation covered 85 to 95 percent of the surface of each site and effectively decreased sediment production.

Golden and others (1984) summarized studies on erosion rates from site preparation (Figure 3-39). The rates reflect soil movement measured at the bottom of a slope, not the quantity of sediment actually reaching streams. Therefore, the numbers estimate the worst-case erosion if a stream is located directly at the toe of a slope with no intervening vegetation. Rates are averages for 3- to 4-year recovery periods.

Dissmeyer (1980) showed that discing produced more than twice the erosion rate of any other method (Figure 3-40). Bulldozing,

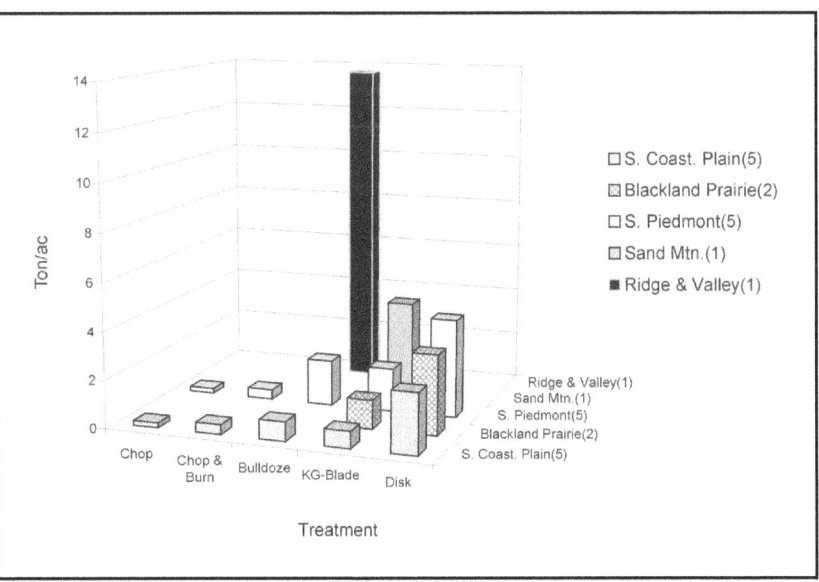

Figure 3-39. Predicted erosion rates using various site preparation techniques for physiographic regions in the southeastern United States (after Golden et al., 1984). Numbers in parentheses indicate number of predictions for the region.

shearing, and sometimes grazing were associated with relatively high rates of erosion, and chopping or chopping and burning produced moderate erosion rates. Logging also produced moderate erosion rates in this study when the effect of skid and spur roads was included. The lowest rate of erosion was associated with burning.

Beasley and Granillo (1985) compared storm flow and sediment losses from mechanically and chemically prepared sites in southwest Arkansas over a 4-year period. Mechanical preparation (clear-cutting followed by shearing, windrowing, and replanting with pine seedlings) increased sediment losses in the first 2 years after treatment. A subsequent decline in sediment losses in the mechanically prepared watersheds was attributed to rapid growth of ground cover. Windrowing brush into ephemeral drainages and leaving it unburned effectively minimized soil losses by trapping sediment on the site and reducing channel scouring. Chemical site preparation (using herbicides) had no significant effect on sediment losses.

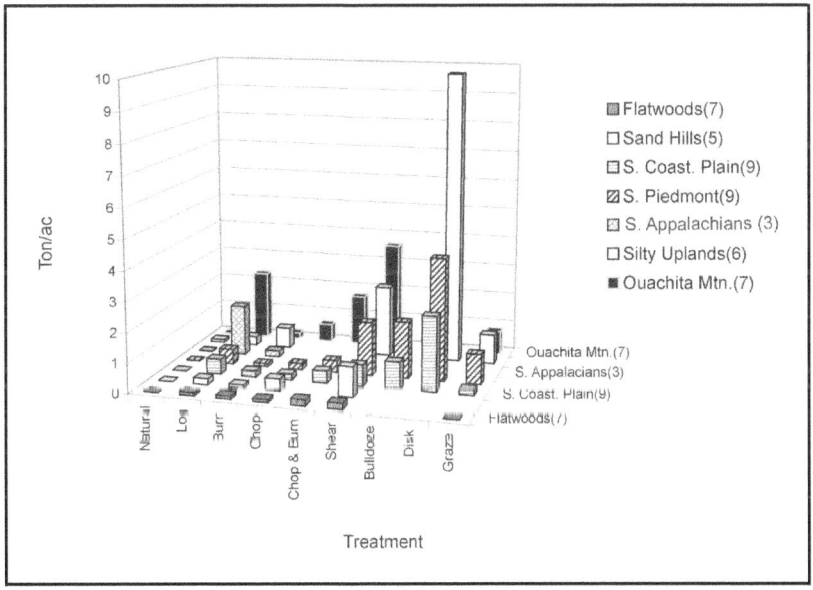

Figure 3-40. Erosion rates for site preparation practices in selected land resource areas in the Southeast (after Dissmeyer, 1980). Numbers in parentheses indicate the number of sites in the region.

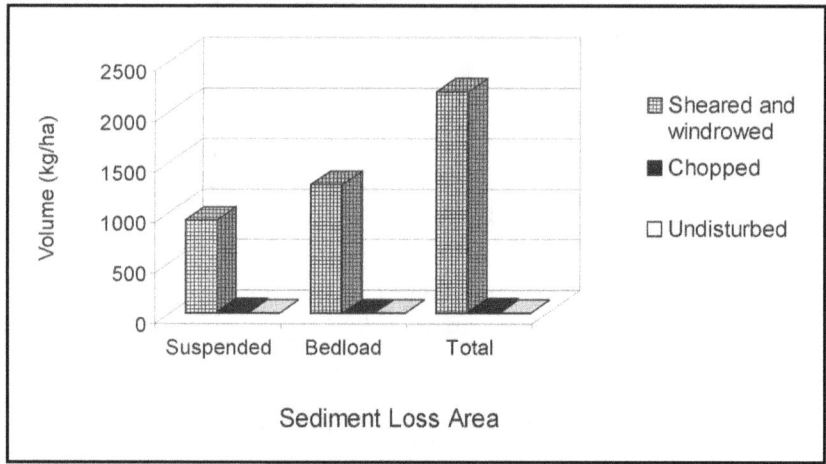

Figure 3-41. Sediment loss (kg/ha) in stormflow by site treatment from January 1, to August 31, 1981 (TX) (after Blackburn et al., 1982).

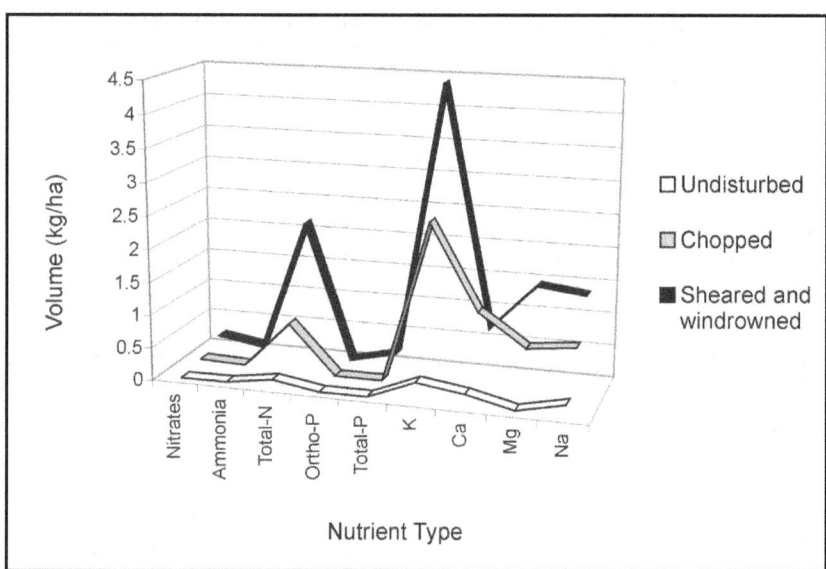

Figure 3-42. Nutrient loss (kg/ha) in stormflow by site treatment from January 1 to August 31, 1981 (TX) (after Blackburn et al., 1982).

Blackburn and others (1982) studied water quality changes associated with two site preparation methods in Texas. Figure 3-41 shows that shearing and windrowing (which exposed 59 percent of the soil) produced 400 times more sediment loading than chopping (which exposed 16 percent of the soil) during site preparation in this study. The authors also found that total nitrogen losses from sheared and windrowed watersheds were nearly 20 times greater than those from undisturbed watersheds and three times greater than those from chopped watersheds (Figure 3-42).

Mechanical Site Preparation in Wetlands

Under certain circumstances, a permit is needed for mechanical forestry site preparation activities when used for the establishment of pine plantations in the Southeast. EPA and the U.S. Army Corps of Engineers recently issued a memorandum to clarify the applicability of forested wetlands BMPs to these circumstances. Refer to the Wetlands Forest Management Measure for a discussion of permitting requirements in forested wetlands.

Benefits of Site Preparation Practices

Three studies summarized here compare the costs and benefits of different site preparation methods. Dissmeyer and Foster (1987) estimated the long-term costs and benefits of light and heavy site preparation in the Southeast. They concluded that light site preparation would yield more wood production and a higher internal rate of return on investment (Table 3-30). Heavy site preparation methods involve a greater initial investment than light site preparation methods but did not yield more wood per unit area.

Table 3-30. **Analysis of Two Management Schedules Comparing Cost and Site Productivity in the Southeast (Dissmeyer and Foster, 1987)**

| Year | Silviculture Treatment | Light Site Preparation[a] | | Heavy Site Preparation[b] | |
		Investment Per Hectare[c]	Wood Produced M³/ha	Investment Per Hectare[c]	Wood Produced M³/ha
1984	Site Prep/Tree Planting	$297		$420	
1999	Thinning	$252	64.2 pulpwood	$180	46.0 pulpwood
2010	Thinning	$256	22.3 saw timber 33.3 pulpwood	$331	5.3 saw timber 22.0 pulpwood
2020	Final Harvest	$2,422	133.5 saw timber 15.2 pulpwood	$2,071	112.3 saw timber 22.0 pulpwood
Present Net Value (at 4%)		$623		$304	
Internal Rate of Return		12.4%[d]		10.1%	

[a] Light site preparation includes chop and light burn or chop with herbicides, and reduces soil exposure and erosion.
[b] Heavy site preparation includes bulldozing or windrowing or shearing and windrowing, and increases erosion and sediment yields over those for light site preparation.
[c] 1984 dollars.
[d] Based on 4% inflation rate assumed.
Source: Adapted from Patterson, 1984. Dollars in Your Dirt. Alabama's Treasured Forests. Spring: 20-21

Dissmeyer (1986) analyzed the economic benefits of controlling erosion during site preparation. Site preparation methods that increased soil exposure, displacement, and compaction increased site preparation costs and erosion from the site prepared (Table 3-31) and decreased timber production. Using light site preparation techniques such as a single chop and burn reduced erosion, increased timber production on the site, and cost less per unit area treated than more intensive site preparation methods. Heavy site preparation techniques such as shearing and windrowing removed nutrients, compacted soil, increased erosion and site preparation costs, and resulted in a lower present net value of timber.

The U.S. Forest Service (1987) examined the costs of three alternatives to slash treatment: (1) broadcast burn and protection of streamside management zones, (2) yarding of unmerchantable material (YUM) of 15 inches in diameter or more, and (3) YUM of

Table 3-31. **Site Preparation Comparison (VA, SC, NC) (Dissmeyer, 1986)**

Treatment	Treatment Cost ($/acre)	Erosion Index[a]
No site preparation	$59	1.0
Burn only	$67	1.1
Single chop and burn	$119	2.3
Double chop and burn	$178	3.0
Single shear and burn	$216	4.3
Shear twice and burn	$253	5.1
Rootrake and disk and burn	$253	16.0
Rootrake and burn	$253	16.0

Note: All costs updated to 1998 dollars
[a] The index is an expression of relative erosion potential resulting from each treatment.

8 inches in diameter or more (Table 3-32). The two YUM alternatives cost approximately $625-$1,180/acre, in comparison to broadcast burning at $1,300/acre (1998 dollars). In addition, the YUM alternatives protected highly erodible soils from direct rainfall and runoff effects, reduced fire hazards, resulted in meeting air and water quality standards, and allowed for the rapid establishment of seedlings on clear-cut areas.

Table 3-32.   Comparison of Costs for Yarding Unmerchantable Material (YUM) vs. Broadcast Burning (OR) (USDA-FS, 1987)

Activity	Broadcast Burn and Protect SMA	YUM 15" in Diameter and No Burn	YUM 8" in Diameter and No Burn
Broadcast burn	$502/acre	N/A	N/A
SMA protection	$646/acre	N/A	N/A
YUM, fell hardwood, lop and scatter	N/A	$438/acre	$1,004/acre
Planting cost	$143/acre	$187/acre	$172/acre
Totals	$1,291/acre	$624/acre	$1,177/acre

Note: All costs updated to 1998 dollars.

Best Management Practices

Site Preparation Practices

◆ *Do not conduct mechanical site preparation, except for drum chopping, on slopes greater than 30 percent.*

On sloping terrain greater than 10 percent, or on highly erosive soils, operate mechanical site preparation equipment on the contour.

◆ *Do not conduct mechanical site preparation in SMAs.*

◆ *Do not place slash in perennial or intermittent drainages, and remove any slash that accidentally enters drainages.*

Slash can clog the channel and cause alterations in drainage configuration and increases in sedimentation. Extra organic material can lower the dissolved oxygen content of the stream. Slash also allows silt to accumulate in the drainage and to be carried into the stream during storm events.

◆ *Provide SMAs of sufficient width to protect streams from sedimentation by the 10-year storm.*

◆ *Locate windrows a safe distance from drainages to avoid material movement into the drainages during high-runoff conditions.*

Locating windrows above the 50-year floodplain usually prevents windrowed material from entering floodwaters.

◆ *Avoid mechanical site preparation operations during periods of saturated soil conditions, which might cause rutting and accelerate soil erosion.*

◆ *Minimize soil movement when shearing, piling, or raking.*

◆ *Minimize incorporation of soil material into windrows and piles during their construction.*

This can be accomplished by using a rake or, if using a blade is unavoidable, keeping the blade above the soil surface and removing only the slash. This helps retain nutrient-rich topsoil, which promotes rapid site recovery and tree growth and increases the effectiveness of the windrow in minimizing sedimentation.

Forest Regeneration Practices

◆ *Distribute seedlings evenly across the site.*

◆ *Order seedlings well in advance of planting time to ensure their availability.*

◆ *Hand plant highly erodible sites, steep slopes, and lands adjacent to stream channels (SMAs).*

◆ *Operate planting machines along the contour to avoid ditch formation.*

  • Ensure that soil conditions (slope, moisture conditions, etc.) are suitable for machine operation.

  • Close slits or drilling furrows periodically to avoid channeling flow.

3G: FIRE MANAGEMENT

Management Measure for Fire Management

Prescribe fire for hazardous fuel reduction and control or suppression of wildfire in a manner that reduces potential nonpoint source pollution of surface waters:

(1) Prescribed fire should not cause excessive sedimentation due to the combined effect of partial or full removal of canopy and removal of ground fuels, litter layer and duff.

(2) Prescriptions for wildland fire use should protect against excessive erosion or sedimentation to the extent practicable.

(3) All bladed firelines, for prescribed fire and wildfire, should be stabilized with water bars and/or other appropriate techniques if needed to control excessive sedimentation or erosion of the fireline.

(4) Wildfire suppression and rehabilitation should consider possible NPS pollution of watercourses, while recognizing the safety and operational priorities of fighting wildfires.

Management Measure Description

The goal of this management measure is to minimize nonpoint source pollution and erosion resulting from prescribed fire used for site preparation, fuel hazard reduction, and activities associated with wildfire control or suppression. Studies have shown that pre-scribed burning, if carefully planned and done using appropriate BMPs, has no signifi-cant effect on water quality (South Carolina Forestry Commission, 2000).

Prescribed burning reduces hazardous fuels. Where tree species are ecologically depen-dent on fire for regeneration or maintenance of healthy stands, fire is an essential forest management tool. Particularly in the interior west and much of the south, ecosystems developed in the presence of frequently-occurring, low-intensity ground fires. Returning these stands to a structure that more closely resembles that which occurred under these frequent fire regimes requires the use of prescribed fire. Because fire suppression has contributed to increased levels of fuels, wildland fires occurring in these areas burn quite hot and consume a lot of material (live and dead).

The severity of burning and the proportion of the watershed burned are the major factors that affect the influence of prescribed burning on streamflow and water quality. Fires that burn severely on steep slopes close to streams and that remove most of the forest floor and litter down to the mineral soil are most likely to adversely affect water quality. The amount of erosion following a fire depends on

- The amount of ground cover remaining on the soil
- The steepness of the slope
- The time, amount, and intensity of subsequent rainfall
- The severity of fire
- The erodibility of the soil and soil type

- How rapidly a site revegetates
- The type of vegetation

Periodic, low-intensity prescribed fires usually have little effect on water quality, and revegetation of burned areas reduces sediment yield from prescribed burning and wildfires.

Cost of Prescribed Burning

Costs associated with prescribed fire depend on the size of the fire crew, the amount of heavy equipment needed at the site to control the burn, the areal extent and intensity of the burn, and the topography of the area being burned. Table 3-33 provides a range of costs associated with prescribed burning (Hansit, personal communication, 2000; Holburg, personal communication, 2000).

Table 3-33.   Range of Prescribed Fire Costs

Topography	Crew Cost[a]	Heavy Equipment Cost[a]
Mountainous	$50 to $100 per acre	$200 to $400 per acre
Flat land	$3 to $60 per acre	$75 to $300 per acre

[a] Hansit, personal communication, 2000; Holburg, personal communication, 2000.

Best Management Practices

Prescribed Fire Practices

◆ *Plan burning to take into account weather, time of year, and fuel conditions so that these help achieve the desired results and minimize effects on water quality.*

Evaluate ground conditions to control the pattern and timing of the burn.

◆ *Execute the prescribed burn with an agency-qualified crew and burn boss.*

◆ *Do not conduct intense prescribed fire for site preparation in the SMA.*

◆ *Do not pile and burn for slash removal purposes in the SMA.*

◆ *Avoid construction of fire lines in the SMA.*

◆ *Avoid conditions that require extensive blading of fire lines by heavy equipment when planning burns.*

◆ *Use handlines, firebreaks, and hose lays to minimize blading of fire lines.*

◆ *Avoid burning on steep slopes in high-erosion-hazard areas or areas that have highly erodible soils.*

Prescribed Fire in Wetlands

◆ *Whenever possible, conduct burns in wetlands in a manner that does not completely remove the organic layer of the forest floor.*

Prescribed burns conducted in wetlands have the potential to be the most severe due to the increased fuels available. Conduct the fire to minimize the potential to increase surface runoff and soil erosion.

◆ *When conducting prescribed fire to regenerate fire-dependent species, such as aspen, minimize consumption of the organic layer and openings in the vegetation to that which is necessary to obtain adequate regeneration.*

◆ *Do not construct firelines that could drain wetlands.*

◆ *Avoid intense burning.*

Intense burning can accelerate erosion by consuming more organic cover than desired.

Wildfire Practices

Wildfire can change erosion rates on the burned area in two ways. First, fire eliminates vegetative soil cover. Second, chemical changes in the soil following fire may create an increased resistance to water infiltration in the upper soil layer, and this can increase surface runoff and sheet erosion (Elliot et al., 1998). The magnitude of these effects depends on how hot a fire burns, slope, vegetation type, and soil resistance to erosion. Erosion following fire is greatest where a fire has burned most severely and the fire is followed by a strong storm, a year of moderately high rainfall, or a spring with a large volume of snowmelt.

◆ *Whenever possible leave a 300-foot buffer on both sides of a waterway when using aerially applied fire retardants. If necessary to apply retardant within the 300-foot zone, used the application method that will most accurately keep the retardant from entering the stream.*

The U.S. Forest Service will stop purchasing fire retardant chemicals that contain sodium ferrocyanide. A recent study revealed that mixtures with the chemical can decompose to produce amounts of cyanide that exceed EPA water quality guidelines for freshwater organisms.

◆ *Do not clean application equipment in watercourses or locations that drain into watercourses.*

◆ *Close water wells and temporary water catchments excavated for wildfire-suppression activities as soon as practical following fire control.*

◆ *During wildfire emergencies, firelines, road construction, and stream crossings are unrestricted by BMPs when necessary for health and safety of firefighters and the public and protection of resources from greater damage due to wildfire. However, use BMPs whenever possible and begin remediation as soon as possible after the emergency is controlled.*

Fireline Practices

Fireline construction is an integral part of both wildfire suppression and preparation for prescribed burning. Because of the possibility of water quality degradation following fireline construction, however, precautions are necessary to ensure that water quality is not impaired when firelines are constructed (Florida Department of Agriculture and Consumer Services, 1993). Fireline construction involves removing all organic material to expose mineral soil, and this can result in excessive erosion and water quality degradation. In wetland systems, firelines can function as drainage corridors, resulting in excessive drainage and converting a wetland to a non-wetland system. Implementation of one or more of the following practices can minimize water quality effects from fireline construction.

◆ *Use natural or in-place barriers (e.g., roads, streams, and lakes) to minimize the need for fireline construction in situations where artificial construction of firelines could result in excessive erosion and sedimentation.*

◆ *Avoid placing firelines through sensitive areas such as wetlands, marshes, prairies, and savannas unless absolutely necessary.*

◆ *When crossing water bodies with plowing equipment, raise the plow to prevent connecting the fireline directly to the water body. Water bodies can be used as firelines to avoid unnecessarily disturbing riparian zones.*

◆ *Construct firelines with the minimum disturbance possible that still allows for safe and effective firefighting, for instance handline rather than cat line when possible.*

◆ *Construct firelines in a manner that minimizes erosion and sedimentation and prevents runoff from directly entering watercourses.*

◆ *Avoid constructing firelines in SMAs. When necessary to construct line in SMAs, use appropriate strategies following direction in Land Management Plans for protection of resources*

◆ *Minimize construction of fireline straight up and down hill. Balance location of fireline with potential for larger fire consuming greater amounts of material.*

The following minimum impact suppression techniques (MIST) for firelines are recommended to minimize water quality impacts (http://www.nps.gov/crmo/firemp/crmofmp_aj.htm).

- Minimize fireline construction by taking advantage of natural barriers, rock outcrops, trails, roads, streams, and other existing fuel breaks.
- Construct firelines to be as narrow as necessary to halt the spread of the fire and place then to avoid impacts to water resources.
- Leave unburned material within the final line.
- Minimize clearing and scraping.
- Flag the route to the fire from the nearest trail or road to minimize off-road travel and soil disturbance.

Fireline Rehabilitation

◆ *Where possible, use alternatives to plowed lines such as harrowing, foam lines, wet lines, or permanent grass.*

◆ *Get cover on the site as soon as possible after the fire is out to maintain erosion control measures on firelines.*

◆ *Revegetate firelines with native species.*

◆ *Install grades, ditches, and water bars as soon as it is safe to begin rehabilitation work.*

◆ *Install water bars on any fireline running up and down the slope, and direct runoff onto a filter strip or sideslope, not into a drainage.*

3H: REVEGETATION OF DISTURBED AREAS

Management Measure for Revegetation of Disturbed Areas

Reduce erosion and sedimentation by rapid revegetation of areas disturbed by harvesting operations or road construction:

(1) Revegetate disturbed areas (using seeding or planting) promptly after completion of the earth-disturbing activity. Local growing conditions will dictate the timing for establishment of vegetative cover.

(2) Use mixes of species and treatments developed and tailored for successful vegetation establishment for the region or area.

(3) Concentrate revegetation efforts initially on priority areas such as disturbed areas in SMAs or the steepest areas of disturbance (e.g., on roads, landings, or skid trails) near drainages.

Management Measure Description

Revegetating disturbed areas restabilizes the soil in these areas, reduces erosion, and helps to prevent sediment and pollutants associated with sediment (such as phosphorus and nitrogen) from entering into nearby surface waters. Vegetation controls soil erosion by dissipating the impact force of raindrops, reducing the velocity of surface runoff, trapping dry sediment and preventing it from moving farther downslope, stabilizing the soil with roots, and contributing organic matter to the soil, which increases soil infiltration rates.

Nutrient and soil losses to streams and lakes are reduced by revegetating harvested, burned, or other disturbed areas. In some cases, planting early to establish erosion protection quickly and then again later to provide more permanent protection is necessary and advisable to prevent excessive erosion.

Good ground cover is key to reducing erosion. Good ground cover is defined as living plants within 5 feet of the ground and litter or duff with a depth of 2 inches or more (Kuehn and Cobourn, 1989).

Benefits and Costs of Revegetation Practices

The effectiveness of revegetation for controlling erosion, particularly on steep slopes and road fills, depends on protecting the slope until vegetative growth can take hold and grow enough to serve as a soil stabilizer. Straw mulch and netting are common ways to protect a newly seeded and fertilized slope. Adding straw mulch can reduce erosion by one-eighth to one-half. Adding netting with mulch can reduce erosion by nearly 100 percent to negligible levels (Figure 3-43) (Bethlahmy and Kidd, 1966).

Megahan (1987) estimated that the cost of seeding with plastic netting placed over the seeded area (approximately $8,200 per acre) is almost 50 times more than the cost of dry

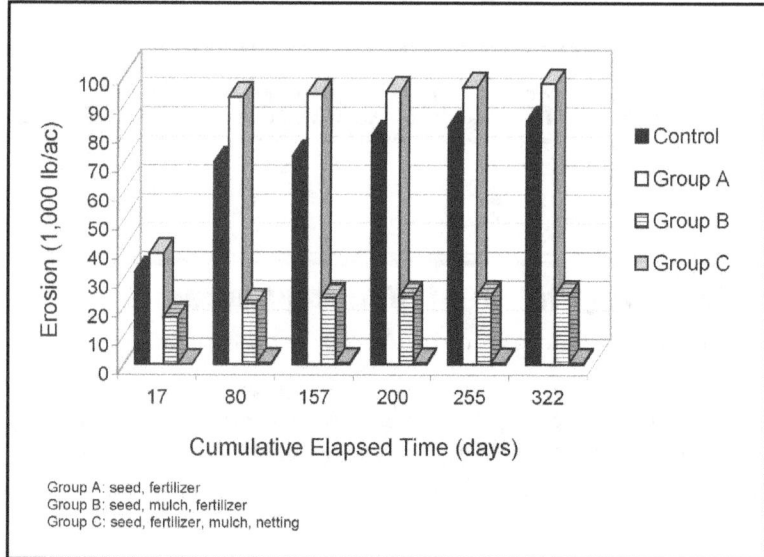

Group A: seed, fertilizer
Group B: seed, mulch, fertilizer
Group C: seed, fertilizer, mulch, netting

Figure 3-43. Comparison of the effectiveness of seed, fertilizer, mulch, and netting in controlling cumulative erosion from treated plots on a steep road fill in Idaho (after Bethlahmy and Kidd, 1966).

seeding alone (approximately $180 per acre). Other cost estimates related to practices for forest regeneration are presented in Tables 3-34 to 3-36. Dubensky (1991) estimated the economic effect of regeneration practices on the overall cost of a harvesting operation (Table 3-34). Lickwar (1989) compared revegetation costs for disturbed areas of various slope gradients in the Southeast (Table 3-35). Minnesota's Stewardship Incentives Program estimated the costs of reestablishing permanent vegetation with native and introduced grasses (Table 3-36).

Table 3-34. Economic Effect of Implementation of Proposed Management Measures on Road Construction and Maintenance (Dubensky, 1991)[a]

Management Practice	Increased Cost
Fiber for road and landing construction/maintenance	$5.00/ton
Ripping, shaping, and seeding log decks	$214/deck
Seeding firelines or rough logging roads	$24/100 ft
Construction and seeding of water bars	$15 each
Construction of rolling dips on roads	$24 each

All costs updated to 1998 dollars
[a] Public comment information provided by the American Paper Institute and the National Forest Products Association.

Table 3-35. Cost Estimates (and Cost as a Percent of Gross Revenues) for Seed, Fertilizer, and Mulch (1987 Dollars) (Lickwar, 1989)

Practice Component	Steep Sites[a]		Moderate Sites[b]		Flat Sites[c]	
Seed, fertilizer, and mulch	$19,950	(3.41%)	$18,438	(2.72%)	$17,590	(1.36%)

Note: All costs updated to 1998 dollars.
[a] Based on a 1,148-acre forest and gross harvest revenues of $399,685. Slopes average over 9 percent.
[b] Based on a 1,104-acre forest and gross harvest revenues of $473,182. Slopes ranged from 4 percent to 8 percent.
[c] Based on a 1,832-acre forest and gross harvest revenues of $899,491. Slopes ranged from 0 percent to 3 percent.

Table 3-36.   Estimated Costs for Revegetation (1991 Costs) (Minnesota DNR, 1991)

Practice	Total Cost[a]
Establishment of permanent vegetative cover (includes seedbed preparation, fertilizer, chemicals and application, seed, and seeding as prescribed in the plan)	
Introduced grasses	$96/acre
Native grasses	$176/acre

Note: All costs updated to 1998 dollars.
[a] The costs shown represent the total cost of the practice. Calculations were made by dividing the maximum Federal cost share by 0.75 to obtain the total cost.

Best Management Practices

◆ *Use mixtures of seeds adapted to the site, and avoid the use of invasive species. Choose annuals to allow natural revegetation of native understory plants, and select species that have adequate soil-binding properties.*

The selection of appropriate grasses and legumes is important for vegetation establishment. Grasses vary as to climatic adaptability, soil chemistry, and plant growth characteristics. USDA Natural Resources Conservation Service technical guides at the statewide level are excellent sources of information about seeding mixtures and planting prescriptions. The U.S. Forest Service, state foresters, and county extension agents can also provide helpful suggestions.

Using native species is both important and practical, and plenty of hardy native species are usually available. Nonnative species can outcompete and eliminate native vegetation, and the use of nonnative species often results in increased maintenance activities and expense.

Seeding rates (e.g., pounds per 1,000 square feet) are generally recommended for individual seed varieties and seed mixtures. Following such recommendations usually provides adequate cover and soil protection, whereas overseeding can create seedling overcrowding and subsequent failure.

◆ *On steep slopes, use native woody plants planted in rows, cordons, or wattles.*

These species may be established more effectively than grass and are preferable for binding soils.

◆ *Seed as soon as practicable after soil disturbance, preferably before rain, to increase the chance of successful vegetation establishment.*

Timing depends on the species to be planted and the schedule of operations, which determines when protection is needed.

◆ *Mulch as needed to hold seed, retard rainfall impact, and preserve soil moisture.*

Critical, first-year mulch applications provide the necessary ground cover to curb erosion and aid plant establishment. Various materials, including straw, bark, and wood chips, can be used to temporarily stabilize fill slopes and other disturbed areas and to improve conditions for germination immediately after construction. In most cases, mulching is done together with seeding and planting to establish stable banks. Both the type and the

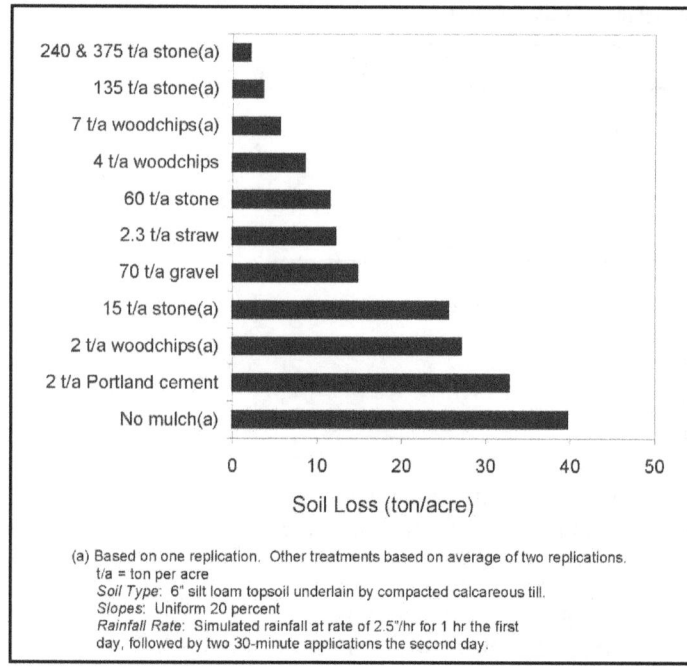

Figure 3-44 shows a bar chart of Soil Loss (ton/acre) for various treatments:

- 240 & 375 t/a stone(a)
- 135 t/a stone(a)
- 7 t/a woodchips(a)
- 4 t/a woodchips
- 60 t/a stone
- 2.3 t/a straw
- 70 t/a gravel
- 15 t/a stone(a)
- 2 t/a woodchips(a)
- 2 t/a Portland cement
- No mulch(a)

Soil Loss (ton/acre), axis from 0 to 50.

(a) Based on one replication. Other treatments based on average of two replications.
t/a = ton per acre
Soil Type: 6" silt loam topsoil underlain by compacted calcareous till.
Slopes: Uniform 20 percent
Rainfall Rate: Simulated rainfall at rate of 2.5"/hr for 1 hr the first day, followed by two 30-minute applications the second day.

Figure 3-44. Soil losses from a 35-foot-long slope (after Hynson et al., 1972).

amount of mulch applied vary considerably between regions and depend on the extent of the erosion potential and the available materials (Hynson et al., 1982). Figure 3-44 summarizes the effectiveness of various types of mulch (including Portland cement) for reducing erosion.

◆ *Fertilize according to site-specific conditions.*

Fertilization is often necessary for successful grass establishment because road construction commonly results in the removal or burial of fertile topsoil. To determine fertilizer formulations, it is best to compare available nitrogen, phosphorus, potassium, and sulphur in the soils to be treated with the requirements of the species to be sown. It might be necessary to refertilize periodically after vegetation establishment to maintain growth and erosion control capabilities. Fertilizer and other chemical management techniques are covered in depth in section 3I of the document.

◆ *Use biosolids as an alternative to commercial fertilizers.*

Biosolids is the name given to the solid material remaining after raw sewage has been treated. Biosolids can be used for forest regeneration efforts as a viable alternative to using commercial fertilizers. Biosolids are rich in nitrogen, as well as other nutrients essential for plant growth, including phosphorus, zinc, boron, manganese, and chromium (King County, Washington, 1999). The nutrients in biosolids are mostly in an organic form, so the biosolids act like a slow-release fertilizer, releasing only 15-20 percent of their nutrients during the first year after an application (Meyers, 1998). They also have a high content of organic matter, which increases soil infiltration rates and helps improve the ability of the soil to retain water, making it available for trees during dry periods. Biosolids can increase the growth rate of trees growing on relatively infertile soils to match that of trees growing on fertile soils.

Biosolids that are applied to the forest are delivered to the forest as a semisolid product with a content of approximately 20 percent solids and 80 percent water. The biosolids can be dispersed using a device that propels them aerially over an area, or they can be applied using a high-pressure hose. From a single point, they can be spread to a 250-foot radius or more across young tree growth and to a 60-foot radius in thinned timber stands.

The application rate (in ton/acre) of biosolids can be determined based on the nitrogen content of the biosolids. Specific amounts of nitrogen can be specified for each area to be treated based on soil testing and the nutrient requirements of the species involved. In the Northwest, application rates vary from 3 dry ton/acre of biosolids for timber to 7 dry ton/acre for young plantations, which corresponds to 150 to 350 pounds of plant-available nitrogen per acre (King County, Washington, 1999).

Streams and other water bodies are protected during biosolids applications by 33-foot buffer areas that are not fertilized. States regulate the use and application of biosolids, and obtaining a permit is usually necessary before biosolids may be used.

The potential for long-term effects from metals and pathogens in biosolids has been raised as a concern, but biosolids that meet EPA and state standards pose very little environmental threat (USEPA, 1994).

◆ *Protect seeded areas from grazing and vehicle damage until plants are well established.*

◆ *Inspect all seeded areas for failures, and make necessary repairs and reseed within the planting season.*

◆ *During non-growing seasons, apply interim surface stabilization methods to control surface erosion.*

Possible methods include mulching (without seeding) and installation of commercially produced matting and blankets. Alternative methods for planting and seeding include hand operations, the use of a wide variety of mechanical seeders, and hydroseeding.

3I: FOREST CHEMICAL MANAGEMENT

Forest Chemical Management

Use chemicals when necessary for forest management in accordance with the following to reduce nonpoint source pollution effects due to the movement of forest chemicals off-site during and after application:

(1) Conduct applications by skilled and, where required, licensed applicators according to the registered use, with special consideration given to effects to nearby surface waters.

(2) Carefully prescribe the type and amount of pesticides appropriate for the insect, fungus, or herbaceous species.

(3) Prior to applications of pesticides and fertilizers, inspect the mixing and loading process and the calibration of equipment, and identify the appropriate weather conditions, the spray area, and buffer areas for surface waters.

(4) Establish and identify buffer areas for surface waters. (This is especially important for aerial applications.)

(5) Immediately report accidental spills of pesticides or fertilizers into surface waters to the appropriate state agency. Develop an effective spill contingency plan to contain spills.

Management Measure Description

Chemicals used in forest management are generally pesticides (insecticides, herbicides, and fungicides) and fertilizers. Since pesticides can be toxic, they have to be mixed, transported, loaded, and applied correctly and their containers disposed of properly to prevent potential nonpoint source pollution. Since fertilizers can also be toxic or can shift the ecosystem's energy dynamics, depending on the exposure and concentration, it is important that they be handled and applied properly.

Pesticides and fertilizers are occasionally used in forestry to reduce mortality of and favor desired tree species and improve forest production. Many forest stands or sites never receive chemical treatment, and for those that do receive treatment, typically no more than two or three applications are made during an entire tree rotation (40 to 120 years).

Even though few applications are made, forestry chemicals can enter surface waters and precautions can be taken to prevent water contamination.

A number of studies conducted before 1990 demonstrate the importance of following current state and federal guidelines for forest chemical applications for protecting surface waters and groundwater. Norris and others (1991) compiled information from multiple studies that evaluated the peak concentrations of herbicides, insecticides, and fertilizers in soils, lakes, and streams (see Table 3-37). These studies were conducted from 1967 to 1987. Norris (1968) found that application of 2,4-D to marshy areas led to higher-than-normal levels of stream contamination. When ephemeral streams were treated, residue levels of hexazinone and picloram greatly increased with storm-generated flow. Glyphosate was aerially applied (3.3 kg/hectare) to an 8-hectare forest ecosystem in the Oregon Coast Range. The study area contained two ponds and a small perennial stream. All were unbuffered and received direct application of the herbicide. Glyphosate residues were detected for 55 days after application with peak stream concentrations of 0.27 mg/L. It was demonstrated that the concentration of insecticides in streams was significantly greater when the chemicals were applied without a buffer strip to protect the watercourse. When streams were unbuffered, the peak concentrations of malathion ranged from 0.037 to 0.042 mg/L. When buffers were provided, however, the concentrations of malathion

Table 3-37. Peak Concentrations of Forest Chemicals in Soils, Lakes, and Streams After Application (Norris et al., 1991)

Chemicals[a] and System[b]	Application Rate (kg/hectare)	Concentration (mg/L or mg/kg*)		Time Interval[c]	Time to Non-detection	Source[d]
		Peak	Subsequent			
Herbicides						
2,4-D	2.24	0.001-0.13			1-168 h[e]	17
Marsh	2.24	0.09				17,18
2,4-D BE						
Built pond	23.0					1
Water		3.0	1.0	85 d		
			0.2	180 d		
Sediment		8.0*	4.0*	13+ d		
			0.4–0.6*	82–182 d		
Aquatic plants			206*	7 d		
			8*	82 d	182 d	
2,4-D AS						
Reservoir		3.6	0	13 d		7
Picloram						
Runoff		0.078				19
Runoff		0.038				23
Ephemeral stream	2.8	0.32		157 d	915 d	9
Stream	0.37					3
Hexazinone						
Stream (GA)	1.68	0.044		3–4 m		11
Forest (GA)	1.68					14
Litter		0.177*	<0.01*	60+ d		
Soil		0.108*	<0.01*	90 d		
Ephemeral stream		0.514		3 d		
Perennial stream		0.442		3 d		
Atrazine						
Stream	3.0	0.42	0.02	17 d		16
Built ponds						10
Water		0.50	0.05	14 d		
			0.005	56 d		
Sediments		0.50*	0.9*	4 d		
		0.50*	0.25*	56 d		
Triclopyr						
Pasture (OR)	3.34	0.095*	0.09	5.5 h		20
Glyphosate						
Water	3.3	0.27	<0.01	3 d		15
Dalapon						
Field irrigation water		0.023–3.65	<0.01	Sev h		5

Table 3-37. (continued)

Chemicals[a] and System[b]	Application Rate (kg/hectare)	Concentration (mg/L or mg/kg*) Peak	Concentration (mg/L or mg/kg*) Subsequent	Time Interval[c]	Time to Non-detection	Source[d]
Insecticides						
Malathion						
Streams	0.91					24
Unbuffered		0.037–0.042				
Buffered		0–0.017				
Carbaryl						
Streams & ponds (E)		0–0.03				24
Streams, unbuffered (PNW)		0.005–0.011			48 h	24
Water	0.84	0.026–0.042				8
Brooks with buffer	0.84	0.001–0.008				22
Rivers with buffer	0.84	0.000–0.002				22
Streams, unbuffered	0.84	0.016				22
Ponds	0.84					6
Water		0.254			100-400 d	
Sediment		<0.01–5.0*[f]				
Acephate						
Streams		0.003–0.961		1 d		4
Pond sediment & fish	0.56	0.113–0.135	0.013-0.065	14 d		21
Fertilizers						
Urea	224					
Urea-N						
Forest stream (OR)		0.39	0.39	48 h		12
Dollar Cr (WA)		44.4				13
NH_4^+-N						
Forest stream (OR)		<0.10				12
Tahuya Cr (WA)		1.4				13
NO_3^+-N						
Forest stream (OR)		0.168				12
Elochoman R (WA)		4.0				13

[a] 2,4-D BE = 2,4-D butoxyethanol ester; 2,4-D AS = 2,4-D amine salt + ester.
[b] E = eastern USA; Cr = Creek; GA = Georgia; PNW = Pacific Northwest; OR = Oregon; R = River; WA = Washington; buffer = wooded riparian strip.
[c] d = day; h = hours; m = months; sev h = several hours. Intervals are times from application to measurement of peak or subsequent concentration, whichever is the last measurement indicated.
[d] 1 = Birmingham and Colman (1985); 2 = Bocsor and O'Connor (1975); 3 = Davis et al. (1968); 4 = Flavell et al. (1977); 5 = Frank et al. (1970); 6 = Gibbs et al. (1984); 7 = Hoeppel and Westerdahl (1983); 8 = Hulbert (1978); 9 = Johnsen (1980); 10 = Maier-Bode (1972); 11 = Mayack et al. (1982); 12 = Moore (1970); 13 = Moore (1975b); 14 = Neary et al. (1983); 15 = Newton et al. (1984); 16 = M. Newton (Oregon State University, personal communication, 1967); 17 = Norris (1967); 18 = Norris (1968); 19 = Norris (1969); 20 = Norris et al. (1987); 21 = Rabeni and Stanley (1979); 22 = Stanley and Trial (1980); 23 = Suffling et al. (1974); 24 = Tracy et al. (1977).
[e] Normally less than 48 h.
[f] One extreme case: 23.8 mg/kg peak concentration, 16 months to nondetection.

were reduced to levels that ranged from undetectable to 0.017 mg/L. The peak concentrations of carbaryl ranged from 0.000 to 0.0008 mg/L when watercourses were protected with a buffer, but they increased to 0.016 mg/L when watercourses were unbuffered.

Moore (1971), as cited in Norris et al. (1991), compared nitrogen loss from a watershed treated with 224 kg urea-N per hectare to nitrogen loss from an untreated watershed. The study demonstrated that the loss of nitrogen from the fertilized watershed was 28.02 kg/hectare whereas the loss of nitrogen from the unfertilized watershed was only 2.15 kg/hectare (Table 3-38).

Table 3-38. Nitrogen Losses from Two Subwatersheds in the Umpqua Experimental Watershed (OR) (Norris et al., 1991)

Loss Locus or Statistic	Urea-N	NH$_3$-N	NO$_3$-N	Total
	Absolute loss (kg/hectare)			
Watershed 2 (treated)	0.65	0.28	27.09	28.02
Watershed 4 (untreated)	0.02	0.06	2.07	2.15
Net loss (2-4)	0.63	0.22	25.02	25.87
	Proportional loss			
Percent of total	2.44	0.85	96.71	100.00

Riekerk and others (1989) found that the greatest risk to water quality from pesticide application in forestry operations occurred from aerial application because of drift, wash-off, and erosion processes. They found that aerial applications of herbicides resulted in surface runoff concentrations roughly 3.5 times greater than those for application on the ground.

The Riekerk and others (1989) study results also suggested that tree injection application methods would be considered the least hazardous for water pollution, but would also be the most labor-intensive. Hand application of herbicides usually poses little or no threat to water quality in areas where there is no potential for herbicides to wash into watercourses through gullies. Providing buffer areas around streams and water bodies can effectively eliminate adverse water quality effects from forestry chemicals.

Megahan (1980) summarized data on changes in water quality following the fertilization of various forest stands with urea. The major observations from this research are summarized below:

- Increases in the concentration of urea-N ranged from very low to a maximum of 44 ppm, with the highest concentrations attributed to direct application to water surfaces.

- Higher concentrations occurred in areas where buffer strips were not left beside stream banks.

- Chemical concentrations of urea and its by-products tended to be relatively short-lived due to transport downstream, assimilation by aquatic organisms, or adsorption by stream sediments.

Based on his review, Megahan concluded that the effects of fertilizer application in forested areas could be significantly reduced by avoiding application techniques that could result in direct deposition into the water body and by maintaining a buffer area along the stream bank. Other researchers have presented information supporting Megahan's conclusions (Hetherington, 1985; Malueg et al., 1972).

Cost of Forest Chemical Applications

The cost of chemical management depends on the method of application (Table 3-39). Generally, chemicals are applied by hand, from an airplane or helicopter (aerial spray), or mechanically. When forest chemicals are applied mechanically, it is most common to use a boom sprayer.

Table 3-39.   Average Costs for Chemical Management (Hansit, 2000; Holburg, 2000)

Application Practice	Average Cost
Hand application	$100/acre
Aerial application	$55–$70/acre

Best Management Practices

◆ *For aerial spray applications, mark and maintain a buffer area of appropriate width around all watercourses and water bodies to avoid drift or accidental application of chemicals directly to surface waters (Figure 3-45).*

Buffer width is determined by taking into considerations the altitude of application, weather conditions, and drop size distribution (Ice and Teske, 2000). Careful and precise marking of application areas for aerial applications helps avoid accidental contamination of open waters.

Models are available to help the forest manager calculate pesticide application details. The Spray Drift Task Force, in collaboration with EPA and USDA, co-developed AgDRIFT, a new model, to provide estimates of spray drift deposition under different pesticide application and meteorological conditions (see www.agdrift.com). The Forest Service Cramer-Barry-Grim (FSCBG) spray dispersion model analyzes data on aircraft,

Figure 3-45.   Establish buffer zones of appropriate width during aerial applications of forest chemicals to protect water quality, people, and animals (Washington State DNR, 1997).

meteorology, pesticides, and target areas to predict deposition and drift (see www.fs.fed.us/foresthealth/technology). A personal computer version of the model is available that combines and implements mathematical models to assist forest managers in planning and implementing aerial spray operations.

◆ *Apply pesticides and fertilizers during favorable atmospheric conditions.*

Do not apply pesticides when wind conditions increase the likelihood of significant drift. It is also best to avoid pesticide application when temperatures are high or relative humidity is low because these conditions influence the rate of evaporation and enhance losses of volatile pesticides.

◆ *Ensure that pesticide users abide by the current pesticide label, which might specify whether users be trained and certified in the proper use of the pesticide; allowable use rates; safe handling, storage, and disposal requirements; and whether the pesticide may be used only under the provisions of an approved State Pesticide Management Plan.*

Consistency between management measures and practices for pesticides and those in the approved State Pesticide Management Plan helps ensure consistency in the method and means of use.

◆ *Locate mixing and loading areas, and clean all mixing and loading equipment thoroughly after each use, where pesticide residues will not enter streams or other water bodies.*

◆ *Dispose of pesticide wastes and containers according to state and federal laws.*

◆ *Take precautions to prevent leaks and spills.*

◆ *Develop a spill contingency plan that provides for immediate spill containment and cleanup, and notification of proper authorities.*

Maintain an adequate spill and cleaning kit that includes the following:

- Detergent or soap.
- Hand cleaner and water.
- Activated charcoal, adsorptive clay, vermiculite, kitty litter, sawdust, or other adsorptive materials.
- Lime or bleach to neutralize pesticides in emergency situations.
- Tools such as a shovel, broom, and dustpan and containers for disposal.
- Proper protective clothing.

◆ *Apply slow-release fertilizers when possible.*

This practice reduces potential nutrient leaching to groundwater, and it increases the availability of nutrients for plant uptake.

◆ *Apply fertilizers during maximum plant uptake periods to minimize leaching.*

◆ *Base fertilizer type and application rate on soil and/or foliar analysis.*

Conduct foliar analysis approximately once per year to diagnose nutrient toxicities or deficiencies and to determine the correct fertilization program to follow. Foliar analysis is the process whereby leaves from trees are dried, ground, and chemically analyzed for their nutrient content. Compare the results of foliar analysis to available nitrogen, phosphorus, potassium, and sulphur in the soils to be treated and to the requirements of the species.

◆ *Consider the use of pesticides as only one part of an overall program to control pest problems.*

Integrated Pest Management (IPM) strategies have been developed to control forest pests without total reliance on chemical pesticides. The IPM approach uses all available techniques, including chemical and nonchemical. An extensive knowledge of both the pest and the ecology of the affected environment is necessary for IPM to be effective.

◆ *Base selection of pesticide on site factors and pesticide characteristics.*

These factors include vegetation height, target pest, adsorption (attachment) to soil organic matter, persistence or half-life, toxicity, and type of formulation.

◆ *Check all application equipment carefully, particularly for leaking hoses and connections and plugged or worn nozzles. Calibrate spray equipment periodically to achieve uniform pesticide distribution and rate.*

◆ *Always use pesticides in accordance with label instructions, and adhere to all federal and state policies and regulations governing pesticide use.*

3J: WETLANDS FOREST MANAGEMENT

Management Measure for Wetlands Forest Management

Plan, operate, and manage normal, ongoing forestry activities (including harvesting; road design, construction, and maintenance; site preparation and regeneration; and chemical management) to adequately protect the aquatic functions of forested wetlands.

Management Measure Description

Forested wetlands provide many beneficial functions that need to be protected. Among these are floodflow alteration, sediment trapping, nutrient retention and removal, provision of important habitat for fish and wildlife, and provision of timber products. The extent of wetlands (including forested wetlands) in the continental United States has declined greatly in the past 40 years because of conversion to other land uses. There are currently approximately 100 million acres of wetlands in the 48 contiguous states, or about one-half of their extent at the time of European settlement. Although the rate of wetlands loss has slowed in recent years, the United States continues to sustain a net loss of approximately 58,000 acres per year. Forestry activities are the third leading cause of wetlands loss–behind urban development and agriculture–and accounted for 23 percent of wetland losses from 1986 to 1997 (Dahl, 2000). Given the historic and ongoing losses, it is critical that additional effects to wetlands be avoided and minimized to the maximum extent possible.

Potential effects of forestry operations in wetlands include the following:

- Loss and/or degradation due to discharges of dredged or fill material.

- Sediment production from road construction and use and equipment operation resulting in wetlands filling.

- Drainage alteration as a result of improper road construction and ditching. An excellent discussion of the relationship between forest roads and drainage is contained in the U.S. Forest Service document Water/Road Interaction Technology Series (USDA-FS, 1998b).

- Stream obstruction caused by failure to remove logging debris.

- Soil compaction caused by operation of logging vehicles during flooding periods or wet weather. Skid trails, haul roads, and log landings are areas where compaction is most severe.

- Contamination from improper application or use of pesticides.

- Loss of integrity of whole wetland landscapes (and the functions they serve) as a cumulative effect of incremental losses of small wetland tracts.

Potential adverse effects associated with road construction and maintenance in forested wetlands are alteration of drainage and flow patterns, increased erosion and sedimentation, habitat loss and degradation, and damage to existing timber stands. In an effort to prevent these potential adverse effects, section 404 of the Clean Water Act requires the use of appropriate BMPs for road construction and maintenance in wetlands so that flow and circulation patterns and chemical and biological characteristics are not impaired (see text below).

Harvest planning and selection of the right harvest system are essential in achieving the management objectives of timber production, ensuring stand establishment, and avoiding adverse effects on water quality and wetland functions and values. The potential effects of reproduction methods and cutting practices on wetlands include changes in water quality, water quantity, temperature, nutrient cycling, and aquatic habitat. Streams can also become blocked with logging debris if SMAs are not properly maintained or if appropriate practices are not employed in SMAs.

Site preparation includes but is not limited to the use of prescribed fire, chemicals, and/or mechanical site preparation. Extensive site preparation on bottoms where frequent flooding occurs can cause excessive erosion and stream sedimentation. The degree of acceptable site preparation is governed by the amount and frequency of flooding, soil type, and species suitability and is dependent on the regeneration method used.

Forestry in Wetlands: Section 404

Section 404 establishes a program that regulates the discharge of dredged or fill material into waters of the United States, including wetlands. The Corps and EPA jointly administer the program. The Corps administers the day-to-day program, including permit decisions and jurisdictional determinations; develops policy and guidance; and enforces Section 404 provisions. EPA develops and interprets environmental criteria used in evaluating permit applications; determines the scope of geographic jurisdiction; and approves and oversees state assumption. EPA also identifies activities that are exempt, enforces Section 404 provisions, and has the authority to elevate and/or veto Corps permit decisions. In addition, the U.S. Fish and Wildlife Service, the National Marine Fisheries Service, and state resource agencies have important advisory roles.

Section 404(f) exempts normal forestry activities (for example, bedding, seeding, harvesting, and minor drainage) that are part of an established, ongoing forestry operation. A forest operation ceases to be "established" when the area in which it was conducted has been converted to another use or has lain idle so long that modifications to the hydrological regime are necessary to resume operations (40 CFR Part 232.3(c)(1)(ii)(B)). This exemption does not apply to activities that represent a new use of the wetland and that would result in a reduction in reach or impairment of flow or circulation of waters of the United States, including wetlands. In addition, Section 404(f) provides an exemption of discharges of dredged or fill material for the purpose of constructing or maintaining forest roads, where such roads are constructed or maintained in accordance with BMPs to assure that the flow and circulation patterns and chemical and biological characteristics of the navigable waters are not impaired, that the reach of the navigable waters is not reduced, and that any adverse effect on the aquatic environment will be otherwise minimized. Following are the section 404(f) regulations pertaining to forestry activities, including the BMPs for forest road construction or maintenance.

Code of Federal Regulations, Title 40, section 232.3: Activities Not Requiring a Section 404 Permit

Except as specified in paragraphs (a) and (b) of this section, any discharge of dredged or fill material that may result from any of the activities described in paragraph (c) of this section is not prohibited by or otherwise subject to regulation under this part.

(a) If any discharge of dredged or fill material resulting from the activities listed in paragraph (c) of this section contains any toxic pollutant listed under section 307 of the Act, such discharge shall be subject to any applicable toxic effluent standard or prohibition, and shall require a section 404 permit.

(b) Any discharge of dredged or fill material into waters of the United States incidental to any of the activities identified in paragraph (c) of this section must have a permit if it is part of an activity whose purpose is to convert an area of the waters of the United States into a use to which it was not previously subject, where the flow or circulation of waters of the United States may be impaired or the reach of such waters reduced. Where the proposed discharge will result in significant discernible alterations to flow or circulation, the presumption is that flow or circulation may be impaired by such alteration.

Note: For example, a permit will be required for the conversion of a cypress swamp to some other use or the conversion of a wetland from silvicultural to agricultural use when there is a discharge of dredged or fill material into waters of the United States in conjunction with construction of dikes, drainage ditches or other works or structures used to effect such conversion. A conversion of section 404 wetland to a non-wetland is a change in use of an area of waters of the U.S. A discharge which elevates the bottom of waters of the United States without converting it to dry land does not thereby reduce the reach of, but may alter the flow or circulation of, waters of the United States.

(c) The following activities are exempt from section 404 permit requirements, except as specified in paragraphs (a) and (b) of this section:

* * *

(6) Construction or maintenance of farm roads, forest roads, or temporary roads for moving mining equipment, where such roads are constructed and maintained in accordance with best management practices (BMPs) to assure that flow and circulation patterns and chemical and biological characteristics of waters of the United States are not impaired, that the reach of the waters of the United States is not reduced, and that any adverse effect on the aquatic environment will be otherwise minimized. The BMPs which must be applied to satisfy this provision include the following baseline provisions:

(i) Permanent roads (for farming or forestry activities), temporary access roads (for mining, forestry, or farm purposes) and skid trails (for logging) in waters of the United States shall be held to the minimum feasible number, width, and total length consistent with the purpose of specific farming, silvicultural or mining operations, and local topographic and climatic conditions;

(ii) All roads, temporary or permanent, shall be located sufficiently far from streams or other water bodies (except for portions of such roads which must cross water bodies) to minimize discharges of dredged or fill material into waters of the United States;

(iii)   The road fill shall be bridged, culverted, or otherwise designed to prevent the restriction of expected flood flows;

(iv)   The fill shall be properly stabilized and maintained to prevent erosion during and following construction;

(v)   Discharges of dredged or fill material into waters of the United States to construct a road fill shall be made in a manner that minimizes the encroachment of trucks, tractors, bulldozers, or other heavy equipment within the waters of the United States (including adjacent wetlands) that lie outside the lateral boundaries of the fill itself;

(vi)   In designing, constructing, and maintaining roads, vegetative disturbance in the waters of the United States shall be kept to a minimum;

(vii)   The design, construction and maintenance of the road crossing shall not disrupt the migration or other movement of those species of aquatic life inhabiting the water body;

(viii)   Borrow material shall be taken from upland sources whenever feasible;

(ix)   The discharge shall not take, or jeopardize the continued existence of, a threatened or endangered species as defined under the Endangered Species Act, or adversely modify or destroy the critical habitat of such species;

(x)   Discharges into breeding and nesting areas for migratory waterfowl, spawning areas, and wetlands shall be avoided if practical alternatives exist;

(xi)   The discharge shall not be located in the proximity of a public water supply intake;

(xii)   The discharge shall not occur in areas of concentrated shellfish production;

(xiii)   The discharge shall not occur in a component of the National Wild and Scenic River System;

(xiv)   The discharge of material shall consist of suitable material free from toxic pollutants in toxic amounts; and

(xv)   All temporary fills shall be removed in their entirety and the area restored to its original elevation.

Best Management Practices

Wetland Harvesting Practices

◆   *Conduct forest harvesting according to preharvest planning designs and locations.*

Planning and close supervision of harvesting operations are needed to protect site integrity and enhance regeneration. Harvesting without regard to season, soil type, or type of equipment can damage the site productivity; retard regeneration; cause excessive rutting, churning, and puddling of saturated soils; and increase erosion and sedimentation of streams. Harvesting without regard to other activities occurring in the watershed can cause unacceptable cumulative effects.

◆   *Establish a streamside management area (SMA) adjacent to natural perennial streams, lakes, ponds, and other standing water in the forested wetland following the components of the SMA management measure.*

◆ *Select the harvesting method to minimize soil disturbance and hydrologic effects on the wetland.*

In seasonally flooded wetlands, a guideline is to use conventional skidder logging that employs equipment with low-ground-pressure tires, cable logging, or aerial logging. Comparisons of cable logging and helicopter logging have concluded that helicopter operations cause less site disturbance, are more economical, and provide greater yield. Table 3-40 presents one set of harvesting system recommendations by type of forested wetland (Florida Division of Forestry, 1988). Another alternative is to conduct harvesting during winter months when the ground is frozen (see below).

◆ *Use ultrawide, high-flotation tires on logging trucks and skidders to reduce soil compaction and erosion.*

Using dual-tired skidders and high-floatation tires for log hauling reduces soil damage, soil compaction, surface runoff, and sedimentation (Aust et al., 1994).

◆ *When ground skidding, use low-ground-pressure tires or tracked machines and confine skidding to a few primary skid trails to minimize site disturbance, soil compaction, and rutting. Adjust tire pressure on skidders during wet weather or when conducting forested wetland harvesting (Aust, Virginia Polytechnic Institute and State University, personal communication, 1999).*

Table 3-40. Recommended Harvesting Systems by Forested Wetland Site[a] (Florida Department of Agriculture and Consumer Services, 1988)

Site Type	Conventional	Conventional with Controlled Access[b]	Cable or Aerial	Barge or High Flotation Boom
Flowing Water				
Minoral Soil				
Alluvial River Bottom	B	A	C	C
Organic Soil				
Black River Bottom	B	A	C	C
Branch Bottom	A[c]	B	C	C
Cypress Strand	B	A	A	A
Muck Swamp	C	A	A	A
Nonflowing Water				
Mineral Soil				
Wet Hammock	B	A	C	C
Organic Soil				
Cypress Dome	B	A	A	A
Peat Swamp	C	A	A	A

Note: A = recommended; B = recommended when dry; C = not recommended.
[a] Recommendations include cost considerations
[b] Preplanned and designated skid trails and access roads.
[c] Log from the hill (high ground).

Research conducted by Randy Foltz of the Intermountain Research Station in the Lowell Ranger District of the Willamette National Forest, Oregon (1994), addressed the use of variable tire pressure as a BMP for forest roads. His study showed that by reducing the tire pressure on logging trucks from their highway inflation of 90 psi to between 30 and 70 psi, sediment runoff was reduced on average by 67 percent. The percentage reduction in sediment runoff was directly correlated with the rainfall quantity and traffic volume.

◆ *When soils become saturated, suspend ground skidding harvesting operations. Use of ground skidding equipment during excessively wet periods can result in unnecessary site disturbance and equipment damage.*

Wetland Road Design and Construction Practices

◆ *Locate, design, and construct forest roads according to preharvest planning.*

Forestry activities in wetlands are often subject to municipal, county, state, and federal regulations. Therefore, sufficient time should be set aside to obtain all necessary permits.

Improperly located, designed, or constructed forest roads can cause changes in hydrology, accelerate erosion, reduce or degrade fisheries habitat, and destroy or damage existing stands of timber.

◆ *Use temporary roads in forested wetlands.*

A temporary road in a wetland needs to provide adequate cross-road drainage at all natural drainageways. Temporary drainage structures include culverts, bridges, and porous material such as corduroy or chunkwood.

Construct permanent roads only to serve large and frequently used areas, as approaches to watercourse crossings, or to provide access for long-term fire protection. Use the minimum design standard necessary for reasonable safety and the anticipated traffic volume. Various temporary wetland crossing options are compared in Table 3-41.

Blade the surface of a wetland to be as flat as possible prior to constructing a temporary road (Hislop and Moll, 1996, cited in Blinn et al., 1998). Do not disturb the root mat in any wetland that has grass mounds or other uneven vegetation. Any temporary wetland crossing is enhanced by using a root or slash mat to provide additional support to the equipment.

◆ *Construct fill roads only when absolutely necessary for access since fill roads have the potential to restrict natural flow patterns.*

Where construction of fill roads is necessary, use a permeable fill material (such as gravel or crushed rock) for at least the first layer of fill. The use of pervious materials helps maintain the natural flow regimes of subsurface water. Figure 3-46 demonstrates the different effects of impervious and pervious road fills on wetland hydrology. Permeable fill material is not a substitute for using bridges where needed or for installing adequately spaced culverts at all natural drainageways. Use this practice in conjunction with cross drainage structures to ensure that natural wetland flows are maintained (i.e., so that fill does not become clogged by sediment and obstruct flows).

◆ *Provide adequate cross drainage to maintain the natural surface and subsurface flow of the wetland.*

Table 3-41.   Temporary Wetland Crossing Options (Blinn, 1996)

Crossing Option	Description	Application	Cost
Wood Mats	Individual cants that are strung together using two 3/16-inch galvanized steel cables to make a single-layer crossing.	Wet mineral or sandy soils or existing road beds. Wood mats are not recommended for undisturbed peat or very weak clay soils. They require a relatively level surface with grades up to 4 percent, a fairly straight alignment, and no cross slope.	Approximately $170 to initially construct a 10' x 12' mat
Wood Planks/ Panels	Wood planks or panels are constructed using lumber planking to create a two-layer crossing. Parallel runners are laid down on each side where the vehicle's tires will pass and then lumber is nailed perpendicular to these runners.	Most wetland soils, if sized properly. The surface width needed depends on the soil strength. Wood plank crossings require a relatively level surface with grades up to 4 percent, a fairly straight alignment, and no cross slope.	Approximately $150 to initially construct an 8' x 12' wood plank
Wood Pallets	Wood-pallet crossing mats are sturdy, commercially available, multilayered variation of a three-layer wood pallet (used for shipping or storage) that has been designed specifically for traffic.	Most wetland soils, if sized appropriately. The require a relatively level surface with grades up to 4 percent, a fairly straight alignment, and no cross slope. Most appropriate for hauling or forwarding operations.	Approximately $350 for a commercial 8' x 16' pallet
Bridge Decking	The decking of a timber bridge can be used to cross a small wetland area.	Most wetland soils, if sized properly. Easy to install and remove. Require a relatively level ground surface.	Approximately $6,000 for a 30' x 12' bridge
Expanded Metal Grating	Metal grating is relatively light and the surface is rough enough to provide some traction. Built by hand-placing the grating sections in the wheel paths.	Most shallow wetland soils, sandy soils, or on an existing road. It is not recommended for undisturbed peat or very weak clay soils. Performance is enhanced where there is an adequate root or slash mat to provide additional support.	Approximately $100 for a 4' x 8' grate
PVC or HDPE Pipe and Plastic Road	A PVC and HDPE pipe mat is constructed using 4-inch diameter PVC or HDPE pipes that are tightly connected using galvanized steel cables. Plastic roads are similar to pipe mats except that they are not built to ease the transition of tires between the firm soil and the road.	Most wetland soils, if sized properly. Mat width needed depends on soil strength. Require a relatively level surface with grades up to 4 percent, a fairly straight alignment, and no cross slope.	Approximately $200 for a 4' x 12' pipe mat. Plastic road that is 8' x 40' costs approximately $2,000

Table 3-41.   (continued)

Crossing Option	Description	Application	Cost
Tire Mats	A tire mat or panel of tires created by interconnecting tire sidewalls with corrosion-resistant fasteners. Tire threads are also used in some designs. Mats of varying length and width can be created.	Most wet mineral soils with different designs for distinct soils and situations. Tire mats require a relatively level surface with grades up to 5 percent, a fairly straight alignment, and no cross slope.	Approximately $300 for a 5' x 10' mat
Corduroy	Corduroy is a crossing made of brush, small logs cut from low-value and noncommercial trees on-site, or mill slabs that are laid perpendicular or parallel to the direction of travel.	Most wetland soils. Corduroy crossings require a relatively level surface with grades up to 4 percent, a fairly straight alignment, and no cross slope.	Low
Pole Rails	When attempting to support skidding or forwarding machinery equipped with high flotation or dual tires, one or more straight hardwood poles cut from on-site trees can be laid parallel to the direction of travel below each wheel.	Skidding and felling machinery equipped with wide, high-flotation tires and used across small mineral soil wetlands. Should only be used on relatively level surface with grades up to 4 percent, a fairly straight alignment, and no cross slope.	Low
Wood Aggregate	Wood particles ranging in size from chips to chunks can provide cohesion and support on soft soils. Wood aggregate is used in the same way as gravel, except that it is lighter and temporary due to natural deterioration.	The traffic capability of most wet soils can be improved substantially with the application of wood aggregate. Can be used on a variety of grades, alignments, and cross slopes.	Competitive with local sources of gravel fill.
Equipment with Wide Tires, Duals, Bodies, or Tire Tracks	These mobility options provide a method for increasing the contact area between the equipment and the soil so that the machine's weight is spread over a larger surface area.	Many wetland soils. Performance is enhanced in areas where there is adequate root or slash mat to provide additional support to the equipment.	Wide tires may cost more than $4,000 each, tire tracks may cost approximately $7,000 for a set of two tracks.
Central Tire Inflation (CTI)	CTI is a low-ground-pressure option currently for use on hauling vehicles only, but will likely be available on other equipment in the future.	Many wetland soils. The reduced tire pressure, when used with radial ply tires, results in a larger tire "footprint," which reduces the vehicle pressure applied to the ground.	Cost depends on the number of axles retrofitted. 18 axles = $16,000

This can be accomplished through adequate sizing and spacing of water crossing structures, proper choice of the type of crossing structure, and installation of drainage structures at a depth adequate to pass subsurface flow. Designed and constructed according to these considerations helps ensure that bridges, culverts, and other structures do not perceptibly diminish or increase the duration, direction, or magnitude of the minimum, peak, or mean flow of water on either side of the structure.

◆ *Construct roads at natural ground level to minimize the potential to restrict flowing water.*

Float the access road fill on the natural root mat. If the consequences of the natural root mats' failing are serious, use reinforcement materials such as geotextile fabric, geo-grid mats, or log corduroy. Figure 3-47 depicts a cross section of the practice of floating the road. Protect the root mat beneath the roadway from equipment damage by diverting through traffic to the edge of the right-of-way, shear-blading stumps instead of grubbing, and using special wide-pad equipment. Also, protect the root mat from damage or puncture by using fill material that does not contain large rocks or boulders.

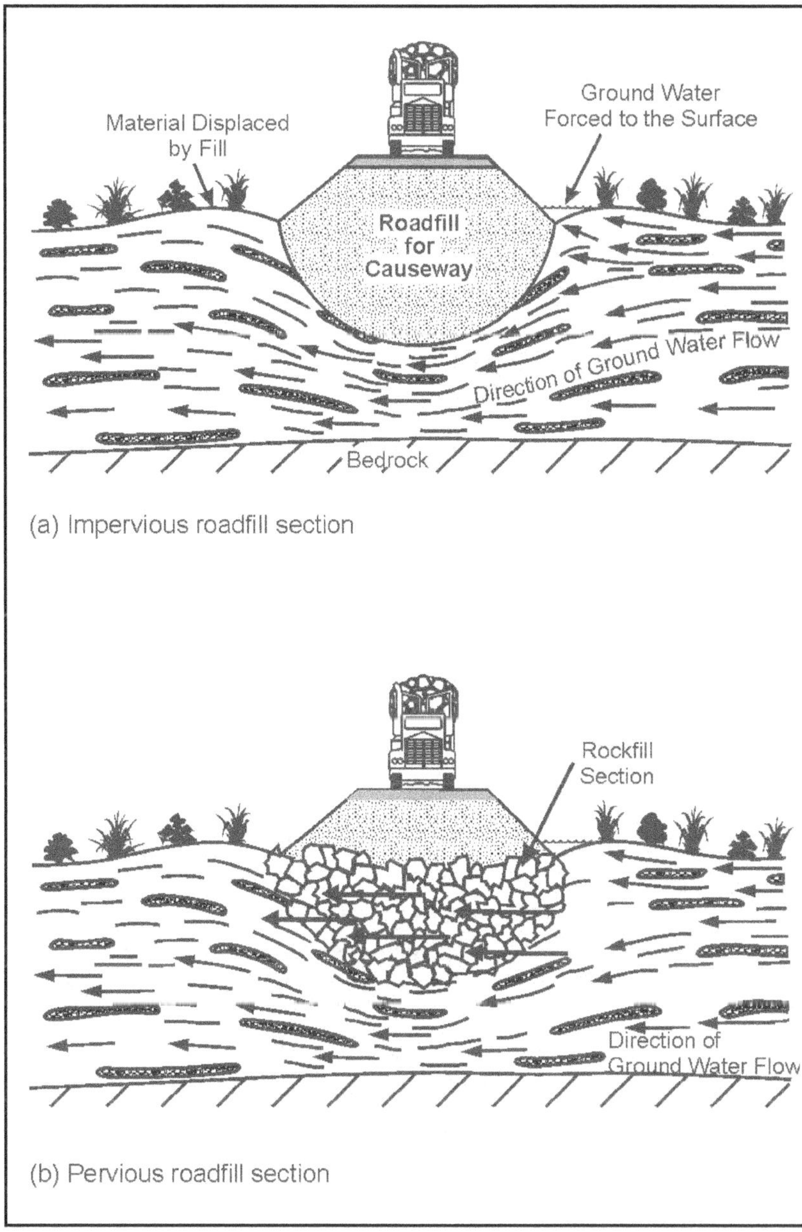

◆ *Discharges of dredged or fill material into wetlands or other waters of the United States must comply with CWA section 404 (see text above).*

Figure 3-46. Comparison of impervious (a) and pervious (b) roadfill sections. Impervious roadfill consolidates natural material and restricts groundwater flow. Pervious roadfill allows movement of groundwater through it and minimizes flow changes (adapted from Thronson, 1979).

Practices for Crossing Wetlands in Winter

Winter provides an opportunity to cross wetlands with little effect. Roads are often constructed across wetlands in winter to take advantage of frozen ground.

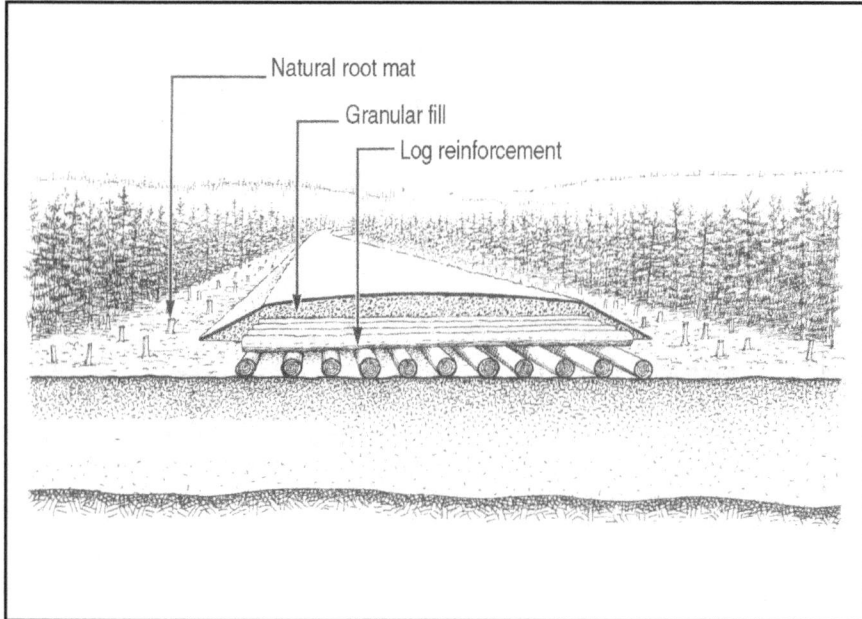

Figure 3-47. Elements of a road crossing through a swamp wetland, cross section (Ontario MNR, 1990).

◆ *The following are recommendations for crossing wetlands in winter, for all wetland types (Minnesota Division of Forestry, 1995):*

• If permanent structures are to be used, follow BMP installation guidelines for permanent roads.

• Select the shortest practical route to minimize potential problems with drifting snow and crossing of open water.

• Avoid crossing open water or active springs. If crossing is unavoidable, temporary crossings are preferred over permanent crossings. These can be ice bridges, temporarily installed bridges, or timber mats.

• Avoid using soil fill.

• Install structures that block water flow so that they can be easily removed prior to the spring thaw. Remove these structures during a winter thaw.

• Use planking, timber mats, or other support alternatives to improve the capability of the road to support heavy traffic. If removal would cause more damage than leaving them in place, these structures can be left as permanent sections on frozen roads. Avoid clearing practices that result in berms of soil or organic material, which can disrupt normal water flow in wetlands.

• Do not operate machinery during a winter thaw. Resume operations only when conditions are adequate to support equipment.

• Remove temporary fills and structures to the extent practical when no longer needed.

• Install buffer strips near open water.

• Anchor temporary structures at one end only to allow them to move aside during high-water flows.

◆ *To avoid excessive damage, equipment operations are best avoided on any portion of a road where ruts are deeper than 6 inches below the water surface for a continuous distance of more than 100 yards (Wiest, 1998).*

Wetland Site Preparation and Regeneration Practices

◆ *Select a regeneration method that meets the site characteristics and management objectives.*

Choice of regeneration method has a major influence on the stand composition and structure and on the forestry practices to be applied over the life of the stand. Natural regeneration may be achieved by clear-cutting the existing stand and relying on regeneration from seed from adjacent stands, the cut trees, or stumps and from root sprouts (coppice). Successful regeneration depends on recognizing the site type and its characteristics, evaluating the stocking and species composition in relation to stand age and site capability, planning regeneration options, and using sound harvesting methods. Schedule harvest during the dormant season to take advantage of seed sources and to favor coppice regeneration. Harvest trees at a stump height of 12 inches or less when practical to encourage vigorous coppice regeneration. Artificial regeneration may be accomplished by planting of seedlings or direct seeding. Table 3-42 presents an example of regeneration system recommendations (Georgia Forestry Association, 1990).

◆ *Conduct mechanized site preparation and planting of sloping areas on the contour.*

◆ *To reduce disturbance, conduct bedding operations in high-water-table areas during dry periods of the year.*

The degree of acceptable site preparation depends on the amount and frequency of flooding, the soil type, and the species suitability.

◆ *Minimize soil degradation by limiting operations on saturated soils.*

Wetland Fire Management Practices

Site preparation burns in wetlands are often the most severe (hottest) and have the most potential to increase surface runoff and soil erosion.

Table 3-42. Recommended Regeneration Systems by Forested Wetland Type (Georgia Forestry Association, 1990)

Type	Natural Regeneration				Artificial Regeneration		
	Clear-cut	Group Selection	Shelter Wood	Seed[a] Tree	Mechanical Site Prep.[b]	Plant	Direct Seed
Flood Plains, Terraces, Bottomland							
Black River	A	B	B	C	D	C	C
Red River	A	B	B	C	D	B	B
Branch Bottoms	A	B	B	C	D	C	C
Piedmont Bottoms	A	B	B	C	D	B	B
Muck Swamps	A	C	C	C	D	C	C
Wet Flats							
Pine Hammocks & Savannahs	A	B	B	B	A	A	B
Pocosins or Bays	A	C	B	B	B	B	B
Cypress Strands	A	C	C	C	D	C	C
Cypress Domes: Peat Swamps							
Peat Swamps	A	C	C	C	C	C	C
Cypress Domes	A	C	C	C	D	C	C
Gulfs, Coves, Lower Slopes	A	B	B	C	C	B	C

Note: A = highly effective; B = effective; C = less effective; D = not recommended.

[a] Seed tree cuts are not recommended on first terraces of floodplains, terraces, and bottomland.

[b] Mechanical site preparation to convert wetlands to pine plantation is regulated by Section 404 of the Clean Water Act and a permit may be required for site preparation to convert some of the wetlands identified in the table, i.e., floodplains, bottomlands, pocosins, bays, cypress strands, peat swamps, cypress domes.

◆ *Conduct site preparation burns in a manner such that they do not completely remove the organic layer from the forest floor.*

◆ *Do not construct firelines for site preparation that will drain wetlands.*

Chemical Management Practices

◆ *Where feasible and applicable, apply herbicides by injection to individual stems.*

◆ *For chemical and aerial fertilizer applications, maintain and mark a buffer area around all surface water to avoid drift or accidental direct application.*

Avoid application of pesticides with toxicity to aquatic life, especially aerial applications. Aerial applications generally require a buffer from water, agricultural lands, and homes. Motorized ground applications require a buffer from water. The first pass of each application is be made parallel to the buffer zone. A buffer is not necessary for hand applications; however, hand-applied forest chemicals have to be applied to specific targets, and chemicals need to be prevented from entering the water. Before any application of a chemical, consult state laws and regulations for chemical application for proper buffer establishment. Have a person licensed in chemical application perform all work (Washington State DNR, 1997).

◆ *Apply slow-release fertilizers when possible.*

This practice reduces the potential of the nutrients leaching to groundwater, and it increases the availability of nutrients for plant uptake.

◆ *Apply fertilizers when leaching will be minimized.*

◆ *Base fertilizer type and application rate on soil and/or foliar analysis.*

To determine fertilizer formulations, it is best to compare available nitrogen, phosphorus, potassium, and sulphur in the soils to be treated with the requirements of the species to be sown.

EPA and Corps of Engineers Memorandum to the Field

Mechanical Site Preparation Activities and CWA Section 404

Under certain circumstances, a CWA section 404 permit is required for mechanical silvicultural site preparation activities in wetlands. In 1995, EPA and the U.S. Army Corps of Engineers issued a memorandum to clarify the applicability of section 404 to mechanical silvicultural site preparation activities in the Southeast.

The memorandum (particularly the descriptions of wetlands, activities, and BMPs in the memorandum) focuses on the southeastern United States. However, the guidance in the memorandum is generally applicable when addressing mechanical silvicultural site preparation activities in wetlands elsewhere in the country.

The memorandum clarifies the applicability of forested wetlands BMPs to silvicultural site preparation activities for the establishment of pine plantations in the Southeast. Mechanical silvicultural site preparation activities conducted in accordance with the

BMPs discussed below, which are designed to minimize effects to the aquatic ecosystem, will not require a Clean Water Act section 404 permit. These BMPs further recognize that certain wetlands should not be subject to unpermitted mechanical silvicultural site preparation activities because of the adverse nature of potential effects associated with these activities on these sites.

EPA and the Corps will continue to work closely with state forestry agencies to promote the implementation of consistent and effective BMPs that facilitate sound silvicultural practices. In those states where no BMPs specific to mechanical silvicultural site preparation activities in forested wetlands are currently in place, EPA and the Corps will coordinate with those states to develop BMPs. In the interim, mechanical silvicultural site preparation activities conducted in accordance with the memorandum will not require a section 404 permit.

Circumstances in Which Mechanical Site Preparation Activities Require a Section 404 Permit

Mechanical silvicultural site preparation activities can have measurable and significant effects on aquatic ecosystems when conducted in wetlands that are permanently flooded, intermittently exposed, or semipermanently flooded, and in certain additional wetland communities that exhibit aquatic functions and values that are more susceptible to effects from these activities. For the wetland types identified below, mechanical silvicultural site preparation activities require a permit so that individual proposals can be evaluated on a case-by-case basis for site preparation and potential associated environmental effects.

A permit will be required in the following areas unless they have been so altered through past practices (including the installation and continuous maintenance of water management structures) as to no longer exhibit the distinguishing characteristics described below (see *Circumstances in which Mechanical Silvicultural Site Preparation Activities Do Not Require a Permit* below). Of course, discharges incidental to activities in any wetlands that convert waters of the United States to non-waters always require authorization under Clean Water Act section 404.

Permanently flooded wetlands, intermittently exposed wetlands, and semipermanently flooded wetlands. Permanently flooded wetland systems are characterized by water that covers the land surface throughout the year in all years. Intermittently exposed wetlands are characterized by surface water that is present throughout the year except in years of extreme drought. Semipermanently flooded wetlands are characterized by surface water that persists throughout the growing season in most years and, even when surface water is absent, a water table usually at or very near the land surface. Examples of these wetlands include cypress-gum swamps, muck and peat swamps, and cypress strands/domes.

Riverine bottomland hardwood wetlands. These are seasonally flooded (or wetter) bottomland hardwood wetlands within the first or second bottoms of the floodplains of river systems. Site-specific characteristics of hydrology, soils, and vegetation and the presence of the alluvial features mentioned in the memorandum determine the boundary of riverine bottomland hardwood wetlands. National Wetlands Inventory maps provide a useful reference for the general location of these wetlands on the landscape.

White cedar swamps. These wetlands are greater than 1 acre in headwaters and greater than 5 acres elsewhere. They are underlain by peat of greater than 1 meter and vegetated by natural white cedar representing more than 50 percent of the basal area, where the total basal area for all tree species is 60 square feet or greater.

Carolina bay wetlands. These are oriented, elliptical depressions with a sand rim that are either underlain by clay-based soils and vegetated by cypress or underlain by peat of greater than 0.5 meter and typically vegetated with an overstory of red, sweet, and loblolly bays.

Nonriverine forest wetlands. The wetlands in this group are rare, high-quality wet forests, with mature vegetation, located on the Southeastern Coastal Plain. Their hydrology is dominated by high water tables. Two forest community types fall into this group: (1) nonriverine wet hardwood forests, poorly drained mineral soil interstream flats (comprising 10 or more contiguous acres), typically on the margins of large peatland areas, seasonally flooded or saturated by high water tables, with vegetation dominated (greater than 50 percent of basal area per acre) by swamp chestnut oak, cherrybark oak, or laurel oak alone or in combination, and (2) nonriverine swamp forests, very poorly drained flats (comprising 5 or more contiguous acres), with organic soils or mineral soils with high organic content, seasonally to frequently flooded or saturated by high water tables, with vegetation dominated by bald cypress, pond cypress, swamp tupelo, water tupelo, or Atlantic white cedar alone or in combination.

Low pocosin wetlands. These are the central, deepest parts of domed peatlands on poorly drained interstream flats, underlain by peat soils greater than 1 meter, typically vegetated by a dense layer of short shrubs.

Wet marl forests. These are hardwood forest wetlands underlain with poorly drained, marl-derived, high-pH soils.

Tidal freshwater marshes. These wetlands are regularly or irregularly flooded by fresh water. They have dense herbaceous vegetation and occur on the margins of estuaries or drowned rivers or creeks.

Maritime grasslands, shrub swamps, and swamp forests. These are barrier island wetlands in dune swales and flats, underlain by wet mucky or sandy soils. They are vegetated by wetland herbs, shrubs, and trees.

Circumstances in Which Mechanical Site Preparation Activities Do Not Require a Section 404 Permit

Mechanical silvicultural site preparation activities in wetlands that are seasonally flooded, intermittently flooded, temporarily flooded, or saturated or are in existing pine plantations and other silvicultural sites (except as listed above) do not require a permit if conducted according to the BMPs listed below in *Best Management Practices.* Of course, silvicultural practices conducted in uplands never require a Clean Water Act section 404 permit (see *Code of Federal Regulations* text above).

Seasonally flooded wetlands are characterized by surface water that is present for extended periods, especially early in the growing season, but is absent by the end of the season in most years. (When surface water is absent, the water table is often near the surface.) Intermittently flooded wetland systems are characterized by substrate that is usually exposed and the presence of surface water for variable periods without detectable seasonable periodicity. Temporarily flooded wetlands are characterized by surface water that is present for brief periods during the growing season, but also by a water table that usually lies well below the soil surface for most of the season. Saturated wetlands are characterized by substrate that is saturated to the surface for extended periods during the growing season, but also by the absence of surface water most of the time. Examples typical of these wetlands include pine flatwoods, pond pine woodlands, and wet flats (e.g., certain pine/hardwood forests).

Best Management Practices

The BMPs below are from a joint EPA and Corps of Engineers *Memorandum to the Field* (see below) on the application of BMPs to mechanical silvicultural site preparation activities for the establishment of pine plantations in the Southeast. The guidance is, however, generally applicable to mechanical silvicultural site preparation activities in wetlands elsewhere in the country. Every state in the Southeast has developed BMPs for forestry to protect water quality, and most have also developed specific BMPs for forested wetlands.

The BMPs listed here are the minimum to be applied for mechanical silvicultural site preparation activities in forested wetlands where these activities do not require a permit (see *Memorandum to the Field* below). In circumstances where a permit is required, BMPs specifically required for the individual operation will be detailed in the permit.

The BMPs below were developed because silvicultural practices have the potential to result in effects on an aquatic ecosystem. Mechanical silvicultural site preparation activities have the potential to cause effects such as soil compaction, turbidity, erosion, and hydrologic modifications if the activities are not effectively controlled by BMPs.

◆ *Position shear blades or rakes at or near the soil surface and windrow, pile, and otherwise move logs and logging debris by methods that minimize dragging or pushing through the soil to minimize soil disturbance associated with shearing, raking, and moving trees, stumps, brush, and other unwanted vegetation.*

◆ *Conduct activities in such a manner as to avoid excessive soil compaction and maintain soil tilth.*

◆ *Arrange windrows in such a manner as to limit erosion, overland flow, and runoff.*

◆ *Prevent disposal or storage of logs or logging debris in SMAs.*

◆ *Maintain the natural contour of the site and ensure that activities do not immediately or gradually convert the wetland to a non-wetland.*

◆ *Conduct activities with appropriate water management mechanisms to minimize off-site water quality effects.*

The full text of the memorandum is available on the Internet at <http://www.epa.gov/owow/wetlands/guidance/silv2.html>.

CHAPTER 4:
USING MANAGEMENT MEASURES TO PREVENT AND SOLVE NONPOINT SOURCE POLLUTION PROBLEMS IN WATERSHEDS

Management measures and associated management practices applied at harvest sites and along roads provide essential control of erosion and sedimentation, and it is important that all management measures and management practices applicable to a harvest site or road be applied to limit as much as possible the amount of soil erosion and the potential for water pollution that can result from forest harvesting activities.

The watershed perspective enables the practitioner to go beyond the effects from a single harvest area or individual road to consider all activities occurring within the watershed that could affect water resources. Each activity can have its own effect on water quality, and the watershed perspective views the effects due to harvesting and road construction within the context of the overall effects of forestry activities together with other activities such as recreational uses and conversions of land use. It is the collective effects of all of these activities that determine how water quality is affected, and these cumulative effects on water quality wouldn't normally be recognized if the effects arising from individual harvesting activities are considered alone.

Research has determined that the use of BMPs on forestland results in smaller increases in nutrients and suspended sediment load after logging than when BMPs are not used. This points to the need for a watershed approach to water quality management, and such an approach within the context of forest harvesting and road construction and use implies, at a minimum, the following:

- Applying management measures and management practices that are appropriate not only to the harvest site, but that take into consideration the current state of water quality in receiving waters, given all that is happening in the watershed, and the effect that forestry activities could have.

- The foreseeable future needs to be considered as well. Some effects of harvesting and road building can last beyond the duration of a harvest or the completion of road construction, and if other activities that could effect water quality are planned in the watershed in the timeframe during which those effects are expected to continue, mitigation of these long-term effects might be necessary.

- Maintenance of older roads built with outdated management practices (those dating from the 1950s to the mid-1970s), which can be significant sources of sediment, is an essential part of forested watershed management. Long-term management plans for forest roads include their inventory, maintenance, and closure; and closure of unused, unneeded, and high-erosion-risk roads.

The EPA Watershed Approach

Watersheds are areas of land that drain to a single stream or other water resource. Watersheds are defined solely by drainage areas and not by land ownership or political boundaries.

Since 1991, the USEPA has promoted the watershed protection approach as a holistic framework for addressing complex pollution problems such as those from nonpoint sources. The watershed protection approach is a comprehensive planning process that considers all natural resources in the watershed, as well as social, cultural, and economic factors. The process tailors workable solutions to ecosystem needs through participation and leadership of stakeholders.

Although watershed approaches may vary in terms of specific objectives, priorities, elements, timing, and resources, all should be based on the following guiding principles.

- *Partnerships*. People affected by management decisions are involved throughout and help shape key decisions. Cooperative partnerships among federal, state, and local agencies and non-governmental organizations with interests in the watershed are formed. This approach ensures that environmental objectives are well integrated with those for economic stability and other social/cultural goals of the area. It also builds support for action among those individuals who are economically dependent upon the natural resources of the area.

- *Geographic focus*. Resource management activities are coordinated and directed within specific geographic areas, usually defined by watershed boundaries, areas overlaying or recharging groundwater, or a combination of both.

- *Sound management techniques based on strong science and data*. Collectively, watershed stakeholders employ sound scientific data, tools, and techniques in an iterative decision-making process. Typically, this includes:

  - Assessment and characterization of the natural resources in the watershed and the people who depend upon them.

  - Goal setting and identification of environmental objectives based on the condition or vulnerability of resources and the needs of the aquatic ecosystem and the people.

  - Identification of priority problems.

  - Development of specific management options and action plans.

  - Implementation, evaluation, and revision of plans as needed.

Operating and coordinating programs on a watershed basis makes good sense for environmental, financial, social, and administrative reasons. For example, by jointly reviewing the results of assessment efforts for drinking water protection, pollution control, fish and wildlife habitat protection, and other resource protection programs, managers from all levels of government can better understand the cumulative effects of various human activities and determine the most critical problems within each watershed. Using this information to set priorities for action allows public and private managers from all levels to allocate limited financial and human resources to address the most critical needs.

Establishing environmental indicators helps guide activities toward solving those high-priority problems and measuring success.

The final result of the watershed planning process is a plan that is a clear description of resource problems. Goals to be attained, and identification of sources for technical, educational, and funding assistance needed. The successful plan provides a basis for seeking support and for maximizing the benefits of that support.

Cumulative Effects

The watershed approach is a useful mechanism for managing the resources within a defined geographical boundary, and it provides a basis for cumulative effects assessment as well. Though it is not a formal analytical framework for the evaluation of cumulative effects, the watershed approach shares with cumulative effects assessment (CEA) a consideration of all relevant activities and influences. Furthermore, a watershed is a natural geographic boundary for the analysis of cumulative effects on water quality because the influences of upstream activities can create a cumulative effect on down-stream water quality.

Definition

Current environmental regulations provide at least two definitions of cumulative effects (CEs):

> Cumulative effect is the effect on the environment which results from the incremental effect of the action when added to other past, present, and reasonably foreseeable future actions regardless of what agency (federal or non-federal) undertakes such other actions. Cumulative effects can result from individually minor but collectively significant actions taking place over a period of time (40 CFR 1508.7).

> Cumulative effects are the changes in an aquatic ecosystem that are attributable to the collective effect of a number of individual discharges of dredged or fill material. Although the effect of a particular discharge may constitute a minor change in itself, the cumulative effect of numerous such piecemeal changes can result in a major impairment of the water resources and interfere with the productivity and water quality of existing aquatic ecosystems (40 CFR 230.11).

CEs can be very difficult to quantify and assess, and they are best understood by focusing on the mechanisms by which watershed processes are affected (Reid, 1993). Watershed processes are affected when a land use activity causes a change in the production and transport of one or more watershed products (water, sediment, organic material, chemicals, or heat). Most land use activities affect only one of four aspects of the environment—vegetation, soils, topography, or chemicals—and other watershed changes result from initial effects on these. Understanding CEs within a watershed context involves: (1) understanding how specific land uses affect vegetation, soils, topography, or chemicals; (2) determining to what extent these changes affect watershed processes; and (3) understanding how changes to vegetation, soils, topography, chemicals, and watershed processes affect particular resources and values.

Cumulative effects can be additive or synergistic (MacDonald, 2000). Additive effects are those in which each land use activity creates a discrete effect on an individual resource or value and the total effect is the sum of the individual effects. Synergistic effects are those in which the combined effect of individual activities on a resource or value are greater than the sum of their individual effects. Synergistic effects can occur through the interaction of different chemicals or types of effects on a single resource. Many times with synergistic effects, each effect is analyzed and determined to individually not be detrimental to a particular resource, but the combined or cumulative effect of the three activities do create a significant impact on a resource.

Assessment of CEs should also take into account whether they are on-site or off-site. On-site CEs can occur if a change persists long enough for later activities to affect the same resource or for the effects of off-site activities to be transported to the site of the change. The temporal dimension of on-site CEs is important to their assessment, while the spatial dimension is limited to the original site of the effect. Off-site CEs occur when a land use activity causes a change in a watershed process such that effects are created at a location other than where the original land use activity occurred. Off-site CEs occur when watershed processes are altered long enough for the off-site effects to accumulate over time; when watershed processes are affected at multiple sites in a watershed and the watershed products that are affected are transported to the same site, or when an off-site effect interacts with an on-site effect. Both the temporal and spatial dimension of off-site CEs are important to consider when analyzing them.

The Importance of Considering and Analyzing Cumulative Effects

Cumulative effects are of concern with respect to forest roads; forest road construction, use, and maintenance; and forest harvesting because the changes that can occur in watershed processes following these activities can persist for many years. This persistence increases the potential for cumulative effects to occur.

Traditionally, effect assessment has evaluated the likely effects of single actions on the environment. But single areas and ecosystems are often affected by more than single actions or projects. The collective effect of numerous small actions can cause serious degradation, though the effects of each small action by itself might be undetectable. Even after an area or ecosystem has been degraded, an analysis of the effects of an additional action might conclude that there would be only minor or no significant effect. An analysis of the additive effect of the single additional action—the cumulative effects—however, might conclude that the action could be detrimental (USEPA, 1992). Cumulative effects analysis also differs from many types of traditional environmental assessment in the need to predict the consequences of "reasonably foreseeable future actions."

The importance of cumulative effects assessment, then, lies in the difference between traditional effect assessment and cumulative effects assessment. Traditional effect assessment is performed with respect to the proposed disturbance, whereas cumulative effects assessment is performed with respect to valued environmental functions (USEPA, 1992). An assessment of an action might have little to no detectable significant effect in terms of pollutant additions or habitat loss, as determined by traditional effect assessment, but might have a clearly disturbing effect on ecosystem functioning as determined by cumulative effects assessment. As more habitat is lost or fragmented and pollutants are generated, environmental stewardship demands that we pay more attention to the collective effects of our actions on ecosystems and their functioning and place less stress on the absolute quantities of pollutants that are generated or habitat lost as a result of each action. Cumulative effects assessment is the means to do this.

Problems in Cumulative Effects Analysis

Cumulative effects analysis, as conceived, is a powerful approach to assessing the overall effect of our actions on the environment and of managing those actions such that species and ecosystems continue to function properly. Unfortunately, many practical problems are associated with performing a cumulative effects analysis, including the following:

- Because total maximum daily load (TMDL) assessments calculate all point source and non-point source pollution for a watershed, a TMDL is essentially a cumulative effects analysis. Agencies responsible for implementing TMDL's have been hesitant to do so because of limitations in personnel, water quality data, and understanding of watershed dynamics. There is also a lack of available methodologies for tracking pollutants such as clean sediment (MacDonald, 2000).

- Ecosystems are complex and our knowledge of their workings is still limited, yet cumulative effects assessment involves identification of the ecosystem components of relevance that will be the focus of the cumulative effects analysis (Berg et al., 1996).

- The boundaries for cumulative effects assessment might be different from those relevant to other analyses, such as nonpoint source pollution or TMDL assessment. A single watershed might be appropriate for assessing nonpoint source pollution, but many watersheds might be involved in cumulative effects analysis for effects on forest conservation (Berg et al., 1996).

- Current guidelines published by the CEQ (1997) do not explicitly address natural processes, spatial variability, and temporal variability within project areas. Natural variability and rates of recovery can affect prediction and detection of cumulative impacts (MacDonald, 2000).

- Effects from individual projects often last for no longer than one human generation, whereas the time frame for changes in ecosystem processes that are the focus of cumulative effects assessment is typically an order of magnitude longer (Berg et al., 1996).

- The effects of most management activities diminish over time, and so then does the magnitude of possible cumulative effects. This leads to a problem of temporal scale related to determining the magnitude of human-induced cumulative effects relative to natural variability over a long time lag (MacDonald, 1997).

- The scale of cumulative effects analysis is very different from that used for traditional effect assessment, and effects due to individual projects might be undetectable using the analytical methods necessary for cumulative effects assessment. For instance, patterns on the landscape, such as whether 10,000 hectares are contiguous or not, are relevant for cumulative effects analysis; a small clear-cut, important at the local scale, might not appear in an analysis at a scale of thousands of hectares (Berg et al., 1996).

- When working at the scale necessary for cumulative effects assessment, areas that contain fragmented jurisdictions with multiple-agency oversight, differences in regulatory structure between jurisdictions and agencies, and conflicting interests and mandates are involved (Berg et al., 1996).

- To adequately assess the future consequences of multiple perturbations in a watershed, the status of ecosystem recovery from past perturbations must be estimated.

Complexity of the analysis increases because recovery times for various components in a system are not necessarily identical, and knowledge is often inadequate to quantify recovery rates. For instance, "recovery" of stream flow magnitude and rate after timber harvest is largely a function of the rate of revegetation of the watershed. Sediment produced by roads associated with the timber harvest will typically take much longer to move through stream channels and "recover" to pre-road levels. Understanding of both types of recovery is needed and they cannot be substituted for each other.

Within the context of forestry activities and forested watersheds, the following difficulties are encountered when attempting to assess cumulative effects (Reid, 1993):

- The effects of forest management activities on streamflow has been studied extensively, yet it remains difficult to determine what effects a management activity will have on a stream because hydrologic response varies greatly with basin size, flow magnitude, season, climate, geology, and type and intensity of forest management activity. The results of studies done in one basin are therefore difficult to extrapolate to other basins. It can be important to determine whether forestry activities will have effects on watershed processes because of the potential consequences if the effects are substantial enough, but such a determination can be costly. It can also be costly, however, to take measures to prevent watershed effects from forestry activities when such effects might not materialize.

- Variability in storm intensity and runoff processes limit the ability to detect human-induced effects on streamflow. Even with years of monitoring data, it can be difficult to distinguish between human-induced effects and natural variability in watershed processes. The process of determining cause and effect is complicated by the fact that different activities can cause similar responses and one activity might not always elicit the same response.

- The dynamics of natural forest communities must be understood to interpret or predict the effects of changes, and natural disturbance frequencies, patterns, characteristics, recovery rates; these are not well understood. Monitoring would be a useful tool to increase our understanding of these dynamics, but the sequences of changes that can lead to CEs, or the combinations of changes that can lead to CEs are varied and can take long periods of time to take effect (e.g., 50 years). Monitoring these effects is often not possible due to the time frame involved.

- If a system responds incrementally, changes can be easily identified; but many changes, such as landslides or floods, do not occur incrementally. Instead, changes, such as loss of vegetation water storage and increased soil compaction, might be relatively benign and accumulate until some event, such as a 50-year storm, triggers a substantial response. These thresholds at which substantial and important CEs occur often cannot be predicted, and knowledge of them is based on studying them after they occur.

- The rate of recovery from land use depends on the type of land use and on the watershed processes that are affected.

Approaches to Cumulative Effects Analysis

Four general approaches for predicting cumulative effects include the use of analytical models, assessments of previous management activities, use of a collection of procedures that address specific anticipated impacts, and use of a checklist to indicate what cumulative effects might be expected to occur because of a land use activity. Models can be used to predict changes to physical or biological aspects of a watershed, or to predict the magnitude of change in a watershed process or characteristic that might trigger a particular type of impact (Reid, 1993). Models are useful because the cumulative effects of repeated timber harvests in a watershed could be estimated or monitored experimentally only in a study lasting several centuries (Ziemer and Lisle, 1991). While modeling does represent a simplification of nature and depends on a modeler's skill, modeling results can represent average conditions and explore the effects of large spatial and temporal scales. They can also be useful for conducting "what if" analyses, where the effects of different sequences of harvesting or precipitation events, for example, are explored. This characteristic of models contrasts sharply with monitoring studies, in which the unique sequence of events that occurs during a monitoring distorts the results.

Many models have been developed for specific locations and cannot easily be applied to other areas. The limitations of the models are stated in user's guides or instructions for use, but the models, nevertheless, are often put into general use regardless of whether the assumptions of the model are valid for a particular application or whether the methods of the model have been tested and validated (Reid, 1993). Many models are meant to be used to predict particular impacts, yet their methods are used to test for the likelihood of a variety of other possible impacts for which the method was not developed. Used properly, however, models can shed light on the importance of processes and variables to watershed behavior and treatment effects, but have limited value for precisely predicting watershed behavior (Reid, 1993). A large amount of data generally is required for modeling, and its acquisition can involve intensive monitoring. Data analysis also can be complex, and these factors have kept the use of models very limited (MacDonald, 1997).

Slightly less complicated than modeling would be an analysis involving a broad-scale assessment of previous management activities. Such a method would use one or more management indices to assess the relative likelihood of a cumulative effect, rather than explicitly modeling cause-and-effect (MacDonald, 1997). The EPA Synoptic Approach and the *Washington State Watershed Analysis Method* (described below) are examples of this level of analysis.

Another approach for assessing cumulative effects consists of a collection of procedures used to evaluate a variety of impacts. A relevant subset of impacts is generally considered. This approach provides flexibility in determining what impacts will be considered, but it provides no guidance on determining which impacts should be evaluated (Reid, 1993). The *Water Resources Evaluation of Non-point Silvicultural Sources* (WRENSS) (described below) method is an example of a procedure-based approach.

A third general approach consists of a checklist of items to consider during an assessment. A checklist provides guidance in determining what impacts to evaluate but does not provide methods for doing so (Reid, 1993). Checklists are useful for (1) identifying which issues to look at in more detail, (2) helping to ensure that a range of issues are considered, (3) providing a simple means to address the issue of cumulative effects assessment. Disadvantages associated with checklists include the strictly qualitative nature of the assessments, their lack of repeatability, and their lack of documentation (MacDonald, 1997). The California Department of Forestry questionnaire (described below) is an example of a checklist assessment method.

Each approach has its strengths and weaknesses, and a workable approach should be a combination of these separate approaches. For example, a checklist or expert system could be used to guide users through a decision tree to identify the impacts to be considered, and then a set of procedures could be selected to address them (Reid, 1993). Modeling could be employed to assess the sensitivities of the watershed to various treatment scenarios.

Five techniques that have been developed for assessing cumulative effects are described below.

1. EPA The Synoptic Approach

The Synoptic Approach was developed by EPA for the evaluation of cumulative effects on wetlands for section 404 permit review. It does not provide a precise, quantitative assessment of cumulative effects, but is used to rate cumulative effects on resources of interest (Berg et al., 1996). The Synoptic Approach has two major steps—definition of the synoptic indices and selection of landscape indicators.

Synoptic Indices

Four synoptic indices are used for assessing cumulative effects and relative risk—function, value, functional loss, and replacement potential. The function index refers to the total amount of a particular function a wetland provides within a landscape subunit without consideration of the ecological or social benefits of that function. Landscape elements function within landscapes through physical, chemical, and biological processes to provide habitat, cleanse water, prevent flooding, and perform other functions. The value index refers to the value of ecological functions with respect to public welfare. Tangible benefits (e.g., hunting, camping, timber, carbon dioxide sequestration) and intangible benefits (e.g., aesthetic, existence value) can both be included, as well as future value as the future benefit of the functions performed. Note that the value index does not represent economic value since market factors are not considered. The functional loss index represents cumulative effects on a particular valued function that have occurred within a landscape subunit. A complete loss, where an ecosystem element is changed into something else entirely, is a conversion. A partial loss, where ecosystem element type is the same but functioning is altered, is degradation. In the course of a cumulative effects assessment, future loss is considered per the Council on Environmental Quality's regulations (40 CFR 1508.7). Functional loss depends on the characteristics of a particular effect, including the type of effect; its magnitude, timing, and duration; and ecosystem resistance, or the sensitivity of the ecosystem element to disturbance. The replacement potential index represents the ability to replace an ecosystem element and its valued functions. Functional replacement through ecological restoration or natural recovery are both considered. Protection of ecosystem elements and functions is critical for risk reduction if their replacement potential is judged to be low (USEPA, 1992).

Landscape Indicators

Landscape indicators are first-order approximations that represent some particular synoptic index. Quantifying specific synoptic indices for large landscape subunits would be difficult if not impossible, so the Synoptic Approach uses landscape indicators of actual functions, values, and effects (USEPA, 1992).

As an example, a particular management concern might be nonpoint source sediment loading to streams. Nonpoint source sediment loading would then be the synoptic index used in the Synoptic Approach. Since it would be difficult to quantify this over a large area, total area harvested might be chosen as a landscape indicator for forest harvesting. Total harvested area would be the data used to determine cumulative nonpoint source sediment loading effects on the area of concern.

The Synoptic Approach is an ecologically based framework in which locally relevant information and best professional judgment are combined to address cumulative effects. It is not, however, meant to be used to assess the cumulative effects of specific actions. Rather, it is really meant to be used to augment site-specific review processes and to improve best professional judgment. It is probably most effectively used at extremely large landscape scales, such as the state level (Berg et al., 1996). The approach is valuable because it is flexible enough to cover a broad spectrum of management objectives and constraints—the specific synoptic indices and landscape indicators used in an application can be chosen based on the particular goals and constraints of the assessment—and it certainly need not be limited to assessing effects on wetlands. The process allows managers to weigh the need for precision against the constraints of time, money, and information (USEPA, 1992).

2. Washington State Watershed Analysis

The Washington State Watershed Analysis method is used to develop forest plans for individual watersheds based on current scientific understanding of the significant links between physical and biological processes and management activities. The first step in use of the method is screening a watershed to qualitatively define and assess areas of sensitivity to environmental change within the watershed. If any area is found to be sensitive, then the area and the causal mechanism must be addressed by a management plan appropriate to the problem. The management plan will define more precisely the potential effects of management actions and management alternatives. The method uses separate assessment modules for mass wasting, surface erosion, hydrologic change, riparian function, stream channel assessment, fish habitat, water supply/public works, and routing through the fluvial system (Berg et al., 1996).

The Washington State Watershed Analysis process is a collaborative one that involves both scientists and managers, and its products generally are area-specific management prescriptions and monitoring recommendations (Berg et al., 1996).

3. Water Resources Evaluation of Nonpoint Silvicultural Sources (WRENSS)

The WRENSS is a process-based approach to evaluating timber management impacts (Reid, 1993). It consists of a series of procedures for evaluating separate impacts, though it is not intended specifically to address CEs. The original focus of the method was water quality and consideration of the effects of timber management and roads. While its procedures do not address resources other than water quality, it would be possible to add additional methods to evaluate impacts on particular resources and to assess the effects of other land uses. Use of the method can be complex and time consuming.

The method is based on computer simulation modeling that delivers graphs and tables as results that are used to estimate changes in evapotranspiration, flow duration, and soil moisture from different logging plans. Temperature changes are incorporated using a separate model, the Brown model, and sediment modules include methods for estimating surface erosion, ditch erosion, landsliding, earthflow activity, sediment yield, and channel stability.

Application of the method to CE analysis would require the identification of likely environmental changes generated by a project, likely downstream impacts, and the mechanisms generating them.

4. California Department of Forestry Questionnaire

The California Department of Forestry and Fire Protection developed a questionnaire for use by registered professional foresters to assess potential cumulative watershed effects (CWE) from timber management. Completion of the questionnaire involves a four-step process: (1) perform a resource inventory in the assessment area; (2) judge whether the planned timber operation is likely to produce changes to each of those resources; (3) identify the effects of past or future projects; and (4) judge whether significant cumulative effects are likely from the proposed operation. Onsite and downstream beneficial uses, existing channel conditions, and adverse effects from past projects are identified and listed during the first step. The area for analysis is one of manageable size relative to the timber harvest—usually an order 3 or 4 watershed. During the assessment, the user rates the magnitude of a variety of potential effects from the proposed and future projects, and combined past, present, and future projects. The assessment serves as an indicator of need for further review.

Responding to the questionnaire relies on the qualitative observations and professional judgment of the person filling out the forms. The questionnaire is designed to be used within the time constraints of the development of timber harvest plans and serves primarily as a checklist to be certain that all important issues have been considered. Its strength lies in its flexibility: the checklist can be easily altered to accommodate a wide variety of situations and harvesting conditions.

The California Department of Forestry questionnaire addresses a wide variety of uses and effects and includes many that are not related to water quality, e.g., recreational, aesthetic, biological, and traffic uses and values, but it provides only qualitative results. The questionnaire is the only CWE evaluation method that uses an assessment of more than one type of effect from more than one type of mechanism, and it is one of few that incorporates an evaluation of effects that accumulate due to past, present, and future actions (Berg et al., 1996).

5. Phased Approach to Cumulative Effects Assessment

MacDonald (2000), put forth a conceptual process for assessing cumulative effects. The process is an attempt to overcome some of the problems with other approaches to cumulative effects analysis (CEA), including problems in defining key issues, specifying the appropriate spatial and temporal scales, and determining the numerous interactions and indirect effects to analyze. The assessment is broken down into three phases: scoping, analysis, and management.

- The scoping phase is further broken down into steps in which the issues, resources, time scale, spatial scale, risk, and assessment effort are identified for the cumulative effects analysis. The analysis phase is likewise subdivided into five substeps.

- In the analysis phase researchers identify and analyze cause-and-effect mechanisms; natural variability and resource condition; past, present and future activities; relative impacts of past, present and future activities; and validity and sensitivity of the overall cumulative effects analysis.

- The management phase identifies possibilities for mitigation and restoration, as well as key data gaps and monitoring needs.

Figure 4-1 illustrates MacDonald's process for assessing cumulative effects.

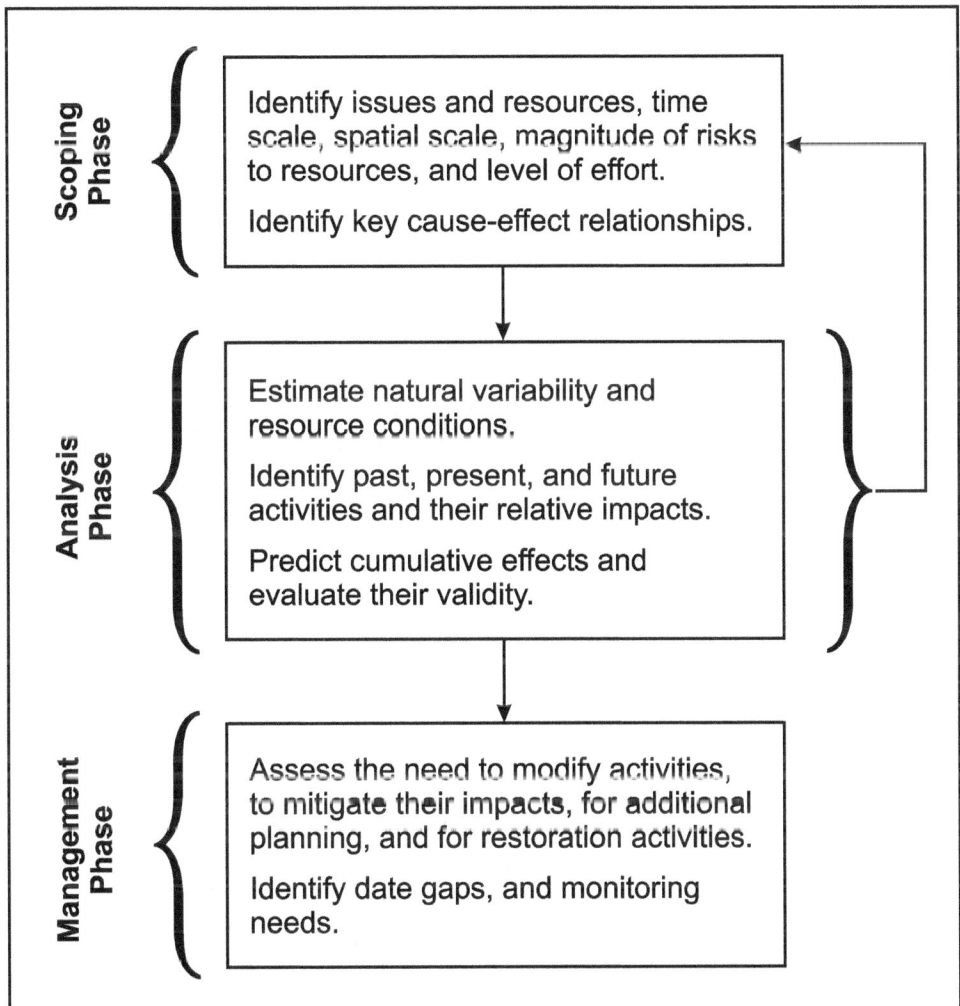

Figure 4-1. Representation of MacDonald's process for assessing cumulative effects (after MacDonald, 2000).

The President's Council on Environmental Quality (CEQ) published guidelines for performing CEA (CEQ, 1997). The CEQ methodology is broken down into three groups of steps that are designed to be integrated into three components of an environmental impact assessment (EIA). The EIA components relevant to CEA are scoping, describing the affected environment, and determining the environmental consequences.

- In the scoping component of an EIA, the CEA steps are to identify significant issues and define assessment goals; establish spatial boundaries of the CEA; establish temporal scale of the CEA; and identify other activities that affect natural and human communities.

- The affected environment component of the EA should incorporate the following CEA steps: characterize the resources, ecosystems and human communities and their resilience to stress; define stresses and regulatory thresholds for measuring stresses; and define baseline conditions for the area defined in the CEA.

- The environmental consequences component of the EIA should identify CEA cause-and-effect relationships between human activities and resources; determine the significance of cumulative effects; develop alternatives to minimize or mitigate significant cumulative effects; monitor cumulative effects and adapt management accordingly.

CEQ lists seven primary methods to develop baseline data and analytical models for cumulative effects analysis (CEA):

- Questionnaires, interviews, and panels to gather initial information
- Checklists to review important activities that may contribute to cumulative effects
- Matrices to tally cumulative effects
- Networks and system diagrams to qualitatively analyze effects of multiple activities on multiple resources in the analysis
- Modeling to quantify the cause-and-effect relationships within the CEA
- Trends analysis to use baseline data to extrapolate future cumulative effects
- Overlay mapping (GIS) to perform spatial analysis and identify areas of high and low impact.

Appendices to the CEQ report provide examples of each method and how it is might be used in CEA. The report is available on the World Wide Web at <http://ceq.eh.doe.gov/nepa/ccenepa/ccenepa.htm>.

The MacDonald (2000) and CEQ (1997) guidelines share many similar components. The spatial and temporal boundaries of the CEA are defined first, along with the resources that will be impacted by cumulative effects. Detailed analysis of cause-and-effect relationships follows, and baseline data is developed to describe present conditions. Both methods include monitoring and mitigation steps toward the end of the process. MacDonald's framework differs from the CEQ methodology by including natural variability in systems, consideration of past and future activities, sensitivity analysis of predictive models, and an up-front determination on the level of effort that is appropriate for the assessment. MacDonald's refinements help address some of the hurdles to CEA implementation that have hampered past efforts.

Forest Watershed Management: An Example

The Umatilla National Forest, located in the Blue Mountains of southeast Washington and northeast Oregon, covers 1.4 million acres of diverse landscapes and plant communities (USDA-FS, 1999). The forest has some mountainous terrain, but mostly consists of V-shaped valleys separated by narrow ridges or plateaus. The landscape also includes heavily timbered slopes, grassland ridges and benches, and bold granite outcroppings. Elevations range from 1,600 to 8,000 feet above sea level.

The Forest is administered by the Forest Supervisors Office in Pendleton, Oregon, along with four Ranger Districts located in Pomeroy and Walla Walla, Washington, and Ukiah and Heppner, Oregon. The actual on the ground management of the forest resources is accomplished at the Ranger District level by the District Ranger and staff, while the Forest Supervisor oversees management and administration. The Forest is challenged daily with protecting both the productivity and the aesthetic values of the land. Managing to provide many resources, benefiting many people "for the long run" is the key principle guiding the Umatilla Management Team.

Because water from the Blue Mountains is important for so many uses, proper management of the watersheds in the Umatilla National Forest is strongly emphasized. The goals of the watershed management program are as follows:

- To maintain streams that are cold, clean, and free of excessive sediments and human-caused pollution.
- To keep stream banks, channels, wetlands, and adjacent floodplains healthy.
- To restore damaged lands to their previous, productive condition.
- To maintain near-natural amounts of runoff water.

The Umatilla National Forest Plan includes important direction for achieving these goals. The plan envisions a basic three-point program for managing forest watersheds:

1. Inventory Basic Watershed Resources

Proper management of a forest watershed demands a good understanding of basic components—soil, water, climate, and vegetation. Managers at the Umatilla National Forest upgrade the resource information base for the forest by conducting the following inventories and surveys:

- Soil
- Water
- Fishery resources
- Potential watershed improvement projects
- Riparian zones (areas adjacent to streams and lakes)

These watershed surveys provide vital information for improving the management of surface water resources.

2. Apply Best Management Practices

The Umatilla National Forest has developed "best management practices"—policies, standards, and methods of operation designed to reduce harmful effects on water while still allowing use of other resources. Maintaining stream surface shading to prevent fish-bearing waters from overheating during the summer is an example of general practices applied throughout the forest. Others are developed specifically for a particular activity.

Forest managers work together in the project planning stages to identify the nature and risk of potential hazards to water resources. As a result, projects can be modified to avoid problem areas and reduce water resource damage.

The forest's watershed management program emphasizes the prevention of problems before they occur. However, it is sometimes necessary to treat watershed problems resulting from past practices. Such treatments might include restoring wet meadows, recontouring gullied lands, or stabilizing eroding stream banks.

Recently, a program to control and treat the acidic wastewater draining into a forest stream where salmon and steelhead spawn was begun in the Umatilla National Forest. These wastes, produced by abandoned gold mines, are now treated in man-made bogs, where toxic metals and other harmful substances are filtered out. Initial results have shown a dramatic recovery in water quality.

3. Monitor and Analyze Results

An extensive water-monitoring program has been developed for the Umatilla National Forest. It measures success in achieving the goal of maintaining healthy and abundant water resources. Monitoring stations are strategically placed at forest management projects to measure

- Stream flow
- Water temperature
- Suspended sediment and turbidity
- Shape and condition of stream channels and riparian areas
- Precipitation, snow pack and other climatic factors
- The soil's ability to infiltrate and hold precipitation
- Physical, chemical and biological components of water quality

These measurements provide a better understanding of how management activities affect water resources and whether our efforts are effective in maintaining high water quality.

CHAPTER 5: MONITORING AND TRACKING TECHNIQUES

This chapter discusses monitoring the implementation and effectiveness of forestry management measures. For the most part, such monitoring is done either for research purposes or to assess compliance with regulatory requirements or recommendations. Therefore, it is usually the domain of universities or government agencies and this chapter is directed primarily at state agencies responsible for compliance with forestry regulations, nonpoint source pollution control regulations, or voluntary forest practice programs. Owners and managers of large forestland tracts are encouraged to work with state officials to develop a means of monitoring the implementation of BMPs on their lands to assess whether they are installed and maintained adequately so that they will protect water quality effectively, regardless of whether the state's program mandates forest practice implementation or encourages voluntary implementation.

Overview

Designing and legally implementing a state program of management practices for forest harvests and forest road construction cannot protect water quality unless the BMPs are implemented by those who actually harvest the timber or manage the land to be harvested. Monitoring the implementation of BMPs is a crucial element of any BMP program. Monitoring provides feedback on whether management practices are implemented per the specifications required or recommended by state and federal governments, on how the forestry practice program is received by harvesters and landowners, and on forestry practice design and use standards and specifications so they can be refined to be more useful and more effective.

Many states have implemented programs to monitor the implementation of forestry practices at harvest sites in conjunction with the passage of forest practice legislation or after a state has established a set of forestry practice recommendations. The end of this chapter provides information about some of these programs. Fewer states monitor the effectiveness of management practices at protecting water quality as part of their BMP implementation monitoring programs. However, even a limited amount of effectiveness monitoring, such as under controlled conditions during experimental harvests, is important to ensure that BMP design specifications and standards are adequate to protect water quality and soils. Once it is determined that BMPs that are installed according to standards and specifications are actually effective, it can be acceptable to monitor only the implementation of BMPs to ensure that they are properly installed, the assumption being that if they are installed adequately, then they effectively protect water quality and forest resources. Such an approach is often necessary because of the difficulty and cost in measuring water quality directly and confounding factors such as upstream pollution sources. Without the initial information that adequately installed BMPs are effective, though, little can be said about the degree of water quality and forest resource protection attained by adequately installing BMPs.

Monitoring Program Fundamentals

The most fundamental step in the development of a monitoring plan is to define the goals and objectives, or purpose, of the monitoring program. In general, monitoring goals are broad statements such as "to measure changes in fish spawning habitat" or "to measure nutrient loading to streams adjacent to harvest sites." Monitoring programs can be grouped according to the following general statements of purpose or expected outcomes:

- Describe status and trend
- Describe and rank existing and emerging problems
- Design management and regulatory programs
- Evaluate program effectiveness
- Respond to emergencies
- Evaluate the implementation of best management practices
- Evaluate the effectiveness of best management practices
- Validate a proposed water quality model
- Perform research

Unlike monitoring goals, monitoring objectives are more specific statements that can be used to add detail, including geographic scale, measurement variables, sampling methods, and sample size, to the monitoring design. Detailed monitoring program objectives enable the designer of the program to define precisely what data will be gathered in order to meet the management goals. Vague or inaccurate statements of objectives lead to program designs that provide too little or too much data, thereby either failing to meet management needs or costing too much.

Numerous guidance documents have been developed, or are in development, to assist resource managers in developing and implementing monitoring programs that address all aspects of monitoring design. Appendix A in *Monitoring Guidance for Determining the Effectiveness of Nonpoint Source Controls* (USEPA, 1997) presents a review of more than 40 monitoring guidances for both point and nonpoint source pollution. These guidances discuss virtually every aspect of nonpoint source pollution monitoring, including monitoring program design and objectives, sample types and sampling methods, chemical and physical water quality variables, biological monitoring, data analysis and management, and quality assurance and quality control.

Once the monitoring goals and objectives have been established, existing data and constraints are considered. A thorough review of literature pertaining to water quality studies previously conducted in the geographic region of interest can help determine whether existing data provide sufficient information to address the monitoring goals and what data gaps exist.

Identification of project constraints address financial, staffing, and temporal elements. Clear and detailed information is obtained on the time frame within which management decisions need to be made, the amounts and types of data that is to be collected, the level of effort needed to collect the necessary data, and equipment and personnel needed to conduct the monitoring. From this information it can be determined whether available personnel and budget are sufficient to implement or expand the monitoring program.

As with monitoring program design, the level of monitoring that will be conducted is largely determined when goals and objectives are set for a monitoring program, although there is some flexibility for achieving most monitoring objectives.

The overall scale of a monitoring program has two components—a temporal scale and a geographic scale. The temporal scale is the amount of time required to accomplish the program objectives. It can vary from an afternoon to many years. The geographic scale can also vary from quite small, such as plots along a single stream reach, to very large, such as an entire river basin. The temporal and geographic scales, like a program's design and monitoring level, are primarily determined by the program's objectives.

If the main objective is to determine the current biological condition of a stream, sampling at a few stations in a stream reach over 1 or 2 days might suffice. Similarly, if the monitoring objective is to determine the presence or absence of a nonpoint source effect, a synoptic survey might be conducted in a few select locations. If the objective is to determine the effectiveness of a watershed forest management program for improving water quality conditions in streams, however, monitoring subwatersheds for 5 years or longer might be necessary. If the objective is to calibrate or verify a model, very intensive sampling might be necessary.

Depending on the objectives of the monitoring program, it might be necessary to monitor only the water body with the water quality problem or it might be necessary to include areas that have contributed to the problem in the past, areas containing suspected sources of the problem, or a combination of these areas. A monitoring program conducted on a watershed scale will include a decision about the watershed's size. The effective size of a watershed is influenced by drainage patterns, stream order, stream permanence, climate, number of landowners in the area, homogeneity of land uses, watershed geology, and geomorphology. Each factor is important because each has an influence on stream characteristics, although no direct relationship exists.

There is no formula for determining appropriate geographic and temporal scales for any particular monitoring program. Rather, once the objectives of the monitoring program have been determined, a combined analysis of them and any background information on the water quality problem(s) being addressed will make it clear what overall monitoring scale is necessary to reach the objectives.

Other factors that can be considered to determine appropriate temporal and geographic scales include the type of water resource being monitored and the complexity of the nonpoint source problem. Some of the constraints mentioned earlier, such as the availability of resources (staff and money) and the time frame within which managers need monitoring information, will also contribute to determination of the scale of the monitoring program.

For additional details regarding nonpoint source monitoring techniques, including chemical and biological monitoring, refer to *Monitoring Guidance for Determining the Effectiveness of Nonpoint Source Controls* (USEPA, 1997). This technical document focuses on monitoring to evaluate the effectiveness of management practices, but also includes approximately 300 references and summaries of more than 40 other monitoring guides. In addition, Chapter 8 of EPA's management measures guidance for section 6217 contains a detailed discussion of monitoring (USEPA, 1993).

Monitoring BMP Implementation

The implementation of management measures and BMPs should be tracked to determine the extent to which the measures are implemented on harvest sites or throughout a watershed. Data on BMP implementation and trends in BMP implementation can be used to address the following goals:

- Determine the extent to which BMPs are implemented in accordance with relevant standards and specifications.

- Determine whether there has been a change from previous years in the extent to which BMPs are being implemented.

- Establish a baseline from which decisions can be made regarding the need for additional incentives for implementation of BMPs.

- Determine the extent to which BMPs are properly maintained and operated.

- Measure the success of voluntary BMP implementation programs.

- Determine how and why BMP use varies from one geographic area to another.

- Support workload and costing analyses for landowner assistance or regulatory programs.

Methods to assess the implementation of management measures are a key focus of the technical assistance to be provided by EPA and NOAA under CZARA section 6217.

Implementation assessments can be done on several scales. Site-specific assessments can be used to assess individual management practices or management measures, and watershed assessments can be used to look at the cumulative effects of implementing multiple management measures. With regard to "site-specific" assessments, it is important to assess individual management practices at the appropriate scale for the practice of interest. For example, to assess the implementation of management measures or management practices for forest roads at harvest sites, only the roads at timber harvesting sites would need to be inspected. In this example, the scale would be a timber harvest area and the sites would be active and inactive roads at the harvest areas. To assess implementation of management measures and practices at streamside management areas, the proper scale might be a harvest area larger than 10 acres and the sites could be areas encompassed by buffer areas for 200-meter stretches of stream. For site preparation and forest regeneration, the scale and site might be an entire harvest site. Site-specific measurements can then be used to extrapolate to a watershed or statewide assessment.

Sampling design, approaches to conducting the evaluation, data analysis techniques, and ways to present evaluation results are described in EPA's *Techniques for Tracking, Evaluating, and Reporting the Implementation of Nonpoint Source Control Measures—Forestry* (USEPA, 1997a), from which much of the text for this chapter has been borrowed. Chapter 8 of EPA's management measures guidance for section 6217 contains a detailed discussion of techniques and procedures to assess implementation, operation, and maintenance of management measures (USEPA, 1993).

Monitoring BMP Effectiveness

By tracking management measures and water quality simultaneously, analysts gain the information necessary to evaluate the performance of the management measures implemented. Management measure tracking provides information on whether pollution controls are being implemented, operated, and maintained adequately. Only with such information is it possible to draw conclusions from water quality monitoring data about the effectiveness of management practices.

A major challenge in attempting to relate implementation of management measures to water quality changes is determining the appropriate land management attributes to track. For example, simply counting the number of management measures implemented in a watershed has little chance of being useful in statistical analyses to relate water quality to land treatment since the count only remotely relates (i.e., a mechanism is lacking) to the measured water quality parameter (e.g., cobble embeddedness). Land treatment monitoring that relates directly to the pollutants or effects monitored at the water quality station is most useful. For example, the spacing of water bars relative to slope might be a more useful parameter to track than the number of miles of road constructed. Since the effect of management measures on water quality might not be immediate or implementation might not be sustained, information on other relevant watershed activities (e.g., urbanization, wildfire frequency and extent) is essential for the final analysis.

Management practice effectiveness has not been well documented on a watershed scale, particularly for watersheds with mixed land uses. Studies of management practice effectiveness have been done at the plot and field scales where specific treatments are used and compared to a control situation. Extrapolations from these data and studies using nonpoint source pollution models constitute most of the information available on a watershed scale. Actual data collection and management practice effectiveness determination on a watershed scale is more complex and, because of natural variability, it requires long periods of monitoring before management practice implementation so that a statistical minimum detectable change level can be established. The minimum detectable change is the minimum measurable change in a water quality parameter over time that is statistically significant, and it is a function of statistical tests, the number of samples taken per year, the number of years of monitoring, and the variates and covariates used in the analyses. Dissmeyer (1994) provides detailed information on monitoring forestry BMPs to evaluate their effectiveness in meeting water quality goals. An approach for watershed monitoring of management practice effectiveness, and the problems associated with the approach and with such studies in general, is discussed in Park and others (1994).

Appropriately collected water quality information can be evaluated with trend analysis to determine whether pollutant loads have been reduced or whether water quality has improved. Valid statistical associations drawn between implementation and water quality data can be used to indicate the following:

- Whether management measures have been successful in improving water quality in a watershed or recharge area.

- The need for additional management measures to meet water quality objectives in the watershed or recharge area.

Greater detail regarding methods to evaluate the effectiveness of land treatment efforts is provided in EPA's nonpoint source monitoring guidance (USEPA, 1997) and management measures guidance for section 6217 (USEPA, 1993).

Importance of BMP Monitoring

Researchers with the U.S. Forest Service reviewed state BMP implementation and monitoring programs and the results from those programs in 1994. At the time, twenty-one states were assessing BMP effectiveness. They found that the states had generally concluded that carefully developed and applied BMPs can prevent serious deterioration of water quality, and that most water quality problems were associated with poor BMP implementation. Water quality monitoring was determined to be essential to understanding the relationship between land disturbance and water quality, as it leads to improved understanding of the interaction of soils and topography with BMP implementation. BMP guidelines can be reassessed continually to make them more cost effective, and the more they can be specified, used, monitored, and fine tuned for specific circumstances, the more cost-effectively they can be used to protect water quality.

Quality Assurance and Quality Control

Quality assurance (QA) and quality control (QC) are commonly thought of as procedures used in the laboratory to ensure that all analytical measurements made are accurate. But QA and QC extend beyond the laboratory and are essential components of all phases and all activities within each phase of a nonpoint source monitoring project.

Definitions of Quality Assurance and Quality Control

Quality assurance is an integrated management system designed to ensure that a product or service meets defined standards of quality with a stated level of confidence. Quality assurance activities involve planning quality control, quality assessment, reporting, and quality improvement.

Quality control is the overall system of technical activities designed to measure quality and limit error in a product or service. A quality control program manages quality so that data meet the needs of the user as expressed in a quality assurance project plan.

Quality control procedures include the collection and analysis of blank, duplicate, and spiked samples and standard reference materials to ensure the integrity of analyses, as well as regular inspection of equipment to ensure it is operating properly. Quality assurance activities are more managerial in nature and include assignment of roles and responsibilities to project staff, staff training, development of data quality objectives, data validation, and laboratory audits. Such procedures and activities are planned and executed by diverse organizations through carefully designed quality management programs that reflect the importance of the work and the degree of confidence needed in the quality of the results.

Importance of Quality Assurance and Quality Control Programs

Although the value of a QA/QC program might seem questionable while a project is under way, its value will be quite clear after a project is completed. If the objectives of the project were used to design an appropriate data collection and analysis plan, all QA/QC procedures were followed for all project activities, and accurate and complete records were kept throughout the project, the data and information collected from the project should be adequate to support a choice from among alternative courses of action. In addition, the course of action chosen should be defensible based on the data and information collected. Development and implementation of a QA/QC program can require up to 10 to 20 percent of project resources (Cross-Smiecinski and Stetzenback, 1994), but this cost can be recaptured in lower overall costs due to the project's being well planned and executed. Likely problems are anticipated and accounted for before they arise, eliminating the need to spend countless hours and dollars resampling, reanalyzing data, or mentally reconstructing portions of the project to determine where an error was introduced. QA/QC procedures and activities are cost-effective measures used to determine how to allocate project energies and resources toward improving the quality of research and the usefulness of project results.

EPA Quality Policy

EPA has established a QA/QC program to ensure that data used in research and monitoring projects are of known and documented quality to satisfy project objectives. The use of different methodologies, lack of data comparability, unknown data quality, and poor coordination of sampling and analysis efforts can delay the progress of a project or render the data and information collected from it insufficient for decision making. QA/QC practices are best used as an integral part of the development, design, and implementation of a nonpoint source monitoring project to minimize or eliminate these problems.

Additional information on QA/QC can be found in Chapter 5 of EPA's nonpoint source monitoring guide (USEPA, 1997) and in EPA documents on QA/QC.

Review of State Management Practice Monitoring Programs

Objectives of the Audits

In general, state audits of harvest sites or other types of forestry operations have as their primary objectives to assess compliance with BMP implementation guidelines and/or the effectiveness of BMPs at preventing soil erosion and protecting water quality. Additionally, because the process of collecting BMP implementation and effectiveness information lends itself well to the collection of related information that can be quite useful to a state forestry department, states also collect information that will help them to

- Identify problem areas where additional landowner training and education is needed to improve BMP implementation.
- Determine which BMP implementation standards and specifications need revision.
- Identify necessary improvements in the BMP monitoring program.

Information on landowner training is easily gathered during the audits if the landowner on whose property a harvest was done is present during the audit or contacted as part of the audit. Landowners can be contacted before the audit in most instances to obtain permission to enter their property, and they can be asked to be present either during the audit, when they can perhaps offer valuable information about the harvest, or after an audit during a discussion of the results.

Analysis of BMP implementation standards and specifications can be done effectively during an audit, or during an analysis of audit results after an annual audit has been completed, by comparing the implementation and effectiveness information gathered during the audit with state implementation specifications. For example, specifications may call for a recommended maximum distance between culverts on forest roads of a given slope. During the audits it might be noticed that, even where these specifications have been adhered to, erosion is unacceptable. It may then be recommended to lower the maximum distance, or it might be noticed that excessive erosion is related to a particular soil type, and a shorter distance might be recommended where this soil type occurs.

Audits can provide valuable information about the monitoring program, too. It might be discovered during the course of audits that instances of particular types of effects to soils or water resources are increasing over the years. Or it might be recognized that certain forestry operations (e.g., prescribed burning or site preparation) might not be accounted for in the audits adequately enough to draw conclusions about effects to water resources. Information collected during the audits can be used to adjust the monitoring program to actual information needs.

Audits conducted by some states serve specific objectives beyond assessments of BMP implementation and effectiveness. A good example is South Carolina, which has designed the data collection aspect of its BMP implementation survey to permit the state to determine the effect of a number of variables on compliance with BMP standards. The variables investigated include

- Physiographic region in which the harvest occurred
- Occurrence of a stream on the harvest site
- Percent slope at the harvest site
- Type of terrain at the harvest site
- Category to which the landowner belonged
- Use of cost share assistance for the harvest
- Landowner's familiarity with state BMPs
- Use of a site preparation contract
- Written requirement for the use of BMPs
- Involvement of a forester in the prescription and supervision of site preparation
- Size of the area being site-prepared for reforestation

Criteria Used to Choose the Audit Sites

States use a number of criteria to select sites for inclusion in BMP audits. Generally, the criteria exclude from the audits those sites where BMPs of interest would not likely have been used, where the types of effects of interest (e.g., impacts to water quality) would be difficult to detect or nonexistent, and sites where detecting whether BMPs had been implemented would be difficult due to changes in site characteristics since their implementation. Other criteria ensure that sites from different topographic or vegetative community areas or administrative jurisdictions (e.g., counties or state forest service regions) are included in the audits.

The use of criteria result in a biased sample of audit sites, and thus the conclusions from the audits cannot be used to draw conclusions about all harvest sites in a state. But complete random sampling of harvest sites would limit the usefulness of the results more than biasing the selection of sites by the use of criteria. Not limiting the sites chosen for the audits would result in the inclusion of sites where harvests had occurred many years previously and physical evidence of BMP implementation would be undetectable, sites in areas where BMPs of interest (such as those related to SMAs) would not have been used, and would possibly result in not including portions of the state of interest to the state forestry agency. Therefore, it is important to use criteria to ensure that audit sites provide the information of interest.

The following are some of the criteria used in state audits.

Geographic Distribution

Generally, an entire state is included in an audit by choosing a minimum number of sites per county. A minimum of one site per county is a common criterion, though if timber harvesting is limited to certain areas, a state might include only those counties in which timber was harvested during the time period of interest (see second criterion). The geographical distribution of audit sites might be related to the quantity of timber harvested in a county by ensuring that the latter is proportional to the number of sites chosen for the county. Depending on the purpose of the audit, some other potential site selection criteria are

- Sites within a specific watershed.
- The geographic distribution of audit sites reflects the distribution of timber harvest ownership group.
- All physiographic regions of the state are represented.

Time Since Harvest

The timber harvest or other management activity of interest (e.g., site preparation, road construction) is to have occurred within a specific period of time, typically 1 to 2 years, prior to the audit. There are two good reasons to conduct audits as soon as possible after a harvest. First, the longer the delay between a harvest and an audit, the more difficult it will be to determine the adequacy of BMP implementation. With the passage of time natural vegetation growth can hide evidence of the adequacy of soil conservation measures, storms can obliterate evidence of the adequacy of erosion control methods, and the like. Second, most erosion and sedimentation caused by a harvest activity occurs during and shortly after the harvest, and the longer the time between a harvest and an audit of the harvest, the less likely it is that the audit results will be able to help correct BMP implementation problems and, therefore, minimize water quality impacts. Ideally, BMP implementation and effectiveness audits should occur during harvest-related activity.

Minimum Size

Audit sites are generally no less than 5 to 10 acres, which ensures that BMP use would have been called for. A minimum volume of harvested timber is another way of ensuring the same.

Proximity to Watercourse

Most states insist that harvest sites have a stream (perennial or intermittent), lake, wetland, or pond of a certain size on or near them. The criterion might be that the watercourse is on the audit site, especially if a primary goal of the audit is to assess implementation of SMA rules or guidelines, or within 200 to 500 feet of the audit site if water quality effects of harvest operations are of particular concern. States that are interested in overall BMP implementation might not care that audit sites be associated with surface waters.

Representation of Ownership

Inclusion of all ownership groups (private nonindustrial, industrial, federal, state, and local) can be a criterion for choosing sites, though generally audit sites are not specifically chosen to represent the ownership groups. If all ownership groups are to be included, states might use this criterion only if a minimum number of sites per ownership group is not reached using the other criteria. When this happens, sites from the over-represented ownership group or groups are randomly deselected and sites from the under-represented group are randomly selected from those of the desired ownership group.

Randomness

Although, as stated above, simple randomness is not an overriding concern in the design of BMP audits, many states do ensure that once the criteria are met, sites are then selected randomly, resulting in a stratified random sampling design.

Audit Focus: BMP Implementation and BMP Effectiveness

Surveys are geared toward investigating either BMP implementation or BMP effectiveness or both of these. The nature of the forestry activity at any given site that is investigated determines which BMPs are appropriate for implementation at the site or required to be used, depending on whether BMP use is mandatory or voluntary. Sites are generally rated based on the BMPs that should have been used at the site. If a timber harvest plan was prepared prior to the harvest, or a road construction plan prepared prior to construction of a road and BMPs were included in the plan(s), then the survey might investigate whether the BMPs included in the plan were actually implemented.

Number of Sites Investigated

The number of sites investigated varies widely and depends on survey design, amount of silviculture activity in the state, and availability of resources (staff and money). If the results of the survey are to be analyzed statistically, then the number of sites investigated must be sufficient for this purpose. See EPA's *Techniques for Tracking, Evaluating, and Reporting the Implementation of Nonpoint Source Control Measures—Forestry* (USEPA, 1997a) for guidance on selecting a sufficient number of sites for statistical analysis purposes. A difficulty for many states is ensuring that the number of harvest sites inspected is adequate to draw meaningful conclusions about overall BMP implementation. The number of sites harvested within the audit timeframe (e.g., 2 years if the audit includes sites harvested within the 2 years prior to the audit) is often not known. Many states do not require preharvest notification, or that a landowner inform the state department of forestry that a harvest will occur and where it will occur. Without this information, a state cannot know with certainty what percentage of harvest sites are included in an audit and finding sites to audit can be a difficult, costly, time-consuming task. Even if a state has a policy of voluntary implementation of its forestry BMPs or guidelines, simply requiring that landowners report to the state department of forestry when and where a harvest will occur and the acreage to be harvested, the state's ability to audit BMP implementation in a timely manner, track BMP implementation trends, assist landowners with proper BMP implementation, and maintain accurate statistics about forestry activity in the state can be greatly improved.

Number of BMPs Evaluated

The number of BMPs investigated at each site varies depending on the objectives of the survey and the number and types of BMPs recommended or required by the state. Surveys that target specific types of operations or locations, such as road construction or SMAs, generally involve investigations of fewer BMPs than surveys to assess the use of BMPs for all aspects of forest harvesting, from temporary road construction to site preparation for reforestation.

Composition of the Investigation Teams

An investigation "team" can range from one person to a team of 5 to 7 people with different specialties. Again, the composition of the survey team depends on the objectives of the survey. If BMP implementation is the only thing being investigated, then a state forester alone might be capable of conducting the survey. If, on the other hand, soil characteristics, erosion hazard, improvements in road construction techniques, water quality effects, or other more complex issues are also being investigated, then a team of individuals that represent the appropriate disciplines is generally used.

When one person conducts the surveys, generally the person is a state forester who is familiar with BMP standards for both implementation and effectiveness. When teams are used for the surveys, the state forester is accompanied by one or more specialists that represent fields such as watershed science, soil science, wildlife biology, hydrology, fisheries, and road engineering. Separate organizations might also be represented, such as environmental or conservation organizations and the logging industry. Where possible, the survey team is accompanied by the landowner on whose property the survey is being conducted, the logger who conducted the harvest, and the state forester who prepared the harvest plan, if applicable. Examples of who might be included on an audit "team" are

- A county or state forester
- A watershed specialist
- A forestry industry representative
- A member of the environmental community
- A nonindustrial private landowner
- A member of a local or regional planning and development board
- A wildlife biologist
- A hydrologist
- A soil conservationist or soil scientist
- A fisheries biology

- A road engineer
- A logging professional

BMP Implementation and Effectiveness Rating Systems

The implementation of individual BMPs is rated in one of two ways. A scale of implementation, usually from 0 to 5 or 0 to 3, is used to rate not only whether a BMP was implemented but also the quality of implementation. Alternatively, BMPs are rated simply as having been implemented, not implemented, or not applicable to the particular site.

Generally, all BMPs applicable to a site are rated individually and the site then receives an overall BMP implementation rating. The latter rating might be made using one of the two rating systems mentioned above or using a 3-tiered rating system of excellent, adequate, or inadequate. The overall site rating is usually derived as an average of the individual BMP ratings at the site. Low ratings for overall BMP implementation—for example zero to two on a 0-to-5 scale, zero on a 0-to-3 scale, and inadequate on a 3-tiered rating system—are indications that follow-up with the landowner or harvester is necessary or that further education and training might be helpful.

Even when only BMP implementation is being assessed, BMP effectiveness is often rated on a qualitative basis as an onsite assessment of whether, in the case of a low score or inadequate BMP implementation, there was a resultant risk to water quality. Risks to water quality are generally rated as simply being present or not. If it is apparent that water quality was affected by inadequate BMP implementation, this is also noted.

When more than one team is responsible for the assessments and where teams are composed of many people, assessment training or a mock assessment is performed prior to the actual assessments to establish a degree of consistency in the ratings among members and teams. Assessments of adequacy of BMP implementation and risk to water quality can involve many subjective judgements, and going through a mock assessment prior to the actual assessments gives all team members a chance to discuss what constitutes adequate or proper implementation for the different BMPs. In addition, in many states, after a site assessment and while the assessment team is still on the site the team gathers to discuss the ratings of the individual team members and to arrive at an overall site rating. If any discrepancies or differences of opinion cannot be settled through discussion alone, the individual BMPs are revisited.

Audit Results

Successful implementation of BMPs by landowners and harvesters, as indicated by audits with high compliance rates, depends on many factors, such as whether a state's BMP program is mandatory or voluntary, how long a state has had a BMP program, how long a state has been monitoring BMP implementation, and the effectiveness of a state's education and training outreach program for BMP implementation.

Results of many state audits for BMP implementation and effectiveness indicate that BMPs are being implemented and, where implemented, they are effective in protecting soil from erosion and water quality. Results are generally reported in one of two ways: an overall compliance rate, in which all ratings for compliance with individual BMPs or groups of BMPs are averaged into a single number, and compliance rates for individual

BMPs or groups of BMPs. A group of BMPs might be all those required for SMAs, for instance.

An overall compliance rate can be misleading because it is essentially an average of averages. That is, an overall compliance rate is generally obtained by averaging the compliance ratings for separate groups of BMPs, and then those averages are averaged. Instances where such a rating would be misleading include where most groups of BMPs are rated to have high compliance while one important group of BMPs, say those for SMAs or stream crossings, has a much lower compliance rate. The compliance information for the latter group is lost in the overall compliance rating. Of course, a low overall compliance rating, caused by low compliance ratings for many groups of BMPs, can hide a high compliance rating for another group of BMPs as well. Similarly, a single or a few high or low ratings for individual BMPs within a group of BMPs can be hidden by averaging together the compliance ratings for a whole group of BMPs. Generally, states gain far more information useful to them and to the public for improving and reporting BMP compliance if ratings for individual BMPs are kept separate. Trend analyses for implementation of individual BMPs are also much more meaningful than reports of changes in overall compliance for BMPs from one audit to the next. Of course, it is very important to keep data relevant to the effectiveness of individual BMPs, such as that on the slopes of roads where failure occurs or the amount of cover retained in SMAs where sediment reaches streams, separate for each BMP so that improvements can be made to state BMP specifications.

EPA Recommendations for Forestry Practice Audits

Implement a preharvest notification system to assist in selecting an adequate and unbiased sampling population of harvest sites, to reduce the cost of site selection, and to help determine, prior to a site visit, that selected sites meet many of the selection criteria such as time since harvest and size of harvest.

If feasible, conduct audits soon after harvests are completed so that improvements can be made to BMPs found to be inadequately implemented and the water quality impacts of those BMPs can be minimized.

Ensure that harvest sites are chosen randomly. Stratification based on desired characteristics of sites is perfectly acceptable, but if this is done then sampling within the strata must be random to ensure the validity of results.

If the geographic extent of an audit includes a critical watershed, create a separate statistically valid sample population for the watershed and do not group information from harvests within the watershed with information from other harvests. It is important to maintain separate information for watersheds that have been designated "critical" and to sample them separately if the information obtained is to be related to and useful for programs instituted to protect the watersheds.

Have a clearly defined process for or means of determining whether a BMP implementation is acceptable or not. Audits may be conducted with teams of experts or by individuals working at different harvest sites. The subjectivity of BMP ratings can be reduced and their objectivity increased by clearly defining what standards and quality of implementation constitute each rating level in the rating scale being used. Auditors well trained to recognize these standards and quality criteria will provide the most objective, consistent, meaningful, and comparable ratings.

Ensure that BMP implementation according to state standards reflects protection of water quality by collecting data that is sufficient to determine the effectiveness of BMPs under specific circumstances, such as different soil types, topographies, and rainfall patterns. Modify state standards if the data collected indicate that existing standards are insufficient under certain circumstances.

If forest practice implementation or effectiveness ratings are to be grouped for reporting purposes, maintain separate groupings for functionally different BMPs. For instance, create separate group ratings for road erosion BMPs, stream crossing BMPs, SMA BMPs, etc., so that an average compliance rating will not hide important information about which BMPs are not being implemented adequately.

Volunteer Water Monitoring

The information presented below is available from the USEPA Web site (http://www.epa.gov/owow/monitoring/volunteer/startmon.html) and as a published brochure (United States Environmental Protection Agency; Office of Water (4503F), Washington, DC 20460; EPA 841-B-98-002; July 1998).

Volunteer water monitoring is monitoring done by local citizens rather than agency personnel. In every state, volunteers monitor the condition of streams, rivers, lakes, reservoirs, estuaries, coastal waters, wetlands, and wells. Volunteers who monitor are people who want to help protect a stream, lake, bay or wetland near where they live, work, or play. Their efforts are of particular value in providing quality data and building stewardship of local waters.

Volunteers make visual observations of habitat, land uses, best management practices used to protect soil and water resources; and the impacts of storms; measure the physical and chemical characteristics of waters; and assess the abundance and diversity of living creatures–aquatic insects, plants, fish, birds, and other wildlife. Volunteers also clean up garbage-strewn waters, count and catalog beach debris, and become involved in restoring degraded habitats. The number, variety, and complexity of these projects are continually on the rise.

Volunteer monitoring programs are organized and supported in many different ways. Projects may be entirely independent or may be associated with state, interstate, local, or federal agencies; with environmental organizations; or with schools and universities. Financial support may come from government grants, partnerships with business, endowments, independent fundraising efforts, corporate donations, membership dues, or a combination of these sources.

Many volunteer groups collect data that supplements the information collected by state and local resource management or planning agencies. These agencies might use the data to

- Evaluate the success of best management practices designed to mitigate problems.
- Screen water for potential problems, for further study or for restoration efforts.
- Establish baseline conditions or trends for waters that would otherwise go unmonitored.

In general, a volunteer monitoring program should work cooperatively with state and local agencies in developing and coordinating its technical components. To ensure that its data are used, the monitoring program also develops a strong quality assurance project plan that governs how volunteers are trained, how samples are collected and analyzed, and how information is stored and disseminated.

By educating volunteers and the community about the value of local waters, the kinds of pollution threatening them, and how individual and collective actions can help solve specific problems, volunteer monitoring programs can

- Make the connection between watershed health and our individual and collective behaviors (cumulative impacts).
- Build bridges among various agencies, businesses, and organizations.
- Create a constituency for local waters that promotes personal and community stewardship and cooperation.

Information on volunteer monitoring efforts locally and nationwide can be found through USEPA. The *National Directory of Volunteer Environmental Monitoring Programs*, published by USEPA, provides information on existing groups around the country and the kinds of monitoring taking place. In addition, USEPA's *Adopt Your Watershed* site on the World Wide Web (http://www.epa.gov/adopt/) provides information on active volunteer groups on a watershed basis.

Local or state environmental protection, natural resource, parks, or fish and game agencies might also be good sources of information. Even if the agency does not sponsor a volunteer program, it might be aware of other programs or groups that are active. Other potential sponsors or sources of information include

- Local community-based groups such as civic or watershed associations, garden clubs, universities, and activist organizations
- Chapters of national environmental organizations
- Regional offices of federal agencies such as USEPA, the US Department of Agriculture's Extension Service, the U.S. Park Service, and the U.S. Fish and Wildlife Service

Volunteer Monitoring Resources

USEPA supports volunteer monitoring by sponsoring national conferences, publishing methods manuals, producing a nationwide directory of volunteer programs, and funding a national newsletter, *The Volunteer Monitor*. Volunteer coordinators in the 10 EPA Regional offices provide some technical assistance for local programs and help coordinate regionwide conferences. The Regions are also responsible for grants to the states that can be used, in part, to support volunteer monitoring programs that help assess nonpoint sources of pollution or that serve to educate the public about nonpoint source issues.

Some USEPA resources on the World Wide Web

Volunteer Monitoring Homepage	http://www.epa.gov/owow/monitoring/volunteer/
Monitoring Water Quality Homepage	http://www.epa.gov/owow/monitoring/
Surf Your Watershed	http://www.epa.gov/surf/
Adopt Your Watershed	http://www.epa.gov/adopt/
Index of Watershed Indicators	http://www.epa.gov/iwi/

Documents on volunteer monitoring published by USEPA are listed below. Copies can be obtained by contacting the Volunteer Monitoring Coordinator, USEPA (4503F), 401 M Street SW, Washington, DC 20460.

National Directory of Citizen Volunteer Environmental Monitoring Programs, Fifth Edition. EPA 841-B-98-009, November 1998.

Proceedings of the Fifth National Citizen's Volunteer Water Monitoring Conference. EPA 841-R-97-007, October 1997.

Proceedings of the Fourth National Citizen's Volunteer Water Monitoring Conference. EPA 841/R-94-003, February 1995.

Proceedings of the Third National Citizen's Volunteer Water Monitoring Conference. EPA 841/R-92-004, September 1992.

Volunteer Estuary Monitoring: A Methods Manual. EPA 842-B-93-004, December 1993.

Volunteer Lake Monitoring: A Methods Manual. EPA 440/4-91-002, December 1991.

Volunteer Monitor's Guide to Quality Assurance Project Plans. EPA 841-B-96-003, September 1996.

Volunteer Stream Monitoring: A Methods Manual. EPA 841-B-97-003, November 1997.

Volunteer Water Monitoring: A Guide for State Managers. EPA 440/4-90-010, August 1990.

The Volunteer Monitor, published semiannually, is the national newsletter of volunteer water monitoring. The newsletter facilitates the exchange of ideas, monitoring methods, and practical advice among volunteer monitoring groups across the country. Subscriptions are free. Address all correspondence to Eleanor Ely, Editor, 1318 Masonic Avenue, San Francisco, CA 94117; phone 415/255-8049; fax 415/255-0199.

Best Management Practices Evaluation Program: U.S. Forest Service, Pacific Southwest Region

The USDA Forest Service Pacific Southwest Region has published *Investigating Water Quality in the Pacific Southwest Region: Best Management Practices Evaluation Program (BMPEP) User's Guide* (USDA-FS, Pacific Southwest Region, 2002). The guide continues an effort begun in 1992 to monitor and evaluate BMP implementation and effectiveness (USDA-FS, Pacific Southwest Region, 1992). The Best Management Practices Evaluation Program, or BMPEP, was developed to facilitate evaluation of BMPs through the generation and analysis of data to assess the efficacy of the Region's water quality program, and identify program shortcomings and initiate corrective actions (USDA-FS, Pacific Southwest Region, 2002).

There are three types of BMP evaluations, Administrative, In-Channel, and On-Site. Individuals or teams of reviewers conduct the evaluations using Forest Service forms. *Administrative Evaluations* involve assessing all BMPs for a project, including procedural BMPs (such as the Timber Sale Planning Process). *In-Channel Evaluations* assess the effectiveness of a set of BMPs applied to a project area for protecting beneficial uses of water. All BMPs prescribed for a project for water quality protection are evaluated by establishing study sites to assess effects on beneficial uses over time. *On-Site Evaluations* involve assessing both the implementation and effectiveness of specific practices (individual or groups of similar BMPs). The BMPs are assessed at the site of implementation and evaluated relative to attainment of each BMP's stated objectives.

For in-channel evaluations, sites are selected on the basis of their being representative of management activities common to the forest being evaluated (e.g., timber, mineral extraction, developed recreation, range use) and located in watersheds that are representative of the forests' dominant landforms and geologic types. Streams selected for project evaluation have a suitable control (or comparison stream) nearby or have established desired future condition criteria that can serve as the basis of comparison. A monitoring plan is also developed for each in-channel evaluation. The monitoring plan describes the location, beneficial uses to be protected, evaluation objectives, data collection parameters and methods, timing/frequency and duration of collection, analytical techniques, and the decision criteria to be used to determine whether the beneficial uses were protected. A follow-up investigation is conducted when data from an in-channel evaluation indicates that beneficial use protection objectives were not met and to identify causes of nonpoint source degradation.

On-site evaluations focus on the implementation and effectiveness of individual BMPs applied on project sites. These evaluations are essentially used to answer the implementation question "Did we do what we said we were going to do to protect water quality?" and the effectiveness question "How well did we protect water quality?" There are 29 different evaluation procedures, each designed to assess a specific BMP or set of closely related BMPs. For example, one procedure evaluates SMAs; another evaluates grazing; and another evaluates recreational facilities. Each evaluation procedure has its own form where ratings and comments are recorded, and each form has an electronic counterpart in database software. The evaluations are completed by those persons responsible for the execution of the practices being evaluated. For example, a Range Conservationist or Resource Officer would conduct the on-site evaluation of grazing, a Sale Administrator or Planner would conduct the evaluation of SMAs, and an Engineer would conduct the evaluation of road drainage control.

Sites to be evaluated are either selected randomly or selected. Randomly identified sites allow for drawing statistical conclusions on the implementation and effectiveness of BMPs. Random sites are picked from a pool of projects that meet specified criteria. Selected sites are identified in various ways, such as from a monitoring plan prescribed in an EA, EIS or LMP; as part of a routine site visit; as part of a follow-up evaluation to an in-channel evaluation to discover sources of problems; or selected for a particular reason specific to local needs. Note that for statistical analysis, only randomly identified sites are used to develop statistical inferences. Selected sites are clearly identified and kept separate from the random sites during data storage and analysis.

When problems in implementation are discovered during an audit, the probable cause and recommended corrective actions to prevent recurrence are noted. Reviewer comments are extremely valuable in this regard. Effectiveness evaluations are made using specific indicators of the success of the BMPs observed or measured on-site. When effectiveness problems are noted, observers comment on the extent, duration, and magnitude of effects on beneficial uses. In addition to describing the effects, observers use the following system to rate the effects:

Extent:

- Pollutant has been mobilized off-site, but <u>does not reach the stream channel</u>; effects are evident <u>near the site</u> of the activity.

- Pollutant has been mobilized off-site and <u>reaches the stream channel</u>; effects are evident at the <u>stream reach scale</u> (<20 channel widths downstream).

- Pollutant has been mobilized off-site and <u>reaches the stream channel</u>; effects are evident at the <u>drainage scale</u> (>20 channel widths downstream), effects typically extending downstream and are expressed in larger order channels.

Duration:

- The pollutant or its effects dissipate within a very <u>short</u> (<5 day) period; they are typically associated with a single activity or precipitation event.

- The pollutant or its effects are observable for an <u>intermediate</u> (<1 season) duration; effects are typically expressed intermittently during high flow or precipitation events, dissipating to near background levels by the next wet season.

- The pollutant or its effects are observable for a <u>long</u> (>1 season) duration; effects are typically chronic and persist beyond the next wet season.

Magnitude:

- Effects to beneficial uses <u>insignificant with no measurable</u> water quality impairment; pollutant may be visible, but not likely detectable by compared measurements above and below the site.

- Effects to beneficial uses are <u>minor with measurable</u> water quality impacts the pollutant or its effects may be measurable up to the reach scale, but with no likely effect on biological or economic values.

- Effects to beneficial uses are <u>significant with measurable</u> water quality impacts resulting in degradation to biological or economic values.

The *User's Guide* (USDA-FS, Pacific Southwest Region, 2002) includes detailed instructions for completing each of the 29 on-site evaluation procedures. Included for each procedure is information on developing the sample pool; selecting evaluation sites; timing the evaluation; filling in the form; and the method used to do the observations, measurements, and recording for all the implementation and effectiveness criteria. Also included are hypothetical examples of a completed form for each procedure.

Important Points to Note About the BMPEP

Effectiveness criteria focus on site-specific indicators, which in most cases represent potential effects to water quality rather than actual effects. For example, rill erosion observed on a road would be listed as poor effectiveness, though any sediment from the erosion site that does reach a stream might have anywhere from a negligible to serious effect.

Observations could indicate that a BMP has been implemented but was not effective. Such results are useful as they indicate shortcomings of BMPs, that a BMP might be inappropriate for a particular area, or that the BMP was implemented poorly. Some form of improvement to the BMP is definitely needed in such a case.

BMPs with a high number of comments about the effects on water quality (potential or real) and/or high ratings of "implemented–not effective" are often those implemented close to water courses. Because of the greater potential of practices near water courses to affect water quality, it is prudent to prescribe conservative BMPs in these locations to provide adequate water quality protection.

It is important for foresters in a particular area to review the specific results from that area and not to rely solely a the regional summary that is generated from the individual evaluations. A BMP found to be effective in one area is not guaranteed have the same effectiveness whenever and wherever it is applied. Forest-specific results are more indicative of the changes that can be made to improve BMP effectiveness in a particular locality.

REFERENCES

Adams, P.W., and J.O, Ringer. 1994. The effects of timber harvesting & forest roads on water quantity & quality in the Pacific northwest: Summary & annotated bibliography. Oregon Forest Resources Institute. April.

Adams, T., and D. Hook. 1993. *Implementation and effectiveness monitoring of forestry best management practices on harvested sites in South Carolina. Monitoring Report Number BMP-1.* South Carolina Forestry Commission, Columbia, South Carolina. March.

Alabama Forestry Commission. 1989. *Water Quality Management Guidelines and Best Management Practices for Alabama Wetlands.*

Alden, A.M., C.R. Blinn, P.V. Ellefson, and P.G. Nordin. 1996. *Timber harvester perceptions of benefits and costs of applying water quality best management practices in Minnesota.* Draft. Staff Paper Series Number 108, University of Minnesota, College of Natural Resources and the Agricultural Experiment Station, Department of Forest Resources, St. Paul, Minnesota. March.

APA. 1995. *Statewide model logger training and education program.* American Pulpwood Association Inc., Washington, DC. April.

Appelbloom, T.W., G.M. Chescheir, R.W. Skaggs, and D.L. Hesterberg. 1998. *Evaluating management practices for reducing sediment production from forest roads.* Presented at the 1998 ASAE Annual International Meeting, Orlando, Florida, July 12-16. American Society of Agricultural Engineers, St. Joseph, Michigan.

Arthur, M.A., G.B. Coltharp, D.L. Brown. 1998. Effects of best management practices on forest streamwater quality in eastern Kentucky. *JAWRA* 34(3):481-495.

Ashton, W.S., and Carlson, R.F. *Determination of Seasonal, Frequency and Durational Aspects of Streamflow with Regard to Fish Passage Through Roadway Drainage Structures.* Fairbanks, Alaska:Institute of Water Resources, University of Alaska, Fairbanks, 99701. AK-RD-85-06:1-51, 1984.

Aust, W.M. 1994. Best management practices for forested wetlands in the southern appalacian region. *Water, Air and Soil Pollution* 77:457-468.

Aust, W. Michael, Professor of Forestry, Virginia Polytechnic Institute and State University, Blacksburg, Virginia. 1999. Personnel communication, March 25, 1999.

Aust, W.M., R.M. Shaffer, and J.A. Burger. 1996. Benefits and costs of forestry best management practices in Virginia. *Southern Journal of Applied Forestry* 20(1):23-29.

Baker, M.B. 1990. *Hydrologic and Water Quality Effects of Fire.* USDA Forest Service, Rocky Mountain Forest and Range Experiment Station. General Technical Report RM-191, pp. 31-42.

Baker, C.O., and F.E. Votapka. 1990. *Fish Passage Through Culverts*. San Dimas, CA:USDA, Forest Service Technology and Development Center. FHWA-FL-90-006:1-67.

Ballard, T.M. 2000. Impacts of forest management on northern forest soils. *Forest Ecology and Management* 133:37-42.

Barber, M.E., and R.C. Downs. 1996. *Investigation of Culvert Hydraulics Related to Juvenile Fish Passage*. Pullman,Washington:Washington State Transportation Center (TRAC), Washington State University. WA-RD-388.1:1-54.

Bates, K. 1994. *Fishway Design Guidelines for Pacific Salmon*. Olympia, Washington. Washington Department of Fish and Wildlife. Working Paper.

Beasley, R.S. 1979. Intensive Site Preparation and Sediment Loss on Steep Watersheds in the Gulf Coast Plain. *Soil Science Society of America Journal*, 43(3):412-417.

Beasley, R.S., and A.B. Granillo. 1985. Water Yields and Sediment Losses from Chemical and Mechanical Site Preparation. In *Forestry and Water Quality - A Mid-South Symposium*, Arkansas Cooperative Extension Service, pp.106-116.

Beasley, R.S., and A.B. Granillo. 1988. Sediment and Water Yields from Managed Forests on Flat Coastal Plain Soils. *Water Resources Bulletin*, 24(2):361-366.

Beasley, R.S., E.L. Miller, and S.C. Gough. 1984. Forest Road Erosion in the Ouachita Mountains. In *Mountain Logging Symposium Proceeding*, June 5-7, 1984, ed. P.A. Peters and J. Luckok, pp.203-213. West Virginia University.

Belt, G.H., J. O'Laughlin, and T. Merrill. 1992. *Design of forest riparian buffer strips for the protection of water quality: Analysis of scientific literature*. Report No. 8. Idaho Forest, Wildlife and Range Policy Analysis Group. June.

Berg, N.H., K.B. Roby, and B.J. McGurk. 1996. Cumulative watershed effects: Applicability of available methodologies to the Sierra Nevada. In *Sierra Nevada Ecosystem Project: Final Report to Congress*, Vol. III, Assessments, Commissioned Reports, and Background Information. University of California, Davis, Centers for Water and Wildland Resources.

Berglund, E.R. 1978. *Seeding to Control Erosion Along Forest Roads*. Oregon State University Extension Service, Extension Circular 885.

Best, D.W., H.W. Kelsey, D.K. Hagans, and M. Alpert. 1995. Role of fluvial hillslope erosion and road construction in the sediment budget of Garret Creek, Humbolt County, California. In K.M. Nolan, H.M. Kelsey, and D.C. Marron (eds), *Geomorphic Processes and Aquatic Habitat in the Redwood Creek Basin, Northwestern California*. U.S. Geological Survey Professional Paper 1454.

Bethlahmy, N., and W.J. Kidd, Jr. 1966. *Controlling Soil Movement from Steep Road Fills*. USDA Forest Service Research Note INT-45.

Bilby, R.E. 1984. Removal of Woody Debris May Affect Stream Channel Stability. *Journal of Forestry*, 609-613.

Birch, T.W. 1996a. *Private Forest-land Owners of the United States, 1994*. Resource Bulletin NE-134. U.S. Department of Agriculture, Forest Service, Northeastern Forest Experiment Station, Radnor, Pennsylvania. August.

Birch, T.W. 1996b. *Private Forest-land Owners of the Northern United States, 1994.* Resource Bulletin NE-136. U.S. Department of Agriculture, Forest Service, Northeastern Forest Experiment Station, Radnor, Pennsylvania. November.

Birch, T.W. 1997a. *Private Forest-land Owners of the Western United States, 1994.* Resource Bulletin NE-137. U.S. Department of Agriculture, Forest Service, Northeastern Forest Experiment Station, Radnor, Pennsylvania. March.

Birch, T.W. 1997b. *Private Forest-land Owners of the Southern United States, 1994.* Resource Bulletin NE-138. U.S. Department of Agriculture, Forest Service, Northeastern Forest Experiment Station, Radnor, Pennsylvania. March.

Biswell, H.H., and A.M. Schultz. 1957. Surface Runoff and Erosion as Related to Prescribed Burning. *Journal of Forestry*, 55:372-374.

Blackburn, W.H., M.G. DeHaven, and R.W. Knight. 1982. Forest Site Preparation and Water Quality in Texas. In *Proceedings of the Specialty Conference on Environmentally Sound Water and Soil Management*, ASCE, Orlando, Florida, July 20-23, 1982, ed. E.G. Kruse, C.R. Burdick, and Y.A. Yousef, pp. 57-66.

Blinn, C., R. Dahlman, L. Hislop, and M. Thompson. 1998. *Temporary stream and wetland crossing options for forest management.* NC-202. U.S. Department of Agriculture, Forest Service, North Central Research Station, St. Paul, Minnesota.

Bohn, A.B. 1998. *Designing forest stream crossings using bankfull dimensions and the computer, and the computer program XSPRO.* Kootenai National Forest, Montana.

Bolstad, P.V., and W.T. Swank. 1997. Cumulative impacts of landuse on water quality in a southern Appalachian watershed. *J. Amer. Water Resources Assoc.* 33(3):519-533.

Brazier, J.R., and G.W. Brown. 1973. *Buffer Strips for Stream Temperature Control.* Oregon State University School of Forestry, Forest Research Laboratory, Corvallis, OR, Research Paper 15.

Brett, J.R. 1952. Temperature tolerance in young Pacific salmon, genus Oncorhynchus. *J. Fish. Res. Board of Canada* 9(6):265-323.

Brown, L., and P.B. Moyle. *Eel River Survey: Final Report.* Performed under contract to Calif. Dept. of Fish and Game. Dept. of Wildlife and Fisheries Biology, University of California at Davis.

Brown, G.W. 1972. Logging and Water Quality in the Pacific Northwest. In Watersheds in *Transition Symposium Proceedings*, Urbana, IL, pp. 330-334. American Water Resources Association.

Brown, G.W. 1974. *Fish Habitat.* USDA Forest Service. General Technical Report PNW-24, pp. E1-E15.

Brown, G.W. 1985. Controlling Nonpoint Source Pollution from Silvicultural Operations: What We Know and Don't Know. In *Perspectives on Nonpoint Source Pollution*, pp. 332-333. U.S. Environmental Protection Agency.

Brown, G.W., and J.T. Krygier. 1970. Effects of Clearcutting on Stream Temperature. *Water Resources Research* 6(4):1133-1140.

Brown, G.W., and J.T. Krygier. 1971. Clear-cut Logging and Sediment Production in the Oregon Coast Range. *Water Resources Research* 7(5):1189-1199.

Brown, T.C., and D. Binkley. 1994. *Effect of management on water quality in North American forests*. General Technical Report RM-248. U.S. Department of Agriculture, Forest Service. Rocky Mountain Forest and Range Experiment Station. Fort Collins, Colorado. June.

Cafferata, P.H., and T.E. Spittler. 1998. *Logging impacts of the 1970's vs. the 1990's in the Caspar Creek watershed*. General Technical Report PSW-GTR-168. U.S. Department of Agriculture Forest Service.

Cannon, S. 2002. Post-wildfire landslide hazards. U.S. Geological Survey, National Landslide Hazards Program. <http://landslides.usgs.gov/html_files/landslides/frdebris/cannon/cannon.html>. Accessed 25 February 2003.

Cariboo Forest Region. 1999. Cariboo Region Research Section. British Columbia Ministry of Forests. <http://www.for.gov.bc.ca/rsi/>. Viewed June 23.

Carr, W.W., and T.M. Ballard. 1980. Hydroseeding Forest Roadsides in British Columbia for Erosion Control. *Journal of Soil and Water Conservation* 35(1):33-35.

Carraway, B., L. Clendenen, and D. Work. 1999. *Voluntary Compliance with Forestry Best Management Practices in East Texas*. Texas Forest Service.

CDF. 1991. *California Forest Practice Rules*. California Department of Forestry and Fire Protection.

CDF. 1998. Information obtained from Internet site. California Department of Forestry and Fire Protection.

Chamberlin, T.W., R.D. Harr, and F.H. Everest. 1991. Timber harvesting, silviculture, and watershed processes. In Influences of forest and rangeland management on salmonid fishes and their habitats. *American Fisheries Society Special Publication* 19: 181-205.

Clairain, E.J., and B.A. Kleiss. 1989. Functions and Values of Bottomland Hardwood Forests Along the Cache River, Arkansas: Implications for Management. In *Proceedings of the Symposium: The Forested Wetlands of the Southern United States*, Orlando, Florida, July 12-14, 1988. USDA Forest Service General Technical Report SE-50, pp. 27-33.

Clayton, J.L. 1981. *Soil Disturbance Caused by Clearcutting and Helicopter Yarding in the Idaho Batholith*. USDA Forest Service Research Note INT-305.

Coats, R.N., and T.O. Miller. 1981. Cumulative Silvicultural Impacts on Watersheds: A Hydrologic and Regulatory Dilemma. *Environmental Management* 5(2):147-160.

Code of Federal Regulations (CFR). Part 232: 404 Program Definitions; Exempt Activities Not Requiring 404 Permits; Sec. 232.3–Activities not requiring permits.

Connecticut Resource Conservation and Development Forestry Committee. 1990. *A Practical Guide for Protecting Water Quality While Harvesting Forest Products*.

Conner, W.H., and J.W. Day, Jr. 1989. Response of Coastal Wetland Forests to Human and Natural Changes in the Environment With Emphasis on Hydrology. In *Proceedings of the Symposium: The Forested Wetlands of the Southern United States*, Orlando, Florida, July 12-14, 1988. USDA Forest Service General Technical Report SE-50, pp. 34-43.

Copstead, R. 1997. Water/Road Interacion Series: Summary of Historic and Legal Context for Water/Road Interaction. USDA Forest Service, *Water/Road Interaction Technology Series*, Sam Dimas Technology and Development Center, Sam Dimas, California. December.

Copstead, R., and D. Johansen. 1998. Water/Road Interacion Series: Examples from Three Flood Assessment Sites in Western Oregon. USDA Forest Service, *Water/Road Interaction Technology Series*, Sam Dimas Technology and Development Center, Sam Dimas, California.

Corbett, E.S., and J.A. Lynch. 1985. Management of Streamside Zones on Municipal Watersheds. In *Conference on Riparian Ecosystems and their Management: Reconciling Conflicting Uses*, April 16-18, Tucson, Arizona, pp. 187-190.

Coweeta. 1999. Coweeta Long-Term Ecological Research World Wide Web Site. <http://sparc.ecology.uga.edu/>. Viewed July 7.

CRIS. 1999. Current Research Information System (CRIS), U.S. Department of Agriculture World Wide Web page. <http://cris.csrees.usda.gov/>. Viewed June 12.

Crumrine, J.P. 1977. Best Management Practices for the Production of Forest Products and Water Quality. In *"208" Symposium on Non-Point Sources of Pollution from Forested Land*, ed. G.M. Aubertin, Southern Illinois University, Carbondale, IL, pp. 267-274.

Cubbage, F.W., W.C. Siegel, and P.M. Lickwar. 1989. State Water Quality Laws and Programs to Control Nonpoint Source Pollution from Forest Lands in the South. In *Water: Laws and Management*, ed. F.E. Davis, pp. 8A-29 to 8A-37. American Water Resources Association.

Cullen, J.B. 1998. *Best Management Practices for Erosion Control on Timber Harvesting Operations in New Hampshire, Resource Manual.* New Hampshire Department of Resources and Economic Development, Division of Forests and Lands, Forest Information and Planning Bureau.

Curtis, J.G., D.W. Pelren, D.B. George, V.D. Adams, and J.B. Layzer. 1990. *Effectiveness of Best Management Practices in Preventing Degradation of Streams Caused by Silvicultural Activities in Pickett State Forest, Tennessee.* Tennessee Technological University, Center for the Management, Utilization and Protection of Water Resources.

CWP. 2000. Center for Watershed Protection web site. <http://www.cwp.org/>. Accessed December 20.

Dahl, T.E. 2000. *Status and trends of wetlands in the conterminous United States 1986 to 1997.* U.S. Department of the Interior, Fish and Wildlife Service, Washington, DC.

Delaware Forestry Association. 1982. *Forestry Best Management Practices for Delaware.*

Desbonnet, A., P. Pogue, V. Lee, and N. Wolff. 1994. *Vegetated Buffers in the Coastal Zone. A Summary Review and Bibliography.* Coastal Resources Center Technical Report No. 2064. University of Rhode Island Graduate School of Oceanography, Narragansett, Rhode Island. July.

Dickerson, B.P. 1975. *Stormflows and Erosion after Tree-Length Skidding on Coastal Plains Soils.* Transactions of the ASAE, 18:867-868,872.

Dissmeyer, G.E. 1980. Predicted Erosion Rates for Forest Management Activities and Conditions in the Southeast. In *U.S. Forestry and Water Quality: What Course in the 80s?* Proceedings, Richmond, VA, June 19-20, 1980, pp. 42-49. Water Pollution Control Federation.

Dissmeyer, G.E. 1986. Economic impacts of erosion control in forests. In *Proceedings of the Southern Forestry Symposium*, November 19-21, 1985, Atlanta, GA, ed. S. Carpenter, Oklahoma State University Agricultural Conference Series, pp. 262-287.

Dissmeyer, G.E. 1994. *Evaluating the Effectiveness of Forestry Best Management Practices in Meeting Water Quality Goals or Standards.* Miscellaneous Publication 1520. U.S. Department of Agriculture, Forest Service, Southern Region, Atlanta, Georgia. July.

Dissmeyer, G.E., ed. 2000. *Drinking Water from Forests and Grasslands. A Synthesis of the Scientific Literature.* General Technical Report SRS-39. U.S. Department of Agriculture, Forest Service, Southern Research Station, Asheville, North Carolina. September.

Dissmeyer, G.E., and B. Foster. 1987. Some Economic Benefits of Protecting Water Quality. In *Managing Southern Forests for Wildlife and Fish: A Proceedings.* USDA Forest Service General Technical Report SO-65, pp. 6-11.

Dissmeyer, G.E., and B.B. Foster. 1990. Economics of forest soil resource managment. In *Sustained Productivity of Forest Soils: Seventh North American Forest Soils Conference.* University of British Columbia, Facilty of Forestry Publication, Vancouver, British Columbia, pp. 515-525.

Dissmeyer, G.E., E.R. Frandsen, R. Solomon, K.L. Roth, M.P. Goggin, S.R. Miles, and B.B. Foster. 1987. *Soil and Water Resource Management: A Cost or a Benefit? Approaches to Watershed Economics through Examples.* U.S. Department of Agriculture, Forest Service, Washington, DC. December.

Dissmeyer, G.E. and E. Frandsen. 1988. *The Economics of Silvicultural Best Management Practices.* American Water Resources Association, Bethesda, MD. pp. 77-86.

Dissmeyer, G.E., and R. Miller. 1991. *A Status Report on the Implementation of the Silvicultural Nonpoint Source Program in the Southern States.*

Dissmeyer, G.E., and R.F. Stump. 1978. *Predicted Erosion Rates for Forest Management Activities in the Southeast.* USDA Forest Service.

Doolittle, G.B. 1990. *The Use of Expert Assessment in Developing Management Plans for Environmentally Sensitive Wetlands: Updating A Case Study in Champion International's Western Florida Region.* Best Management Practices for Forested

Wetlands: Concerns, Assessment, Regulation and Research. NCASI Technical Bulletin No. 583, pp. 66-70.

Dorn, F., Forester, USDA Forest Service, Gifford Pinchot Nation Forest, personal communication, January 13, 2000.

Douglass, J.E., and W.T. Swank. 1975. Effects of Management Practices on Water Quality and Quantity: Coweeta Hydrologic Laboratory, North Carolina. In: *Municipal Watershed Management Symposium Proceedings*. USDA Forestry Service. General Technical Report NE-13, pp. 1-13.

Dubensky, M.M. 1991. Public comment information provided by the American Paper Institute and National Forest Products Association.

Dubois, M., W.F. Watson, T.J. Straka, and K.L. Belli. 1991. Costs and cost trends for forestry practices in the South. *Forest Farmer* 50(3):26-32.

Dunford, E.G. 1962. Logging Methods in Relation to Stream Flow and Erosion. In *Fifth World Forestry Congress 1960 Proceedings*, 3:1703-1708.

Dykstra, D.P., and Froehlich, H.A. 1976a. Costs of Stream Protection During Timber Harvest. *Journal of Forestry* 74(10):684-687.

Dykstra, D.P., and H.A. Froehlich. 1976b. Stream protection: What does it cost? In *Loggers Handbook,* Pacific Logging Congress, Portland, OR.

Dyrness, C.T. 1963. Effects of Burning on Soil. In *Symposium on Forest Watershed Management*, Society of American Foresters and Oregon State University, March 25–28, 1963, pp. 291-304.

Dyrness, C.T. 1967. *Mass Soil Movements in the H.J. Andrews Experimental Forest.* USDA Forest Service, Pacific Northwest Forest and Range Experiment Station. Research Paper PNW-42.

Dyrness, C.T. 1970. *Stabilization of Newly Constructed Road Backslopes by Mulch and Grass-Legume Treatments.* USDA Forest Service, Pacific Northwest Forest and Range Experiment Station. PNW-123.

Eaglin, G.S., and W.A. Hubert. 1993. Effects of logging and roads on substrate and trout in streams of the Medicine Bow National Forest, Wyoming. *North American Journal of Fisheries Management* 13: 844-846.

Ellefson, P.V., A.S. Cheng, and R.J Moulton. 1995. *Regulation of private forestry practices by state governments.* Station Bulletin 605-1995. University of Minnesota, Minnesota Agricultural Experiment Station, St. Paul, Minnesota.

Ellefson, P.V., and P.D. Miles. 1984. Economic Implications of Managing Nonpoint Forest Sources of Water Pollutants: A Midwestern Perspective. In *Mountain Logging Symposium Proceedings*, June 5-7, 1984, West Virginia University, ed. P.A. Peters and J. Luchok, pp. 107-119.

Ellefson, P.V., and R.E. Weible. 1980. *Economic Impact of Prescribing Forest Practices to Improve Water Quality: A Minnesota Case Study Minnesota.* Forestry Research Notes.

Erman, D.C., J.D. Newbold, and K.B. Roby. 1977. *Evaluation of Streamside Buffer Strips for Protecting Aquatic Organisms.* California Water Resources Center, University of California, Davis, CA.

Eschner, A.R., and J. Larmoyeux. 1963. Logging and Trout: Four Experimental Forest Practices and their Effect on Water Quality. *Progress in Fish Culture,* 25:59-67.

Essig, D.A. 1991. *Implementation of Silvicultural Nonpoint Source Programs in the United States,* Report of Survey Results. National Association of State Foresters.

Everst, F.H., and W.R. Meehan. 1981. Forest Management and Anadromous Fish Habitat Productivity. In *Transactions of the 46th North American Wildlife and Natural Resources Conference,* pp. 521-530. Wildlife Management Institute, Washington, DC.

Federal Register. 1993. Clean Water Act Regulatory Programs; Final Rule 58 FR 45008 August 25, 1993.

Feller, M.C. 1981. Effects of Clearcutting and Slash Burning on Stream Temperature in Southwestern British Columbia. *Water Resources Bulletin* 17(5):863-866.

Feller, M.C. 1989. Effects of Forest Herbicide Applications on Streamwater Chemistry in Southwestern British Columbia. *Water Resources Bulletin* 25(3):607-616.

Firth, J. 1992. Sediment reduction in road maintenance. In *Ecosystem Road Management,* USDA Forest Service, Technology & Development Program, San Dimas Technology and Development Center, San Dimas, California. April.

Fischenich, J.C., and J.V. Morrow, Jr. 2000. *Streambank Habitat Enhancement with Large Woody Debris.* U.S. Army Corps of Engineers, Ecosystem Management and Restoration Research Program. May.

Flanagan, S., and M.J. Furniss. 1997. Field Indicators of Inlet Controlled Road Stream Crossing Capacity. In *Water/Road Interaction Technology Series.* USDA Forest Service, Sam Dimas Technology and Development Center, Sam Dimas, California.

Florida Department of Agriculture and Consumer Services, Division of Forestry and Florida Forestry Association. 1988. *Management Guidelines for Forested Wetlands in Florida.*

Florida Department of Agriculture and Consumer Services. 1993. *Silviculture Best Management Practices.* Florida Department of Agriculture and Consumer Services, Division of Forestry.

Florida DEP. 1997. *Biological assessment of the effectiveness of forestry best management practices.* Okaloosa, Gadsen, Taylor and Clay counties, sampled winter of 1996 and 1997. Florida Department of Environmental Protection, Bureau of Laboratories, Division of Administrative and Technical Services. December.

Foltz, R.B. 1994. *Reducing tire pressure reduces sediment.* USDA Forest Service, Intermountain Research Station. Moscow, Idaho. <http://forest.moscowfsl.wsu.edu/engr/library/>.

Forman, R.T.T., and L.E. Alexander. 1998. Roads and their major ecological effects. Annual *Reviews of Ecology and Systematics* 29: 207-231.

Fortunate, N., P. Heffernan, K. Snager, and C. Tootell. 1998. *The 1998 forestry BMP audits report*. Montana Department of Natural Resources and Conservation, Forestry Division. Missoula, Montana.

Frayer, W.E., T.J. Monahan, D.C. Bowden, and F.A. Graybill. 1983. *Status and Trends of Wetlands and Deepwater Habitats in the Conterminous United States, 1950's to 1970's*. Colorado State University Department of Forest and Wood Sciences, Fort Collins, CO.

Fredriksen, R.L., and R.N. Ross. 1974. Timber Production and Water Quality — Progress in Planning for the Bull Run, Portland Oregon's Municipal Watershed. In *Proceedings of the Society of American Foresters*, pp. 168-186.

Fredriksen, R.L., D.G. Moore, and L.A. Norris. 1973. The Impact of Timber Harvest, Fertilization, and Herbicide Treatment on Streamwater Quality in *Western Oregon and Washington. In Forest Soils and Forest Land Management*, Proceedings of the Fourth North American Forest Soils Conference, ed. B. Bernier and C.H. Winget, pp. 283-313.

Frissell, C.A. 1992. *Cumulative effects of land use on salmonid habitat on southwest Oregon streams*. Ph.D. thesis, Oregon State University, Corvalis, OR.

Fryer, J.L., K.S. Pilcher, J.E. Sanders, J. Rohovec, J.L. Zinn, W.J. Groberg and R.H. McCoy. 1976. *Temperature, infectious diseases, and the immune response of salmonid fish*. Funded by U.S. EPA Research Services. EPA-600/3-76-21. 70 p.

Froehlich, H.A. 1973. Natural and man-caused slash in headwater streams. *Loggers Handbook*, Pacific Logging Congress, Vol. XXXIII.

Furniss, M.J., M. Love, and A.S. Flanagan. 1997. Water/road interaction: Diversion potential at road-stream crossings. In *Water/Road Interaction Technology Series*. U.S. Department of Agriculture, Forestry Service, Technology and Development Program, Washington, DC.

Furniss, M.J., T.D. Roelofs, and C.S. Yee. 1991. *Road Construction and Maintenance. Influences of Forest and Rangeland Management on Salmonid Fishes and Their Habitats*. American Fisheries Society Special Publication 19, pp. 297-324.

Furniss, M.J., T.S. Ledwith, M.A. Love, B.C. McFadin, and S.A. Flanagan. 1998. Response of road-stream crossings to large flood events in Washington, Oregon, and northern California. In *Water/Road Interaction Technology Series*. U.S. Department of Agriculture, Forestry Service, Technology and Development Program, Washington, DC. September.

Gambles, B., Forester, Mead Paper Company, personal communication, January 13, 2000.

Gardner, R.B. 1967. Major Environmental Factors That Affect the Location, Design, and Construction of Stabilized Forest Roads. *Loggers Handbook*, vol. 27. Pacific Logging Congress, Portland, OR.

Georgia Forestry Association, Wetlands Committee. 1990. *Best Management Practices for Forested Wetlands in Georgia.*

Georgia Forestry Commission. 1999. *Georgia's best management practices for forestry.* Georgia Environmental Protection Division, Georgia Forestry Commission. January.

Gibson, H.E., and C.J. Biller. 1975. A Second Look at Cable Logging in the Appalachians. *Journal of Forestry* 73(10):649-653.

Glasgow, C. 1993. Six Rivers National Forest watershed management and road restoration. In *Riparian management: Common threads and shared interests.* A western regional conference on river management strategies, February 4-6, Albuquerque, New Mexico. USDA Forest Service General Technical Report RM-226. Pp. 195-197.

Golden, M.S., C.L. Tuttle, J.S. Kush, and J.M. Bradley. 1984. *Forestry Activities and Water Quality in Alabama: Effects, Recommended Practices, and an Erosion-Classified System.* Auburn University Agricultural Experiment Station, Bulletin 555.

Gosselink, J.G., G.P. Shaffer, L.C. Lee, D.M. Burdick, D.L. Childers, N.C. Leibowitz, S.C. Hamilton, R. Boumans, D. Cushman, S. Fields, M. Koch, and J.M. Visser. 1990. Landscape conservation in a forested wetland watershed. Can we manage cumulative impacts? *Bioscience* 40(8):588-600.

Grace, J.M., III. 1998. *Sediment export from forest road turn-outs: A study design and preliminary results.* Presented at the 1998 ASAE Annual International Meeting, Orlando, Florida, July 12-15.

Grayson, R.B., S.R. Haydon, M.D.A. Jayasuriya, and B.L. Finlayson. 1993. Water quality in mountain ash forests—separating the impacts of roads from those of logging operations. *J. of Hydrology* 150:459-480.

Greene, J.L., and W.C. Siegel. 1994. *The status and impact of state and local regulation on private timber supply.* General Technical Report RM-255. USDA Forest Service, Rocky Mountain Forest and Range Experiment Station, Fort Collins, Colorado. September.

Groberg, W.J., R.H. McCoy, K.S. Pilcher and J.L. Fryer. 1978. Relation of water temperature to infections of coho salmon, chinook salmon and steelhead trout with Aeromonas salmonicida and A. hydrophila. *J. Fish. Res. Board of Canada* 35(1): 1-7.

Groberg, W.J., J.S. Rohovec, and J.L. Fryer. 1983. The effects of water temperature on infection and antibody formation induced by Vibrio anguillarum in juvenile coho salmon. *J. World Mariculture* 14:240-248.

Gucinski, H., M.J. Furniss, R.R. Ziemer, and M.H. Brookes (eds). 2001. *Forest Roads: A Synthesis of Scientific Information.* General Technical Report PNW-GTR-509. U.S. Department of Agriculture Forest Service, Pacific Northwest Research Station, Corvallis, Oregon. May.

Hall, J.D., G.W. Brown, and R.L. Lantz. 1987. The Alsea Watershed Study - A Retrospective. In *Managing Oregon's Riparian Zone for Timber, Fish and Wildlife*, NCASI Technical Bulletin No. 514, pp. 35-40.

Hamilton, R.A. undated. *Forest*A*Syst.* National Farm*A*Syst/Home*A*Syst Program, Madison, Wisconsin.

Hansit, K., Forester, Minnesota Department of Natural Resources, Division of Forestry, personal communication, January 14, 2000.

Harr, R.D., and R.A. Nichols. 1993. Stabilizing forest roads to help restore fish habitats: A northwest Washington example. *Fisheries* 18(4):18-22.

Hartman, G., J.C. Scrivener, L.B. Holtby, and L. Powell. 1987. Some Effects of Different Streamside Treatments on Physical Conditions and Fish Population Processes in Carnation Creek, A Coastal Rain Forest Stream in British Columbia. In *Streamside Management: Forestry and Fishery Interactions*, ed. E.O. Salo and T.W. Cundy. College of Forest Resources, University of Washington, Seattle, WA, pp. 330-372.

Haussman, R.F., and E.W. Pruett. 1978. *Permanent Logging Roads for Better Woodlot Management.* USDA Forest Service, State and Private Forestry, Eastern Region.

Henly, R. 1992. *Updated cost study of small landowner timber harvesting plans.* FRRAP Staff. June.

Henly, R.K., and P.V. Ellefson. 1987. State-administered Forestry Programs: Current Status and Prospects for Expansion. *Renewable Resources Journal* 5(4):19.

Hetherington, E.D. 1985. Streamflow Nitrogen Loss Following Forest Fertilization in a Southern Vancouver Island Watershed. Canadian *Journal of Forestry Research* 15(1).34-41.

Hockman-Wert, D. undated. *Landslide Research: How Much is Enough?* Logging and Landslides: A Clear-cut Controversy, http://www.wildfirenews.com/forests/forest/research.html. Viewed July 2002.

Holburg, M., Forester, Virginia Department of Forestry, personal communication, January 14, 2000.

Hornbeck, J.W., and K.G. Reinhart. 1964. Water Quality and Soil Erosion as Affected by Logging in Steep Terrain. *Journal of Soil and Water Conservation* 19(1):23-27.

Hornbeck, J.W., C.W. Martin, and C.T. Smith. 1986. Protecting Forest Streams During Whole-Tree Harvesting. *Northern Journal of Applied Forestry* 3:97-100.

Huff, J.L., and E.L. Deal. 1982. *Forestry and Water Quality in North Carolina.* North Carolina Agricultural Extension Service, North Carolina State University.

Hulet, B., Forester, USDA, Forest Service, Forest Development, Mt. Baker National Forest., personal communication, January 31, 2000.

Humboldt State University. 1999. *Shovel logging 2000.* World Wide Web site. <http://www.humboldt.edu>. Viewed July 28.

Hynson, J., P. Adamus, S. Tibbetts, and R. Darnell. 1982. *Handbook for Protection of Fish and Wildlife from Construction of Farm and Forest Roads.* U.S. Fish and Wildlife Service. FWS/OBS 82/18.

Ice, G. 1985. The Status of Silvicultural Nonpoint Source Programs. In *Perspectives on Nonpoint Source Pollution.* U.S. Environmental Protection Agency, pp. 223-226.

Illinois Department of Conservation. 1990. Forestry Development Cost-Share Program. *Illinois Administrative Code,* Title 17, Chapter I, subcapter d, Part 1536.

Indiana DNR. 1998. *Logging and forestry BMP's for water quality in Indiana.* Field Guide. Indiana Department of Natural Resources. January.

Jackson, C.R. 2000. *Integrated headwater stream riparian management study progress report #4*. Prepared for the National Council for Air and Stream Improvement, Inc.

Jenson, L., Forest Engineer, USDA, Forest Service, Tongass National Forest, personal communication, January 18, 2000.

Johnson, D., and D. Ernst. 1997. *Indiana's Forestry Best Management Practices 1996-97 BMP* Implementation Study Report and Findings.

Johnson, R.C., and R.K. Bronsdon. 1995. Erosion of forest roads. *Agricultural Engineer* 50(4):22-27.

Jones, D. 2000. *Implementation Monitoring of Forestry Best Management Practices for Harvesting and Site Preparation in South Carolina 1997-1999*. Best Management Practices Monitoring Report BMP-4. South Carolina Forestry Commission, Columbia, South Carolina. February.

Jones, J.A., and G.E. Grant. 1996. Peak flow responses to clear-cutting and roads in small and large basins, western Cascades, Oregon. *Water Resources Research* 32(4):959-974.

Jones, J.A., F.J. Swanson, B.C. Wemple, and K.U. Snyder. In press. A perspective on road effects on hydrology, geomorphology, and disturbance patches in stream networks. *Conservation Biology*.

Keim, R.F., and S.H. Schoenholtz. 1999. Functions and effectiveness of silvicultural streamside management zones in loessial bluff forests. *Forest Ecology and Management* 118:197-209.

Ketcheson, G.L., and W.F. Megahan. 1996. *Sediment production and downslope sediment transport from forest roads in granitic watersheds*. Research Paper INT-RP-486. U.S. Department of Agriculture, Forest Service. Intermountain Research Station. May.

King, J.G. 1984. *Ongoing Studies in Horse Creek on Water Quality and Water Yield*. NCASI Technical Bulletin 435, pp. 28-35.

King County, Washington. 1999. *Fertilizing Forests with Biosolids*. Department of Natural Resources. Internet site <http://dnr.metrokc.gov/WTD/biosolids/Forest.htm>. Viewed February 16, 2000.

Kochenderfer, J.N. 1970. *Erosion Control on Logging Roads in the Appalachians*. USDA Forest Service, Northeastern Forest Experiment Station, Research Paper NE-158.

Kockenderfer, J.N. 1995. *Using open-top pipe culverts to control surface water on steep road grades*. General Technical Report NE-194. U.S. Department of Agriculture, Forest Service, Northeastern Forest Experiment Station. March.

Kochenderfer, J.N., and J.D. Helvey. 1984. Soil Losses from a "Minimum-Standard" Truck Road Constructed in the Appalachians. In *Mountain Logging Symposium Proceedings*, June 5-7, ed. P.A. Peters and J. Luckok, West Virginia University.

Kochenderfer, J.N., and G.W. Wendel. 1980. *Costs and Environmental Impacts of Harvesting Timber in Appalachia with a Truck-mounted Crane*. USDA Forest Service Research Paper NE-456.

Kochenderfer, J.N., G.W. Wendel, and H.C. Smith. 1984. *Cost of and Soil Loss on "Minimum-Standard" Forest Truck Roads Constructed in the Central Appalachians.* USDA Forest Service Northeastern Forest Experiment Station, Research Paper NE-544.

Kuehn, M.H., and J. Cobourn. 1989. *Summary Report for the 1988 Cumulative Watershed Effects Analyses on the Eldorado National Forest* - Final Draft.

Kundt, J.F., and T. Hall. 1988. *Streamside Forests: The Vital Beneficial Resource.* University of Maryland Cooperative Extension Service and U.S. Fish and Wildlife Service.

Lantz, R.L. 1971. *Guidelines for Stream Protection in Logging Operations.* Oregon State Game Commission.

Larse, R.W. 1971. Prevention and Control of Erosion and Stream Sedimentation from Forest Roads. In *Proceedings of the Symposium of Forest Land Uses and the Stream Environment*, pp. 76-83. Oregon State University.

Leith, T. 2002. Man with a Plan. In *Perspectives On Line*. The Magazine of the College of Agriculture and Natural Resources. North Carolina State University, Spring. <http://www.cals.ncsu.edu/agcomm/magazine/spring02/man.htm>.

Leyba, P., Forest Engineer, USDA, Forest Service, Santa Fe National Forest, personal communication, January 18, 2000.

Lickwar, P.M. 1989. *Estimating the Costs of Water Quality Protection on Private Forestlands in the South.* Master's thesis submitted to the University of Georgia.

Likens, G.E., F.H. Bormann, N.M. Johnson, D.W. Fisher, and R.S. Pierce. 1970. Effects of Forest Cutting and Herbicide Treatment on Nutrient Budgets in the Hubbard Brook Watershed-Ecosystem. *Ecological Monographs* 40(1):23-47.

Logan, B., B. Clinch. 1991. *Montana Forestry BMP's, Forest Stewardship Guidelines for Water Quality.* Montana Department of State Lands, Missoula, Montana.

Lotspeich, F.B., and F.H. Everest. 1981. *A new method for reporting and interpreting textural composition of spawning gravel.* Research note PNW-369. USDA Forest Service Pacific Northwest Forest and Range Experiment Station, Corvalis, OR.

Louisiana Forestry Association. 1988. *Recommended Forestry Best Management Practices for Louisiana.* Louisiana Department of Agriculture and Forestry.

Lynch, J.A., E.S. Corbett, and K. Mussallem. 1985. Best Management Practices for Controlling Nonpoint-Source Pollution on Forested Watersheds. *Journal of Soil and Water Conservation* 41(1):164-167.

Lynch, J.A., and E.S. Corbett. 1990. Evaluation of Best Management Practices for Controlling Nonpoint Pollution from Silvicultural Operations. *Water Resources Bulletin* 26(1):41-52.

MacDonald, L.H. 1997. *Analyzing cumulative effects: Issues and guidelines.* Draft. November 7.

MacDonald, L.H. 2000. Evaluating and managing cumulative effects: process and constraints. *Environ. Management* 26(3):299-315.

MacDonald, L.H., A.W. Smart, and R.C. Wissmar. 1991. *Monitoring Guidelines to Evaluate Effects of Forestry Activities on Streams in the Pacific Northwest and Alaska*. EPA/910/9-91-001. U.S. Environmental Protection Agency, Region 10, Seattle, Washington. May.

Mader, S.F., W.M. Aust, and R. Lea. 1989. Changes in Functional Values of a Forested Wetland Following Timber Harvesting Practices. In *Proceedings of the Symposium: The Forested Wetlands of the Southern United States*, Orlando, Florida, July 12-14, 1988. USDA Forest Service General Technical Report SE-50, pp. 149-154.

Maine Department of Conservation, Forest Service. 1991. *Erosion and Sediment Control Handbook for Maine Timber Harvesting Operations: Best Management Practices*.

Maine Department of Conservation, Forest Service. 1998. Information obtained from Internet site.

Maine Department of Conservation, Forest Service. 1999. *The impact of timber harvesting on nonpoint source pollution*. Report to the 119th Maine legislature, Joint Standing Committee on Agriculture, Conservation and Forestry. Maine Department of Conservation, Maine Forest Service. January 15.

Maine Department of Conservation, Forest Service. 2002. *2000-2001 Maine Forest Service Report on Forestry Best Management Practices Use and Effectiveness in Maine*. Department of Conservation, Maine Forest Service, Augusta, Maine. March.

Malueg, K.W., C.F. Powers, and D.F. Krawczyk. 1972. Effects of Aerial Forest Fertilization with Urea Pellets on Nitrogen Levels in a Mountain Stream. *Northwest Science* 46:52-58.

Maryland Department of the Environment. Undated. *Soil Erosion and Sediment Control Guidelines for Forest Harvest Operations in Maryland*.

Marzac, D., Forester, Georgia Pacific Corporation, personal communication, January 14, 2000.

McClurkin, D.C., P.D. Duffy, and N.S. Nelson. 1987. Changes in Forest Floor and Water Quality Following Thinning and Clearcutting of 20-year-old Pine. *Journal of Environmental Quality* 16(3):237-291.

McGurk, B.J., and D.R. Fong. 1995. Equivalent road area as a measure of cumulative effect of logging. *Environ. Manage.* 19(4):609-621.

McHenry, M.L., D.C. Morrill, and E. Currence. 1994. *Spawning Gravel Quality, Watershed Characteristics and Early Life History Survival of Coho Salmon and Steelhead in Five North Olympic Peninsula Watersheds*. Lower Elwha S'Klallam Tribe, Port Angeles, WA. and Makah Tribe, Neah Bay, WA. Funded by Washington State Dept. of Ecology (205J grant).

McIlroy, A. 2001. *Mercury levels tied to logging practices. Clear-cut forests fingered in study*. Globe and Mail, March 12.

McMinn, J.W. 1984. Soil Disturbance by Fuelwood Harvesting with a Conventional Ground System and a Cable Miniyarder in Mountain Hardwood Stands. In *Mountain Logging Symposium Proceedings*, ed. P.A. Peters and J. Luchok, June 5-7, 1984. West Virginia University, pp. 93-98.

McNulty, S.G., and G. Sun. 1998. The development and use of best practices in forest watersheds using GIS and simulation models. In *Proceedings of the International Symposium on comprehensive watershed management*. International Research and Training Center on Erosion and Sedimentation, Beijing, China, September 7-10. Pp. 391-398.

Megahan, W.F. 1980. Nonpoint Source Pollution from Forestry Activities in the Western United States: Results of Recent Research and Research Needs. In *U.S. Forestry and Water Quality: What Course in the 80s?*, Proceedings of the Water Pollution Control Federation Seminar, Richmond, VA, June 19, 1980, pp. 92-151.

Megahan, W.F. 1981. Effects of Silvicultural Practices on Erosion and Sedimentation in the Interior West—A Case for Sediment Budgeting. In *Interior West Watershed Management Symposium Proceedings*, ed. D.M Baumgartner. Washington State University, Cooperative Extension, pp. 169-182.

Megahan, W.F. 1983. Appendix C: Guidelines for Reducing Negative Impacts of Logging. In *Tropical Watersheds: Hydrologic and Soils Response to Major Uses or Conversions*, ed. L.S. Hamilton and P.N. King. Westview Press, Boulder, CO, pp. 143-154.

Megahan, W.F. 1986. *Recent Studies on Erosion and Its Control on Forest Lands in the United States.*

Megahan, W.F. 1987. Effects of Forest Roads on Watershed Function in Mountainous Areas. In *Environmental Geotechnics and Problematic Soils and Rocks*, ed. Balasubramaniam et al. pp. 335-348.

Mersereau, R.C., and C.T. Dyrness. 1972. Accelerated Mass Wasting after Logging and Slash Burning in Western Oregon. *Journal of Soil and Water Conservation* 27:112-114.

Miller, J.H., and D.L. Sirois. 1986. Soil Disturbance by Skyline Yarding vs. Skidding in a Loamy Hill Forest. *Soil Science Society of America Journal* 50(6):1579-1583.

Mills, K., and J. Hinkle (eds.). 2001. *Forestry, Landslides, and Public Safety*. Landslides and Public Safety Project Team. Prepared for the Oregon Board of Forestry. June.

Minnesota Department of Natural Resources, Division of Forestry. 1989. *Water Quality in Forest Management, "Best Management Practices in Minnesota."*

Minnesota Department of Natural Resources, Division of Forestry. 1991. *Minnesota Forest Stewardship Program.*

Minnesota Department of Natural Resources. 1995. *Protecting Water Quality and Wetlands in Forest Management, Best Management Practices in Minnesota*. Minnesota Department of Natural Resources, Division of Forestry, Minneapolis, Minnesota.

Minnesota Forest Resources Council. 1999. *Sustaining Minnesota Forest Resources: Voluntary Site-Level Forest Management Guidelines for Landowners, Loggers and Resource Managers*. Minnesota Forest Resources Council, St. Paul, Minnesota.

Missouri Department of Conservation. 2000. Information obtained from Internet site, http://www.conservation.state.mo.us/forest/private/mgmtasst.htm.

Mississippi Forestry Commission. 1989. Mississippi's Best Management Practices Handbook.

Moll, J., R. Copstead, and D. Johansen. 1997. Traveled Way Surface Shape. In *Water/ Road Interaction Technology Series*, USDA Forest Service, Sam Dimas Technology and Development Center, Sam Dimas, California.

Montana State University. 1991. *Montana forestry BMPs*. Montana State University Extension Service. July.

Montgomery, D.R. 1994. Road surface drainage, channel initiation, and slope instability. *Water Resources Research* 30: 1925-1932.

Montgomery, D.R., J.M. Buffington, N.P. Peterson, D. Schuett-Hames, and T.P. Quinn. 1996. Stream-bed scour, egg burial depths, and the influence of salmonid spawning on bed surface mobility and embryo survival. *Canadian Journal of Fisheries and Aquatic Sciences* 53: 1061-1070.

Moore, D.G. 1975. Impact of Forest Fertilization on Water Quality in the Douglass Fir Region—A Summary of Monitoring Studies. In *Proceeding Forestry Issues in Urban America*, New York, NY, September 22-26, 1974. Society of American Foresters.

Morman, D. 1993. *Riparian rule effectiveness study*. Oregon Department of Forestry, Forest Practices Program. March.

Murphy, M.L., and A.M. Miller. 1997. Alaska timber harvest and fish habitat. In *Freshwaters of Alaska*. Ecological Synthesis. Pp. 229-263.

Murphy, M.L., K.V. Koski, J. Heifetz, S.W. Johnson, D. Kirchhofer, and J.F. Thedinga. 1984. Role of Large Organic Debris as Winter Habitat for Juvenile Salmonids in Alaska Streams. In *Western Proceedings of the 64th Annual Conference of the Western Association of Fish and Wildlife Agencies*, Victoria, British Columbia, July 16-19, 1984, pp. 251-262.

Narver, D.W. 1971. Effects of Logging Debris on Fish Production. In *Forest Land Uses and Stream Environment*, ed. J.T. Krygier and J.D. Hall, School of Forestry and Department of Fisheries and Wildlife, Oregon State University, October 19-21, pp. 100-111.

Nawa, R.K., and C.A. Frissell. 1993. Measuring scour and fill of gravel stream beds with scour chains and sliding bead monitors. *No. American J. of Fisheries Management* 13: 634-639.

National Council for Air and Stream Improvement, Inc. (NCASI). 2000. *Handbook of control and mitigation measures for silvicultural operations*. Unpublished draft Technical Bulletin. Research Triangle Park, N.C. National Council for Air and Stream Improvement, Inc.

Neary, D.G. 1985. Fate of Pesticides in Florida's Forests: An Overview of Potential Impacts in Water Quality. In *Proceedings Soil and Crop Science Society of Florida*, pp. 18-24.

Neary, D.G., P.B. Bush, J.E. Douglass, and R.L. Todd. 1985. Picloram Movement in an Appalachian Hardwood Forest Watershed. *Journal of Environmental Quality* 14(4):585-591.

Neary, D.G., W.T. Swank, and H. Riekerk. 1989. An Overview of Nonpoint Source Pollution in the Southern United States. In *Proceedings of the Symposium: Forested Wetlands of the Southern United States*, July 12-14, 1988, Orlando, FL. USDA Forest Service. General Technical Report SE-50, pp. 1-7.

Norris, L.A. 1968. Stream contamination by herbicides after fall rains on forest land. *Western Society of Weed Scientist Research Program Report*, 33-34.

Norris, L.A., and D.G. Moore. 1971. The Entry and Fate of Forest Chemicals in Streams. In *Forest Land Uses and Stream Environment - Symposium Proceedings*, ed. J.T. Krygier and J.D. Hall, Oregon State University, Corvallis, OR, pp. 138-158.

Norris, L.A., H.W. Lorz, and S.V. Gregory. 1991. *Forest Chemicals. Influences of Forest and Rangeland Management on Salmonid Fishes and Their Habitats*. American Fisheries Society Special Publication 19, pp. 207-296.

North Carolina Division of Forest Resources. 1989. *Forestry Best Management Practices Manual*. Department of Environment, Health and Natural Resources.

North Dakota Forestry Service. 1999. *Forestry Best Management Practices*. Molberg Forestry Center, Bottineau, North Dakota.

North Dakota State University. 1998. Information obtained from Internet site.

Nutter, W.L., and J.W. Gaskin. 1989. Role of Streamside Management Zones in Controlling Discharges to Wetlands. In *Proceedings of the Symposium: The Forested Wetlands of the Southern United States*, July, 12-14, 1988, Orlando, Florida. USDA Forest Service. General Technical Report SE-50, pp. 81-84.

Ohio Department of Natural Resources. Undated. *BMPs for Erosion Control on Logging Jobs*. Silvicultural Nonpoint Source Pollution Technical Advisory Committee.

Ohio Forestry Association. 1999. *Certified Logging Company Program*. <http://www.ohioforest.org/> (select Services>Ohio Master Logging Company Program).

Olsen, E.D. 1987. A Case Study of the Economic Impact of Proposed Forest Practices Rules Regarding Stream Buffer Strips on Private Lands in the Oregon Coast Range. In *Managing Oregon's Riparian Zone for Timber, Fish and Wildlife*, NCASI Technical Bulletin No. 514, pp. 52-57.

Ontario Ministry of Natural Resources. 1988. *Environmental Guidelines for Access Roads and Water Crossings*. Queen's Printer for Ontario, Ontario, Canada.

Oregon Department of Forestry. 1979a. *Waterbars*. Forest Practices Notes No. 1. Oregon Department of Forestry, Forest Practices Section, Salem, OR.

Oregon Department of Forestry. 1979b. *Reforestation*. Forest Practices Notes No. 2. Oregon Department of Forestry, Forest Practices Section, Salem, OR.

Oregon Department of Forestry. 1981. *Road Maintenance*. Forest Practices Notes No. 4. Oregon Department of Forestry, Forest Practices Section, Salem, OR.

Oregon Department of Forestry. 1982. *Ditch Relief Culverts*. Forest Practices Notes No. 5. Oregon Department of Forestry, Forest Practices Section, Salem, OR.

Oregon Department of Forestry. 1991. *Forest Practices Rules, Eastern Oregon Region.* Oregon Department of Forestry, Forestry Practices Section, Salem, OR.

Oregon Department of Forestry. 1997. Information obtained from Internet site.

Oregon Department of Forestry. 2002. *Executive Summary.* Technical Report 15. Oregon Department of Forestry Best Management Practices Compliance Monitoring Project, Forest Practices Monitoring Program. April.

OSHA, Department of Labor. 1999. World Wide Web site. <http://www.osha.gov/>. Viewed July 17.

Pacific Southwest Region. 1998. *Best Management Practices Evaluation Program annual report.* U.S. Department of Agriculture, Forest Service, Pacific Southwest Region, San Francisco, California. April.

Page, C.P., and A.W. Lindenmuth, Jr. 1971. Effects of Prescribed Fire on Vegetation and Sediment in Oak-Mountain Mahogany Chaparral. *Journal of Forestry* 69:800-805.

Pardo, R. 1980. What is Forestry's Contribution to Nonpoint Source Pollution? In *U.S. Forestry and Water Quality: What Course in the 80s?* Proceedings of the Water Pollution Control Federation Seminar, Richmond, VA, June 19, 1980, pp. 31-41.

Park, S.W., S. Mostaghimi, R.A. Cooke, and P.W. McClellan. 1994. BMP impacts on watershed runoff, sediment, and nutrient yields. *Water Resources Bulletin* 30(6):1011-1023.

Patric, J.H. 1976. Soil Erosion in the Eastern Forest. *Journal of Forestry* 74(10):671-677.

Patric, J.H. 1980. Effects of Wood Products Harvest on Forest Soil and Water Relations. *Journal of Environmental Quality* 9(1):73-80.

Patric, J.H. 1984. Some Environmental Effects of Cable Logging in the Eastern Hardwoods. In *Mountain Logging Symposium Proceedings*, ed. P.A. Peters and J. Luchok, June 5-7, 1984, West Virginia University, pp. 99-106.

Pennsylvania Bureau of Soil and Water Conservation. 1990. *Erosion and Sediment Pollution Control Program Manual.* Pennsylvania Department of Environmental Resources.

Pettit, C. 2000. West Kootenay landslide reports. <http://www.watertalk.org/reports/landslides-koot/index.html>. Accessed 24 February 2003.

Phillips, M.J. 1997. Forestry best management practices for wetlands in Minnesota. In *Northern Forested Wetlands: Ecology and Management.* Chapter 28.

Phillips, M.J., L.W. Swift, Jr., and C.R. Blinn. Best management practices for riparian areas. In E.S. Verry, J.W. Hornbeck, C.A. Dolloff (eds.), *Riparian management in forests of the continental Eastern United States.* Lewis Publishers, Boca Raton, Florida. Pp. 273-286.

Piatek, K.B., and H.L. Allen. 2000. Site preparation effects on foliar N and P use, retranslocation, and transfer to litter in 15-year old *Pinus taeda. Forest Ecology and Management* 129:143-152.

Pope, P.E. 1978. *Forestry and Water Quality: Pollution Control Practices.* Forestry and Natural Resources, FNR 88, Purdue University Cooperative Extension Services.

Powell, D.S., J.L. Faulkner, D.R. Darr, Z. Zhu, D.W. MacCleery. 1994. *Forest resources of the United States, 1992.* General Technical Report RM-234 (Revised). USDA Forest Service, Rocky Mountain Forest and Range Experiment Station, Fort Collins, Colorado.

Powers, P.D. *Culvert Hydraulics Related to Upstream Juvenile Salmon Passage.* Washington State Department of Fish and Wildlife, Habitat Program: 1-15, 1996.

Reid, L.M. 1993. Research and cumulative watershed effects. USDA Forest Service, Pacific Southwest Research Station, General Technical Report GTR-141, 118 pp.

Reid, L.M. 1998. Cumulative watershed effects: Caspar Creek and beyond. In: Ziemer, R.R., Technical Coordinator, *Proceedings of the conference on coastal watersheds: the Caspar Creek story*, Ukiah, California, May 6. Gen. Tech. Rep. PSW GTR-168. U.S. Department of Agriculture, U.S. Forest Service, Pacific Southwest Research Station, Albany, California. Pp. 117-127.

Rice, R.M., J.S. Rothacher, and W.F. Megahan. 1972. Erosional Consequences of Timber Harvesting: An Appraisal. In *Watersheds in Transition Symposium Proceedings*, AWRA, Urbana, IL, pp. 321-329.

Richter, D.D., C.W. Ralston, and W.R. Harms. 1982. Prescribed Fire: Effects on Water Quality and Forest Nutrient Cycling (Hydraulic Systems, Pine Litter, USA). *Science* 215:661-663.

Riekerk, H. 1983. Environmental Impacts of Intensive Silviculture in Florida. In *I.U.F.R.O. Symposium on Forest Site and Continuous Productivity*. USDA Forest Service, Pacific Northwest Forest and Range Experiment Station. General Technical Report PNW-163, pp. 264-271.

Riekerk, H. 1983. Impacts of Silviculture on Flatwoods Runoff, Water Quality, and Nutrient Budgets. *Water Resources Bulletin* 19(1):73-80.

Riekerk, H. 1985. Water Quality Effects of Pine Flatwoods Silviculture. *Journal of Soil and Water Conservation* 40(3):306-309.

Riekerk, H. 1989. Forest Fertilizer and Runoff-Water Quality. In *Soil and Crop Science Society of Florida Proceedings*, September 20-22, 1988, Marco Island, FL, Vol. 48, pp. 99-102.

Riekerk, H., D.G. Neary, and W.J. Swank. 1989. The Magnitude of Upland Silviculture Nonpoint Source Pollution in the South. In *Proceedings of the Symposium: Forested Wetlands of the Southern United States*, July 12-14, Orlando, FL, pp. 8-18.

Robben, J., and L. Dent. 2002. *Final Report.* Technical Report 15. Oregon Department of Forestry Best Management Practices Compliance Monitoring Project, Forest Practices Monitoring Program. April.

Rothwell, R.L. 1978. *Watershed Management Guidelines for Logging and Road Construction in Alberta.* Canadian Forestry Service, Northern Forest Research Centre, Alberta, Canada. Information Report NOR-X-208.

Rothwell, R.L. 1983. Erosion and Sediment Control at Road-Stream Crossings (Forestry). *The Forestry Chronicle* 59(2):62-66.

Rygh, J. 1990. Fisher Creek Watershed Improvement Project Final Report. Payette National Forest.

Salazar, D.J. and F.W. Cubbage. 1990. Regulating Private Forestry in the West and South. *Journal of Forestry* 88(1):14-19.

Sanders, J.E., K.S. Pilcher, and J.L. Fryer. 1978. Relation of water tmperature to bacterial kidney disease in coho salmon, sockeye salmon and steelhead trout. *J. Fish. Res. Board of Canada* 35(1): 8-11.

Schmid, D., Forester, New York State Department of Environmental Conservation, Division of Lands and Forests, personal communication, February 8, 2000.

Sedell, J., M. Sharpe, D.D. Apple, M. Copenhagen, and M. Furniss. 2000. *Water & the Forest Service.* FS-660. U.S. Department of Agriculture, Forest Service, Washington DC. January.

Sidle, R.C. 1980. *Impacts of Forest Practices on Surface Erosion.* Pacific Northwest Extension Publication PNW-195, Oregon State Univ. Extension Service.

Sidle, R.C. 1989. Cumulative Effects of Forest Practices on Erosion and Sedimentation. In *Forestry on the Frontier Proceedings of the 1989 Society of American Foresters,* September 24-27, Spokane, WA, pp. 108-112.

Smith, W.B., J.S. Visage, D.R. Darr, and R.M. Sheffield. 2001. *Forest Resources of the United States, 1997.* U.S. Department of Agriculture, Forest Service, North Central Research Station, St. Paul, Minnesota.

South Carolina Forestry Commission. 1998. Information obtained from Internet site.

Spence, B.C, G.A. Lomnicky, R.M. Hughes and R.P. Novitski. 1996. *An ecosystem approach to salmonid conservation.* TR-4501-96-6057. ManTech Corp, Corvalis, OR. Funded by National Marine Fisheries Service, U.S. Fish and Wildlife Service and the U.S. Environmental Protection Agency. Available from NMFS, Portland, OR.

Stednick, J.D., L.N. Tripp, and R.J. McDonald. 1982. Slash Burning Effects on Soil and Water Chemistry in Southeastern Alaska. *Journal of Soil and Water Conservation* 37(2):126-128.

Stone, E. 1973. *The Impact of Timber Harvest on Soils and Water.* Report of the President's Advisory Panel on Timber and the Environment, Arlington, VA, pp. 427-467.

Stuart, G., 1996. *Instructor's Guide Eastern Forest Wetland & Streamside Forest Management.* National Wetlands Conservation Alliance. <http://users.erols.com/wetlandg/forest.htm>.

Sun, G., and S.G. McNulty. 1998. Modeling soil erosion and transport on forest landscape. In *Proceedings of Conference 29.* International Erosion Control Association, Reno, Nevada, February 16-20. Pp. 187-198.

Swank, W.T., L.W. Swift, Jr., and J.E. Douglass. 1988. Streamflow Changes Associated with Forest Cutting, Species Conversions and Natural Disturbances. In *Forest Hydrology and Ecology at Coweeta*, Chapter 22, ed. W.T. Swank and D.A. Crossley, Jr., pp.297-312. Springer-Verlag, New York, NY.

Swartley, W.A. 2002. *Forestry BMP Implementation Survey*. North Carolina Division of Forest Resources. February.

Swift, L.W., Jr. 1984a. Gravel and Grass Surfacing Reduces Soil Loss from Mountain Roads. *Forest Science* 30(3):657-670.

Swift, L.W., Jr. 1984b. Soil Losses from Roadbeds and Cut and Fill Slopes in the Southern Appalachian Mountains. *Southern Journal of Applied Forestry* 8(4):209-215.

Swift, L.W., Jr. 1985. Forest Road Design to Minimize Erosion in the Southern Appalachians. In *Forestry and Water Quality: A Mid-South Symposium*, May 8-9, 1985, Little Rock, AR, ed. B.G. Blackmon, pp. 141-151. University of Arkansas Cooperative Extension.

Swift, L.W., Jr. 1986. Filter Strip Widths for Forest Roads in the Southern Appalachians. *Southern Journal of Applied Forestry* 10(1):27-34.

Swift, L.W., Jr. 1988. Forest Access Roads: Design, Maintenance, and Soil Loss. In *Forest Hydrology and Ecology at Coweeta*, Chapter 23, ed. W.T. Swank and D.A. Crossley, Jr., pp. 313-324. Springer-Verlag, New York, NY.

Swift, L.W., Jr., and R.G. Burns. 1999. The Three Rs of Roads: Redesign, Reconstruction, and Restoration. *Journal of Forestry* 97(8):40-44.

Taylor, L., Forest Engineer, USDA Forest Service, Salmon-Challis National Forest, personal communication, January 28, 2000.

Tennessee DOA. (undated) *1996 BMP Implementation Survey Report*. Tennessee Department of Agriculture, Forestry Division.

Tennessee DOA. 1998. Information obtained from Internet site.

Tennessee Department of Conservation, Division of Forestry. 1990. *Best Management Practices for Protection of the Forested Wetlands of Tennessee.*

Tennessee Department of Agriculture, Division of Forestry. 1996. *1996 BMP Implementation Survey Report*. Tennessee Department of Agriculture, Division of Forestry.

Texas Forestry Association. 1989. *Texas Best Management Practices for Silviculture.*

Toliver, J.R., and B.D. Jackson. 1989. Recommended Silvicultural Practices in Southern Wetland Forests. In *Proceedings of the Symposium: The Forested Wetlands of the Southern United States*, Orlando, Florida, July 12-14, 1988. USDA Forest Service General Technical Report SE-50, pp. 72-77.

Torgersen, C.E., D.M. Price, H.W. Li, and B.A. McIntosh. 1999. Multiscale thermal refugia and stream habitat associations of chinook salmon in northeastern Oregon. *Ecological Applications* 9: 301-319.

Trimble, G.R., and S. Weitzman. 1953. Soil Erosion on Logging Roads. *Soil Science Society of America Proceedings* 17:152-154.

Trombulak, S.C., and C.A. Frissell. In press. Review of ecological effects of roads on terrestrial and aquatic communities. *Conservation Biology.*

University of Minnesota Extension Service. 1998. *Best Management Practices (BMPs) can prevent or minimize the impact of forestry activities on rivers, lakes, streams, groundwater, wetlands, and visual quality.* Forest Management Practices Fact Sheet Crossing Options Series #4. <http://www.extension.umn.edu/distribution/naturalresources/DD6973.html>.

USDA-FS. 1994. *Sedimentation concerns from Forest Roads 133, 149, and 468 into two tributaries of E. Br. Tionesta Cr.* Unpublished research report. USDA Forest Service, Allegheny National Forest, Warren, Pennsylvania. March 17.

USDA-FS. 1995. *Effectiveness monitoring of the filter strip between FR 375, Hastings Run, and the tributary to Hastings Run.* Unpublished research report. USDA Forest Service, Allegheny National Forest, Warren, Pennsylvania. January 17.

*USDA-FS. 1997. *Ecosystem Road Management.* U.S. Department of Agriculture, Forest Service. San Dimas Technology and Development Center, San Dimas, California. November.

USDA-FS. 1998. *Best Management Practice Evaluation Program (BMPEP).* U.S. Department of Agriculture, Forest Service, Pacific Southwest Region.

*USDA-FS. 1998. *Water/Road Interaction Technology Series.* U.S. Department of Agriculture, Forest Service, San Dimas Technology and Development Center, San Dimas, California.

USDA-FS. 1999. USDA Forest Service Internet Site: <http://www.fs.fed.us>.

USDA-FS, Pacific Region. 1992. *Investigating water quality in the Pacific Southwest Region: Best Management Practices Evaluation Program.* U.S. Department of Agriculture, Forest Service, Pacific Southwest Region, San Francisco, California.

USDOI, USDA. 2004. *Progress reported on implementing President Bush's Healthy Forests Initiative.* U.S. Department of the Interior, U.S. Department of Agriculture, Washington, DC. Fact sheet. March.

USEPA. 1984. *Report to Congress: Nonpoint Source Pollution in the U.S.,* U.S. Environmental Protection Agency, Office of Water Program Operations, Washington, DC.

USEPA. 1991. *Pesticides and Groundwater Strategy.* U.S. Environmental Protection Agency, Office of Prevention, Pesticides, and Toxic Substances, Washington, DC.

USEPA. 1992. *A synoptic approach to cumulative impact assessment. A proposed methodology.* EPA/600/R-92/167. U.S. Environmental Protection Agency, Office of Research and Development, Washington, DC. October.

USEPA. 1992. *Economic analysis of coastal nonpoint source pollution controls: Forestry.* Draft. US Environmental Protection Agency, Office of Water, Washington, DC. June.

USEPA. 1992a. *Managing Nonpoint Source Pollution, Final Report to Congress on Section 319 of the Clean Water Act (1989)*. U.S. Environmental Protection Agency, Office of Water, Washington, DC. EPA-506/9-90.

USEPA. 1993. *Guidance Specifying Management Measures for Sources of Nonpoint Pollution in Coastal Waters*. EPA840-B-92-002. U.S. Environmental Protection Agency, Office of Water, Washington, DC. January.

USEPA. 1994. *Land Application of Sewage Sludge. A Guide for Land Appliers on the Requirements of the Federal Standards for the Use or Disposal of Sewage Sludge, 40 CFR Part 503*. EPA/831-B-93-002b. U.S. Environmental Protection Agency, Office of Enforcement and Compliance Assurance, Washington, DC. December.

USEPA. 1995. *Watershed protection: A project focus*. EPA841-R-95-003. U.S. Environmental Protection Agency, Office of Water, Washington, DC. August.

USEPA. 1996. *Watershed approach framework*. EPA840-S-96-001. U.S. Environmental Protection Agency, Office of Water, Washington, DC. June.

USEPA. 1997. *Memorandum to the Field*—Corps and EPA Regulatory Program Chiefs, Office of Wetlands, Oceans, and Watersheds. <http://www.epa.gov/owow/wetlands/guidance/silv2.html>.

USEPA. 1997. *Monitoring Guidance for Determining the Effectiveness of Nonpoint Source Controls*. EPA841-B-96-004. U.S. Environmental Protection Agency, Office of Water, Washington, DC. September.

*USEPA. 1997. *Techniques for Tracking, Evaluating, and Reporting the Implementation of Nonpoint Source Control Measures—Forestry*. EPA841-B-97-001. U.S. Environmental Protection Agency, Office of Water, Washington, DC.

USEPA. 2000. *National Water Quality Inventory: 1998 Report to Congress*. EPA841-F-00-006. U.S. Environmental Protection Agency, Office of Water, Washington, DC. June.

USEPA. undated a. *Ohio CWSRF provides loans for riparian zone conservation*. Clean Water State Revolving Fund Activity Update. <http://www.epa.gov/owm/cwfinance/cwsrf/ohiobru.pdf>. Accessed December 24, 2002.

USEPA. undated b. *Ohio CWSRF provides loans for development best management practices*. Clean Water State Revolving Fund Activity Update. <http://www.epa.gov/owm/cwfinance/cwsrf/darbycr.pdf>. Accessed December 24, 2002.

USEPA. 2002a. Facimile transmittal from Stephanie von Feck, USEPA. Fax dated December 24, 2002.

USEPA. 2002b. *Ohio's restoration sponsor program integrates point source & nonpoint source projects*. Clean Water State Revolving Fund Activity Update. June. <http://www.epa.gov/owm/cwfinance/cwsrf/ohio_wrrsp.pdf>. Accessed December 24, 2002.

VANR. 1998. *Water quality monitoring and aquatic bioassessment related to logging practices in the Dowsville Brook, Shepard Brook and Mill Brook watersheds*. Vermont Agency of Natural Resources, Department of Environmental Conservation, Biomonitoring and Aquatic Studies Section. October.

Vermont Department of Forests, Parks, and Recreation. 1987. *Acceptable Management Practices for Maintaining Water Quality on Logging Jobs in Vermont.*

Virginia Department of Forestry. (undated). *Forestry Best Management Practices for Water Quality in Virginia.*

Virginia DOF. 1994. *Best management practice implementation and effectiveness.* 1994. Virginia Department of Forestry, Forest Resources and Utilization Branch, Charlottesville, Virginia. February.

Virginia DOF. 1998. *Conclusions suggested by water quality monitoring near private timber harvests: 1989-1996.* Executive Summary. Virginia Department of Forestry, Charlottesville, Virginia.

Virginia DOF. 2001. *BMP Effort, Implementation, and Effectiveness Field Audit.* Virginia Department of Forestry, Charlottesville, Virginia. November.

Vowell, J., and T. Gilpin. 1998. *Silviculture Best Management Practices 1997 Compliance Survey Report.* Florida Department of Agriculture, Division of Forestry. July.

Vowell, J., and R. Lima. 2002. *Results of Florida's 2001 Silviculture BMP Compliance Survey.* Florida Division of Forestry, Tallahassee, Florida. March.

Warring, M., Virginia Department of Forestry. 1999. Personnel communication. July 19, 1999.

Washington State Department of Ecology. 1996. *Water Quality Standards for Aquatic Life.* Wash. Dept. of Ecology, Olympia, WA

Washington State Department of Ecology. 1998. *Preliminary Draft Evaluation Standards for Protection of Aquatic Life in Washington Surface Waters.* Wash. Dept. of Ecology, Olympia, WA. January 1998.

Washington Department of Ecology. 1999. *Forest Practices Code -* Stream Crossing for Fish Streams Guidebook: 1.2 Types of Fish Passage Structures. March.

Washington State Forest Practices Board. 1988. *Washington Forest Practices Rules and Regulations.* Washington Annotated Code, Title 222; Forest Practices Board Manual, and Forest Practices Act.

Washington DOE. 1994. *Effectiveness of forest road and timber harvest best management practices with respect to sediment-related water quality impacts.* Interim Report No. 2. Ecology Publication No. 94-67. TFW-WQ8-94-001. Washington State Department of Ecology, Timber, Fish, and Wildlife. May.

Washington State DNR. 1997. Information obtained from Internet site.

*Weaver, W.E., and D.K. Hagans. 1994. *Handbook for forest and ranch roads.* Prepared for Mendicino County Resource Conservation District, Ukiah, California. June.

Weitzman, S., and G.R. Trimble, Jr. 1952. Skid-road Erosion Can Be Reduced. *Journal of Soil and Water Conservation* 7:122-124.

Wemple, B.C., J.A. Jones, and G.E. Grant. 1996. Channel network extension by logging roads in two basins, western Cascades, Oregon. *Water Resources Bulletin* 32: 1195-1207.

White, D. *Hydraulic Performance of Countersunk Culverts in Oregon.* Oregon State University. 1-95, 1996. Master of Science.

Whitman, R. 1989. Clean Water or Multiple Use? Best Management Practices for Water Quality Control in the National Forests. *Ecology Law Quarterly* 16:909-966.

*Wiest, R. 1998. *A Landowner's Guide to Building Forest Access Roads.* NA-TP-06-98. USDA Forest Service, Northeastern Area State and Private Forestry, Radnor, Pennsylvania. July.

Willingham, P.W. 1989. Wetlands Harvesting Scott Paper Company. In *Proceedings of the Symposium: The Forested Wetlands of the Southern United States*, Orlando, Florida, July 12-14, 1988. USDA Forest Service General Technical Report SE-50, pp. 63-66.

Wilbrecht, S., Forester, Oregon Department of Forestry, personal communication, January 15, 2000.

Wisconsin Department of Natural Resources. 1989. *Forest Practice Guidelines for Wisconsin.* Bureau of Forestry, Madison, WI. PUBL-FR-064-89.

Wisconsin Department of Natural Resources. 2003. *Wisconsin Forest Management Guidelines.* PUB-FR-226 2003. Wisconsin Department of Natural Resources, Division of Forestry, Madison, Wisconsin. October.

Yee, C.S., and T.D. Roelofs. 1980. *Planning Forest Roads to Protect Salmonid Habitat.* USDA Forest Service. General Technical Report PNW-109.

Yoder, B., Forester, USDA, Forest Service, Gifford Pinchot National Forest, personal communication, January 20, 2000.

Yoho, N.S. 1980. Forest Management and Sediment Production in the South A Review. *Southern Journal of Applied Forestry* 4(1):27-36.

Zedaker, S., Virginia Polytechnic Institute and State University. 1999. Personnel communication. August 2, 1999.

Ziemer, R.R., J. Lewis, R. M. Rice, and T. E. Lisle. 1991. Modeling the cumulative watershed effects of forest management strategies. Journal of Environmental Quality 20(1): 36-42.

Ziemer, R.R., and T.E. Lisle. 1998. Hydrology. Chapter 3 in: Naiman, R.J., and R.E. Bilby, eds., *River Ecology and Management: Lessons from the Pacific Coastal Ecoregion.* Pages 43-68. Springer-Verlag, N.Y.

Readers are encouraged to contact their state department of forestry for information pertaining to BMPs for forestry in their state and region. In addition, some of the above guidances that represent a synthesis of current information are recommended for further reading and are marked with an asterisk (*).

Access road: A temporary or permanent road over which timber is transported from a loading site to a public road. Also known as a haul road.

Alignment: The horizontal route or direction of an access road.

Allochthonous: Derived from outside a system, such as leaves of terrestrial plants that fall into a stream.

Angle of repose: The maximum slope or angle at which a material, such as soil or loose rock, remains stable (stable angle).

Apron: Erosion protection placed on the streambed in an area of high flow velocity, such as downstream from a culvert.

Autochthonous: Derived from within a system, such as organic matter in a stream resulting from photosynthesis by aquatic plants.

Bedding: A site preparation technique whereby a small ridge of surface soil is formed to provide an elevated planting or seed bed. It is used primarily in wet areas to improve drainage and aeration for seeding.

Berm: A low earth fill constructed in the path of flowing water to divert its direction, or constructed to act as a counterweight beside the road fill to reduce the risk of foundation failure (buttress).

Borrow pit: An excavation site outside the limits of construction that provides necessary material, such as fill material for embankments.

Broad-based dip: A surface drainage structure specifically designed to drain water from an access road while vehicles maintain normal travel speeds.

Brush barrier: A sediment control structure created of slash materials piled at the toe slope of a road or at the outlets of culverts, turnouts, dips, and water bars.

Buck: To saw felled trees into predetermined lengths.

Buffer area: A designated area around a stream or waterbody of sufficient width to minimize entrance of forestry chemicals (fertilizers, pesticides, and fire retardants) into the waterbody.

Cable logging: A system of transporting logs from stump to landing by means of steel cables and winch. This method is usually preferred on steep slopes, wet areas, and erodible soils where tractor logging cannot be carried out effectively.

Check dam: A small dam constructed in a gully to decrease the flow velocity, minimize channel scour, and promote deposition of sediment.

Chopping: A mechanical treatment whereby vegetation is concentrated near the ground and incorporated into the soil to facilitate burning or seedling establishment.

Clearcutting: A silvicultural system in which all merchantable trees are harvested within a specified area in one operation to create an even-aged stand.

Contour: An imaginary line on the surface of the earth connecting points of the same elevation. A line drawn on a map connecting the points of the same elevation.

Crown: A convex road surface that allows runoff to drain to either side of the road prism.

Culvert: A metal, wooden, plastic, or concrete conduit through which surface water can flow under or across roads.

Cumulative effect: The impact on the environment that results from the incremental impact of an action when added to other past, present, and reasonably foreseeable future actions regardless of what agency or person undertakes such action.

Cut-and-fill: Earth-moving process that entails excavating part of an area and using the excavated material for adjacent embankments or fill areas.

DBH: Diameter at breast height; the average diameter (outside the bark) of a tree 4.5 feet above mean ground level.

Disking (harrowing): A mechanical method of scarifying the soil to reduce competing vegetation and to prepare a site to be seeded or planted.

Diversion: A channel with a supporting ridge on the lower side constructed across or at the bottom of a slope for the purpose of intercepting surface runoff.

Drainage structure: Any device or land form constructed to intercept and/or aid surface water drainage.

Duff: The accumulation of needles, leaves, and decaying matter on the forest floor.

Ephemeral drainage: A natural channel that carries water only during and immediately following rainstorms and whose channel bottom is seldom below the local water table. Sometimes referred to as a dry wash.

Felling: The process of cutting down standing trees.

Fill slope: The surface formed where earth is deposited to build a road or trail.

Firebreak: Naturally occurring or man-made barrier to the spread of fire.

Fire line: A barrier used to stop the spread of fire constructed by removing fuel or rendering fuel inflammable by use of fire retardants.

Foam line: A type of fire line that incorporates the use of fire-resistant foam material in lieu of, or in addition to, plowing or harrowing.

Ford: Submerged stream crossing where the traffic surface is reinforced to bear intended traffic.

Forest filter strip: Area between a stream and construction activities that achieves sediment control by using the natural filtering capabilities of the forest floor and litter.

Forwarding: The operation of moving timber products from the stump to a landing for further transport.

Geotextile: A product used as a soil reinforcement agent and as a filter medium. It is made of synthetic fibers manufactured in a woven or loose nonwoven manner to form a blanket-like product.

Grade (gradient): The slope of a road or trail expressed as a percentage of change in elevation per unit of distance traveled.

Harrowing (disking): A mechanical means to scarify the soil to reduce competing vegetation and to prepare a site to be seeded.

Harvesting: The felling, skidding, processing, loading, and transporting of forest products.

Haul road: See access road.

Intermittent stream: A stream that flows only during the wet periods of the year or in response to snow melt and flows in a well-defined channel. The channel bottom may be periodically above or below the local water table.

Landing (log deck): A place in or near the forest where logs are gathered for further processing, sorting, or transport.

Leaching: Downward movement of a soluble material through the soil as a result of water movement.

Logging debris (slash): The unwanted, unutilized, and generally unmerchantable accumulation of woody material, such as large limbs, tops, cull logs, and stumps, that remains as forest residue after timber harvesting.

Merchantable: Forest products suitable for marketing under local economic conditions. With respect to a single tree, it means the parts of the bole or stem suitable for sale.

Mineral soil: Soil that contains less than 20 percent organic matter (by weight) and contains rock less than 2 inches in maximum dimension.

Mulch: A natural or artificial layer of plant residue or other materials covering the land surface that conserves moisture, holds soil in place, aids in establishing plant cover, and minimizes temperature fluctuations.

Mulching: Providing any loose covering for exposed forest soils, such as grass, straw, bark, or wood fibers, to help control erosion and protect exposed soil.

Muskeg: A type of bog that has developed over thousands of years in depressions, on flat areas, and on gentle to steep slopes. These bogs have poorly drained, acidic, organic soils supporting vegetation that can be (1) predominantly sphagnum moss; (2) herbaceous plants, sedges, and rushes; (3) predominantly sedges and rushes; or (4) a combination of sphagnum moss and herbaceous plants. These bogs may have some shrub and stunted conifers, but not enough to classify them as forested lands.

Ordinary high water mark: An elevation that marks the boundary of a lake, marsh, or streambed. It is the highest level at which the water has remained long enough to leave its mark on the landscape. Typically, it is the point where the natural vegetation changes from predominantly aquatic to predominantly terrestrial.

Organic debris: Particles of vegetation or other biological material that can degrade water quality by decreasing dissolved oxygen and by releasing organic solutes during leaching.

Outslope: To shape the road surface to cause runoff to flow toward the outside shoulder.

Patch cutting method: A silvicultural system in which all merchantable trees are harvested over a specified area at one time.

Perennial stream: A watercourse that flows throughout a majority of the year in a well-defined channel and whose bottom (in rainfall dominant regimes) is below the local water table throughout most of the year.

Persistence: The relative ability of a pesticide to remain active over a period of time.

Pioneer roads: Temporary access ways used to facilitate construction equipment access when building permanent roads.

Prescribed burning: Skillful application of fire to natural fuels that allows confinement of the fire to a predetermined area and at the same time produces certain planned benefits.

Raking: A mechanical method of removing stumps, roots, and slash from a future planting site.

Regeneration: The process of replacing older trees removed by harvest or disaster with young trees.

Residual trees: Live trees left standing after the completion of harvesting.

Right-of-way: The cleared area along the road alignment that contains the roadbed, ditches, road slopes, and back slopes.

Riprap: Rock or other large aggregate that is placed to protect streambanks, bridge abutments, or other erodible sites from runoff or wave action.

Rut: A depression in access roads made by continuous passage of logging vehicles.

Salvage harvest: Removal of trees that are dead, damaged, or imminently threatened with death or damage in order to use the wood before it is rendered valueless by natural decay agents.

Sanitation harvest: Removal of trees that are under attack by or highly susceptible to insect and disease agents in order to check the spread of such agents.

Scarification: The process of removing the forest floor or mixing it with the mineral soil by mechanical action preparatory to natural or direct seeding or the planting of tree seedlings.

Scour: Soil erosion when it occurs underwater, as in the case of a streambed.

Seed bed: The soil prepared by natural or artificial means to promote the germination of seeds and the growth of seedlings.

Seed tree method: Removal of the mature timber in one cutting, except for a limited number of seed trees left singly or in small groups.

Selection method: An uneven-aged silvicultural system in which mature trees are removed, individually or in small groups, from a given tract of forestland over regular intervals of time.

Shearing: A site preparation method that involves the cutting of brush, trees, or other vegetation at ground level using tractors equipped with angles or V-shaped cutting blades.

Shelterwood method: Removal of the mature timber in a series of cuttings that extend over a relatively short portion of the rotation in order to encourage the establishment of essentially even-aged reproduction under the partial shelter of seed trees.

Silt fence: A temporary barrier used to intercept sediment-laden runoff from small areas.

Silvicultural system: A process, following accepted silvicultural principles, whereby the tree species constituting forests are tended, harvested, and replaced. Usually defined by, but not limited to, the method of regeneration.

Site preparation: A silvicultural activity to remove unwanted vegetation and other material, and to cultivate or prepare the soil for regeneration.

Skid: Short-distance moving of logs or felled trees from the stump to a point of loading.

Skid trail: A temporary, nonstructural pathway over forest soil used to drag felled trees or logs to the landing. Skid trails may either be constructed or simply develop due to use depending on the terrain.

Slash: See logging debris.

Slope: Degree of deviation of a surface from the horizontal, measured as a numerical ratio, as a percent, or in degrees. Expressed as a ratio, the first number is the horizontal distance (run) and the second number is the vertical distance (rise), as 2:1. A 2:1 slope is a 50 percent slope. Expressed in degrees, the slope is the angle from the horizontal plane, with a 90 degree slope being vertical (maximum) and a 45 degree slope being a 1:1 slope.

Stand: A contiguous group of trees sufficiently uniform in species composition, arrangement of age classes, and condition to be a homogeneous and distinguishable unit.

Streamside management area (SMA): A designated area that consists of the stream itself and an adjacent area of varying width where management practices that might affect water quality, fish, or other aquatic resources are modified. The SMA is not an area of exclusion, but an area of closely managed activity. It is an area that acts as an effective filter and absorptive zone for sediments; maintains shade; protects aquatic and terrestrial riparian habitats; protects channels and streambanks; and promotes floodplain stability.

Tread: Load-bearing surface of a trail or road.

Turnout: A drainage ditch that drains water away from roads and road ditches.

Water bar: A diversion ditch and/or hump installed across a trail or road to divert runoff from the surface before the flow gains enough volume and velocity to cause soil movement and erosion, and deposit the runoff into a dispersion area. Water bars are most frequently used on retired roads, trails, and landings.

Watercourse: A definite channel with bed and banks within which concentrated water flows continuously, frequently or infrequently.

Windrow: Logging debris and unmerchantable woody vegetation that has been piled in rows to decompose or to be burned; or the act of constructing these piles.

Yarding: Method of transport from harvest area to storage landing.

APPENDIX A:
EPA FORESTRY RESOURCES

Monitoring guidelines to evaluate effects of forestry activities on streams in the Pacific Northwest and Alaska. EPA910991001.

> The above document is available from U.S. EPA Public Information Center - S1043, 1200 Sixth Avenue, Seattle, WA 98101; phone 206-553-1200, fax 206-553-1049.

Summary of current state nonpoint source control practices for forestry. EPA841S93001.

Water quality effects and nonpoint source control for forestry: An annotated bibliography. EPA841B93005.

Nonpoint pointers: Managing nonpoint source pollution from forestry, pointer no. 8. EPA841F96004H.

Techniques for tracking, evaluating, and reporting the implementation of nonpoint source control measures: Forestry. EPA841B97009.

Evaluating the effectiveness of forestry best management practices in meeting water quality goals or standards (bound copy). EPA841B94005B.

> The above publications are out of print, but can be viewed on the Web from the following link: http://www.epa.gov/clariton/clhtml/pubtitleOW.html.

Facts about silvicultural activities in wetlands. EPA904F91100.

> The above is available from U.S. EPA, Region 4, Library, 345 Courtland Street, N.E., Atlanta, GA 30365; phone 404-347-4216.

Evaluating the effectiveness of forestry best management practices in meeting water quality goals or standards (3-hole punch). EPA841B94005A.

EPA Nonpoint Source News-Notes: published by EPA quarterly and available on the Internet. Occasionally has articles of interest to foresters and forest land owners. Articles from the Nonpoint Source News-Notes series can be obtained from the Internet at: http://www.epa.gov/owow/info/NewsNotes/. Forestry-related articles have included:

- Scientist Links Nutrient Runoff with Forest Defoliation (No. 51, April/May 1998)

- New Management Policies Proposed for National Forest Road System (No. 52, July/August 1998)

- Urban Forests Decline; Runoff Increases in Puget Sound Area (No. 53, September/October 1998)

- Working Buffer Strips Provide Profit and Protection (No. 54, November 1998)

- Report Lists Communities Suffering Flood Losses (No. 54, November 1998)

- Watershed Management Helps Lake Quality (No. 54, November 1998)

- Applying a Watershed Model to Reduce Nonpoint Source Runoff (No. 56, February/March 1999)

- Texas Forest Service Teaches Loggers about BMPs and Water Quality (No. 56, February/March 1999)

- Nine Salmon Listed in Urban Pacific Northwest (No. 57, May 1999)

- Riparian Forest Wildlife Guidelines for Landowners and Loggers (No. 58, July 1999)

- Getting Started With TMDLs (No. 59, November 1999)

Other EPA publications related to forests and forestry can be found at the EPA publications Web site by searching on "forest" or "forestry": http://www.epa.gov/ncepihom/.

Resources for Non-Industrial Private Forest (NIPF) Landowners:

The Sustainable Forestry Partnership has a web page devoted to Nonindustrial Private Forest Landowners: http://sfp.cas.psu.edu/nipf.htm.

USDA Forest Service—List of Publications, Resources

The USDA Forest Service, Washington Office and regional offices have a number of publications and other resources related to forestry. Lists of available publications, some of which are available electronically, and ordering information can be viewed at the Internet sites of the respective offices. Access to the Washington, DC office and the regional office Internet sites can be gained through the Internet site for publications for the USDA Forest Service: http://www.fs.fed.us/publications/.

The documents of the *Water-Road Interaction Technology Series*, published by the U.S. Forest Service, San Dimas Technology and Development Center, San Dimas, California, are available at: http://www.stream.fs.fed.us/water-road.

Other resources that will be of interest to forestland owners and that are available electronically include:

- FishXing (software and learning system for fish passage through culverts): http://www.stream.fs.fed.us/fishxing

- Forest Service Roads Analysis Process: http://www.fs.fed.us/news/roads/DOCSroad-analysis.shtml

- Forest Roads Science Synthesis: http://www.fs.fed.us/news/roads/science.pdf

APPENDIX B:
SOURCES OF TECHNICAL ASSISTANCE

U.S. Department of Agriculture
Natural Resources Conservation Service
P.O. Box 2890
Washington, DC 20013

U.S. Department of Interior
Fish and Wildlife Service
Public Affairs Office
18th and C Streets, NW
Washington, DC 20240

U.S. Department of the Interior
Geological Survey
12201 Sunrise Valley Drive
Reston, Virginia 22092

U.S. Forest Service
Office of Information
Room 3238
P.O. Box 2417
Washington, DC 20013

U.S. Department of Commerce
National Climatic Center
Federal Building
Asheville, North Carolina 28801
(Attn: Publications)

American Forest Institute
1619 Massachusetts Ave., NW
Washington, DC 20036

American Forests
P.O. Box 2000
Washington, DC 20013-2000

Association of Consulting Foresters of America
5400 Grosvenor Lane, Suite 300
Bethesda, Maryland 20814

International Society of Arboriculture
P.O. Box 71
5 Lincoln Square
Urbana, Illinois 61801

International Society of Arboriculture
P.O. Box GG
6 Dunlap Court
Savoy, Illinois 61874

National Arbor Day Foundation
100 Arbor Avenue
Nebraska City, Nebraska 68410

National Arborist Association
P.O. Box 1094
Amherst, New Hampshire 03031-1094

National Association of State Foresters
Hall of the States, #526
444 North Capital Street, NW
Washington, DC 20001

National Urban Forest Council
c/o American Forests
P.O. Box 2000
Washington, DC 20013

Soil and Water Conservation Society
7515 Northeast Ankney Road
Ankney, Iowa 50021-9764

American Sod Producers Association, Inc.
9th and Minnesota Streets
Hastings, Nebraska 68901

The IPM Practitioner
P.O. Box 7414
Berkeley, California 94707
510-524-2567
Directory of Least-Toxic Pest Control Products

Pesticide Hot Line (Autovon 584-3773)
U.S. Army Environmental Hygiene Agency
Pest Management and Pesticide
Monitoring Division
Aberdeen Proving Ground, Maryland 21005

The Internet site of the *National Association of State Foresters,* http://www.stateforesters.org/, has links to many forestry resources, including:

- State Forestry Statistics

- State Forester Directory

- State Forester Home Pages

- State and Private Forestry Programs

- Other Forestry Links

APPENDIX C: FOREST MANAGEMENT CERTIFICATION PROGRAMS

Forest Management and Forest Product Certification

In the past 10 years, forest management monitoring has been extended beyond an evaluation of whether best management practices have been implemented according to state or federal specifications for the protection of habitat values and water quality to encompass ecological, social, and economic values. Independent organizations offer certification of forest management and forest products to forestry operations managed according to an internationally accepted set of criteria for sustainable forest management (Crossley, 1996). The principles and criteria of sustainable forestry are general enough to be applicable to tropical, temperate, and boreal forests, but the standards used to certify individual operations are sufficiently site- and region-specific for critical evaluation of individual forests and forestry operations.

To be certified, forest management must adhere to principles of resource sustainability, ecosystem maintenance, and economic and socioeconomic viability. Resource sustainability means that harvesting is conducted such that the forest remains productive on a yearly basis. Large scale clear-cutting, for instance, such that the forest would have to remain idle and unproductive for many years, would generally not be acceptable. Ecosystem maintenance means that the ecological processes operating in a forest continue to operate without interruption and the forest's biodiversity is maintained. The principle implies that harvesting does not fundamentally alter the nature of the forest. Economic and socioeconomic viability incorporate the two previous principles and imply that forest operations are sufficiently profitable to sustain operations from year to year and that social benefits provided by a forest, such as existence and recreational value, are also maintained over the long term. Economic and socioeconomic viability are incentives for local people to sustain the ecosystem and resources of the forest (Evans, 1996).

Development of guidelines for sustainable forest management began with the International Tropical Timber Organization (ITTO). In 1989 the ITTO Council requested that "best practice" guidelines for sustainable management of natural tropical forests be developed. Soon afterward, global efforts to define and implement "sustainable forest management" began with the United Nations Conference on Environment and Development (UNCED), held in Rio de Janeiro, Brazil, in 1992. Non-binding "Forest Principles" were endorsed by more than 170 countries attending that conference, though many attending countries hoped that a binding "Forests Convention," similar to those for biodiversity and ozone layer protection, would be endorsed. Since Rio, dozens of fora, groups, and processes have been developed to define and evaluate sustainable forest management.

The movement to evaluate forest management and forest products based on principles of sustainable management is an expansion of focus as more knowledge is gained about forest ecological processes and the impacts, both local and global, of poorly managed forests on ecological systems and, consequently, on human economic and social systems. The expansion is similar to the natural expansion of EPA's focus in the realm of water pollution control from point sources of pollution to nonpoint sources of pollution to the present focus on watershed processes. Progress gained in overcoming one problem (e.g., point sources of water pollution) highlight the impacts of other problems (e.g., nonpoint sources of water pollution) and the search for overcoming these problems naturally expands to encompass the new problems that are highlighted. As more sources of impact are recognized, the focus must expand to encompass them. Thus, while water pollution control has become focused on watershed processes and activities occurring within watersheds, forest management is naturally expanding to encompass the processes dependent on the forest (i.e., ecological, social, and economic) and which can be severely limited by poor management.

Two steps are involved in certifying wood products. First, forest management is certified as sustainable according to an evaluation based on accepted principles of sustainable forest management. Various organizations refer to this certification process as forest certification, forest management auditing, or timber certification. Evaluations are always conducted by a third, independent party. The second step is wood-product certification, or forest product labeling. Again, a third party follows the harvested wood through the manufacturing and product development processes, a "chain-of-custody" inspection process, to certify and label the products created from wood harvested from a "sustain-able" forestry operation. Both types of certification are currently carried out by both for-profit companies and not-for-profit organizations that are predominantly based in the United States and the United Kingdom.

The Forest Stewardship Council (FSC) accredits regional groups to certify forest operations. Well known examples of FSC-accredited groups are Scientific Certification Systems (SCS) and the Rainforest Alliance's Smart Wood Program (Evans, 1996). These groups and others not associated with FSC are active in the United States and their evaluation processes are described below.

Forest Stewardship Council

The Forest Stewardship Council was formed in 1993 and is a nonprofit organization registered in Mexico. FSC strives to serve as a global foundation for the development of region-specific forest-management standards with its *Principles and Criteria for Forest Management*. Independent certification bodies, accredited by the FSC in the application of these standards, conduct impartial, detailed assessments of forest operations at the request of landowners. If the forest operations are found to be in conformance with FSC standards, a certificate is issued, enabling the landowner to bring product to market as "certified wood" and to use FSC trademark logo. In 1996 the FSC accredited the SmartWood Program, Scientific Certification System (SCS), the SGS Forestry QUALIFOR Programme (based in the United Kingdom), and the Soil Association for worldwide forest management and chain-of-custody certification.

The FSC-U.S. Working Group, Inc., is the U.S. arm of the FSC. FSC-U.S. partners are businesses (wood product distributors such as Home Depot, timber producers such as

Seven Islands Land Company, and certification bodies), foundations, and non-governmental organizations (NGO). Currently there are 40 NGO partners, including the Consumer's Choice Council, Defenders of Wildlife, and Friends of The Earth.

Programs accredited under the FSC provide two types of service, forest management certification and chain-of-custody certification. For forest management certification, a third party evaluation of a forest management operation is conducted in conformity with FSC principles–specific environmental, social, and economic standards. Certification enables an organization to guarantee that its product or service conforms to FSC standards, which could affect product marketability.

To certify a forest management operation, the certification body studies the forest management system and policies and visits the operation for an evaluation. A certified operation must be monitored annually to ensure that the standards of forest stewardship are maintained throughout the period of certification.

The FSC *Principles and Criteria for Forest Stewardship* emerged out of a desire to provide market rewards through the labeling of forest products with a distinct logo derived from lands recognized for "exemplary" forest management. The principles and criteria apply to all tropical, temperate, and boreal forests and must be incorporated into the evaluation systems and standards of all certification organizations seeking accreditation by FSC. More detailed standards may be prepared at national and local levels.

Principle No. 6 in the FSC criteria relates to environmental impact. It does not specify BMPs, but requires the certified body to maintain, enhance, or restore ecological functions and values; protect and record representative samples of existing ecosystems within the landscape; and prepare written documentation on controlling erosion, minimizing forest damage, and protecting water resources.

Many regional standards and policies require that certified bodies meet or exceed the specifications listed in state forest practices:

- 6.5 (Appalachian Region): Harvesting, road construction and other mechanical operations shall meet or exceed state Best Management Practices, whether voluntary or mandatory, and other applicable water quality regulations. In advance of these activities, planning shall be done to minimize damage to the soil, water and forest resources from these activities. A written description of the operational plan, demonstrating how damage will be minimized, shall be incorporated into the management plan or harvesting contract as appropriate.

- 6.5.1 (Southeast Region): Harvesting, road construction, and other mechanical operations shall be designed to meet or exceed state best management practices and applicable water quality regulations.

Forest Conservation Program—Scientific Certification Systems (US)

The Forest Conservation Program (FCP) was established by Scientific Certification Systems (SCS) in 1991 as a certification program for sustainable forestry. SCS has certified forests in California (Collins Pine Almanor Forest), Pennsylvania (Collins Pennsylvania Forest), Wisconsin (Menominee Forest), and Mexico.

The FCP uses an evaluation process based on the program elements mentioned above: resource sustainability, ecosystem maintenance, and economic and socioeconomic viability. Each program element is evaluated according to a set of criteria that best represents appropriate benchmarks of sustainable forest management in the region of interest. Timber resource sustainability is evaluated based on criteria relating to how fully-stocked stands are, growing conditions, age and/or size class distribution (even-aged management or uneven-aged management), and whether management allows for sustained yearly harvests and avoids idle years.

The forest ecosystem maintenance element is evaluated based on criteria relating to whether non-timber resource values are a part of management and the extent to which natural ecosystem conditions and processes are altered by harvests. The economic and socioeconomic element is concerned with the overall economic viability of forest operations and the socioeconomic impacts of operations on harvesters and the local community.

The FCP program is designed to provide a quantitative and qualitative approach to certification. Forest evaluations are based on five sources of information. The landowner; investigations of information related to harvesting operations (e.g. timber inventory data, timber management plans, business management plans, and employee records); field sampling (e.g., wildlife surveys); field reviews; and interviews with employees, contractors, and individuals and organizations from the community.

SCS provides two levels of recognition under the FCP program, "Well-managed" and "State-of-the-Art Well-managed." Well-managed forests meet FCP standards for sustainable management as described below. "State-of-the-Art Well-managed" forests rank in the top 10 percent of all forests evaluated under the FCP program.

Evaluations are conducted by an evaluation team that consists of persons with expertise in relevant disciplines, such as forestry, wildlife biology, ecology, and economics. Persons with local or regional expertise are incorporated into evaluation teams and all evaluations are peer reviewed. Periodic monitoring of the forest after initial evaluation, lasting 1 to 3 years, is required as part of certification. Evaluation criteria are selected and weighted to account for regional circumstances.

Each criterion is given a ranking from 1 to 100 based on its perceived importance to sustainable management of the particular forest. Forest management is then scored by the evaluation team according to the chosen criteria. Sixty points on a normalized 100-point scale is the "failure threshold" for each criterion. Forests that receive 60 points or more in all three categories are designated "Well-managed." Forests among the top 10 percent of all SCS-rated forests are given the "State-of-the-Art" designation. The designation given to the forest management operation is also applied to products from wood harvested from the certified forest.

The program is practical and feasible for forest managers to implement because standards of what constitutes good performance and what leads to failure to attain certification for each criterion are clearly described and adaptable for local or regional circumstances. The credibility of the certification process depends largely on the strength of the evaluation team (Evans, 1996).

Smart Wood Program—Rainforest Alliance (US)

The Rainforest Alliance established Smart Wood as the first independent forestry certification program in the world in 1990. The program initially focused on tropical forests but is now used to certify forests of all types. Forests have been certified in Java, Honduras, Mexico, Brazil, and Papua New Guinea. The Smart Wood program is similar to the FCP.

Under the program, long-term management data is used to demonstrate that a forest can be classified as a "sustainable source". Without long-term data but with demonstration that management has a commitment to sustainability, a forest can be classified as "well-managed".

Smart Wood companies are companies that handle Smart Wood-certified products. Category 1 companies sell products made exclusively from Smart Wood forests, and Category 2 companies sell products made from a mix of certified and noncertified sources. Products from Smart Wood companies carry one of these designations.

Smart Wood certification is based on three broad principles:

- All operations maintain ecosystem functions, including watershed stability and conservation of biological resources.

- Planning and implementation incorporate sustained yield production for all forest products.

- Management activities have a positive impact on local communities.

Smart Wood is developing detailed regional standards with the assistance of local specialists (Evans, 1996).

Sustainable Forestry Initiative (SFI) SM Program of the American Forest & Paper Association

The American Forest & Paper Association (AF&PA) is the national trade association of the forest, pulp, and paper, paperboard, and wood products industry. AF&PA represents approximately 138 member companies and licensees controlling 84 percent of paper production, 50 percent of solid wood production, and 90 percent of the industrial timberland in the United States.

AF&PA member companies, as a condition of membership, must commit to conduct their business in accordance with the principles and objectives of the Sustainable Forestry InitiativeSM program, instituted in October 1994.

The SFISM program is a comprehensive system of principles, objectives and performance measures that integrates the perpetual growing and harvesting of trees with the protection of wildlife, plants, soil and water quality. It is based on the premise that responsible environmental practices and sound business practices can be integrated to the benefit of landowners, shareholders, customers and the people they serve.

Professional foresters, conservationists and scientists developed the SFI program. They were inspired by the concept of sustainability that evolved from the 1987 report of the World Commission on Environment and Development and was subsequently adopted by the 1992 Earth Summit in Rio de Janeiro. The original 1994 SFI Principles and Implementation Guidelines were modified and implemented to become the industry "Standard" in 1999. The standards will continue to be updated periodically to reflect new information concerning forest management and social changes.

SFI State Implementation Committees have formed in 32 states to bring industry representatives together with other stakeholders to support logger-training programs and provide outreach to nonindustrial private landowners and opportunities for public involvement.

In a response to public pressure to broaden the SFI program to include nonmember participation in the SFI, a licensee program has been developed. To date, more than 1.5 million acres have been added to the SFI program through licensee agreements, increasing the total forest acres enrolled in the SFI program to 56.5 million acres.

Member companies and licensees are required to submit annual reports to AF&PA describing progress in implementing the SFI program. Since its inception, member companies of AF&PA have invested more than $247 million on research related to wildlife, biodiversity, ecosystem management and the environment. By 1998 more than 30,000 independent loggers and foresters completed training in sustainable forestry with an additional 20,000 completing partial training. In addition, SFI participants and professional loggers have distributed information regarding the SFI program to approximately 242,000 landowners across the country since 1994.

Summary of Certification Initiatives in the United States

Independent certification programs provide a framework of broad principles and core criteria against which forest management can be assessed. Similar to state forestry programs for best management practice monitoring, forest management under the certification programs is evaluated with field sampling, examinations of documents, and interviews with staff and local stakeholders. Evaluation teams are interdisciplinary and knowledgeable of local conditions, and certification is based on scores for identifiable management actions.

Although many certification programs are international in scope and focus, the flexibility to tailor the evaluation to local circumstances is built into the process, so the programs have credibility and can be practically implemented on a local level. Furthermore, the framework of the certification process is a practical forest management tool as the internationally accepted criteria on which evaluations are based provide guidance to forest managers for managing operations for sustainability.

The credibility of the process depends on the expertise of the evaluation team. Persons with local expertise must be used for evaluations in order for the certification process to be placed within a local context, and a local context is absolutely necessary because of the complex inclusion of social, economic, and ecological dimensions in the certification process. This complexity can lead to inconsistencies in evaluations and certifications, but some certification programs, notably the Smart Wood Program, are providing regional, national, and international consistency with the development of regional-specific standards.

A separate approach, the Canadian Standards Association Sustainable Forest Management Project (CSA SFM), is based on developing a preferred future condition that meets society's goals, developing an action plan to move toward the future condition, monitoring progress toward achieving that condition, and correcting one's course of action based on monitoring results. An essential element missing from this approach, and an element that makes the FCP and Smart Wood programs so powerful, is a set of clear criteria that define sustainable forest management. In the CSA SFM approach, this definition is left for local stakeholders to define. The result is a lack of consistency from operation to operation and certification to certification (Evans, 1996).

APPENDIX D: NONINDUSTRIAL PRIVATE FOREST (NIPF) MANAGEMENT

The approximately 10 million nonindustrial private forest (NIPF) owners in the United States include individuals, partnerships, estates, trusts, clubs, tribes, corporations, and associations (Pennsylvania State University, 2000). NIPF owners control 261 million acres of timberland and 58 percent of the commercial forests in the United States. More than two-thirds of timberland east of the Mississippi River is in NIPF ownership, whereas the majority of timberland in the West is in public ownership. NIPFs protect watersheds, provide wildlife habitat, offer scenic beauty, and supply 49 percent of the timber harvested in the United States (USDA-FS, 1992).

Many NIPF owners are not fully aware of the potential economic value of properly-managed timberland. Some are unaware of how to properly manage their timber resources (Pennsylvania State University, 2000). Proper management might be secondary to avoiding annual property taxes and capital gains taxes for some owners. Some other owners who do not plan properly for the inheritance their timberland might lose ownership upon their death, and still others, unaware of either management techniques or the economic value of the land, might decide to convert the land to other uses, such as development or agriculture. Owners who view harvesting of the timber on their land as a one-time capital gain may not be aware of the long-term economic and environmental benefits of sustainable timberland management. Andrew Egan of West Virginia University and Stephan Jones of the Alabama Cooperative Extension System studied NIPF owners and timberland management, and found that landowners with knowledge of forests and forestry are more likely to manage their forests in a sustainable manner (Pennsylvania State University, 2000).

Forest*A*Syst, by Rick Hamilton, extension forestry specialist with the Department of Forestry, North Carolina State University, is a self-assessment guide directed at encouraging forest owners to manage their forests for recreation and aesthetics, wildlife, and timber production, while protecting water quality. The guide discusses steps in developing a forest management plan and strongly recommends the assistance of a professional forester in this process. Major topics are site preparation, natural regeneration, artificial seeding, tree planting, weed control, and fertilization in young and middle-age stands; harvesting the mature forest; managing for wildlife habitat; enhancing the visual appearance of the site; improving recreational opportunities; and using management practices to protect water quality. A *Forest*A*Syst* guide for western North Carolina has been developed from the national *Forest*A*Syst* prototype developed by Mr. Hamilton. A similar guide is available for eastern North Carolina. Other states' programs have spun off from the national version, as well, including Tennessee and Alaska, Georgia (in process), New England (developing a *Forest*A*Syst* model for the region), and Kentucky and Hawaii (in process) (Leith, 2002). For additional information on distribution of the publication and support for adapting it to state and local conditions, contact Rick Hamilton at

(919) 515-5574 or by e-mail (hamilton@cfr.crf.ncsu.edu) or contact Larry Biles, USDA-CSREES (Cooperative State Research, Education and Extension Service), Washington, DC, at (202) 401-4926.

Proper implementation of forestry management measures can maintain fish and wildlife habitat, clean water, biological diversity, aesthetics, and a buffer from urban sprawl. To maintain these values, it is recommended that NIPF landowners follow the guidance of the management measures for forestry to protect water quality set forth in this guidance. Because some of the management measures and BMPs mentioned in the guidance, however, are more relevant to state, federal, and industrial timberland owners, this appendix is provided to focus on certain aspects of planning and managing timberlands that are especially intended to assist NIPF owners in addressing BMP implementation and forest management.

Individual landowners are encouraged to use this guidance to manage and protect water quality on their private forestland. If you have turned directly to this appendix, thinking perhaps that the main sections of the guidance are meant for state agencies and industrial landowners, please take the time to review the rest of the document, especially Section 3. The management measures and practices described in the guidance are applicable to all forest landowners, whether 10 acres or 10,000 acres are being managed. Some of the management measures will be more applicable to some forest management goals than others, but the concepts contained in them are equally relevant to water quality protection in all managed forests where trees are harvested.

Preharvest Planning:

Below are listed some of the more important management practices for achieving the Management Measure for Preharvest Planning. Complete discussions of these and other management practices for preharvest planning can be found in Section 3A. Additional management practices that are particularly applicable to the NIPF landowner follow this listing.

Harvest Planning Practices

◆ *Use topographic maps, aerial photographs, soil surveys, geologic maps, and rainfall intensity charts to augment site reconnaissance to lay out and map harvest units. Identify and mark, as needed:*

◆ *Consider potential water quality and habitat impacts when selecting the silvicultural system as even-aged (clear-cut, seed tree, or shelterwood) or uneven-aged (group or individual selection). The yarding system, site preparation method, and any pesticides that will be used should also be addressed in preharvest planning. As part of this practice the potential impacts from and extent of roads needed for each silvicultural system should be considered.*

◆ *In high-erosion-hazard areas, trained specialists (geologist, soil scientist, geotechnical engineer, wild land hydrologist) should identify sites that have high risk of landslides or that might become unstable after harvest. These specialists can recommend specific practices to reduce the likelihood of erosion hazards and protect water quality.*

Road System Planning Practices

◆ *Preplan skid trail and landing locations on stable soils and avoid steep gradients, landslide-prone areas, high-erosion-hazard areas, and poor-drainage areas.*

◆ *Identify areas that will require the least modification for use as log landings and use them to reduce the potential for soil disturbance. Use topographic maps and aerial photographs to locate these areas.*

◆ *Plot feasible routes and locations on an aerial photograph or topographic map to assist in the final determination of road locations.*

◆ *Design roads and skid trails to follow the natural topography and contour, minimizing alteration of natural features.*

◆ *In moderately sloping terrain, plan for road grades of less than 10 percent, with an optimal grade of between 3 percent and 5 percent. In steep terrain, short sections of road at steeper grades can be used if the grade is broken at regular intervals. Vary road grades frequently to reduce culvert and road drainage ditch flows, road surface erosion, and concentrated culvert discharges.*

◆ *Plan to surface most forest roads, and select a road surface material suitable for the intended road use.*

◆ *Lay out roads, skid trails, and harvest units to minimize the number of stream crossings.*

◆ *To minimize soil disturbance and road damage, plan to suspend operations when soils are highly saturated. Damage to forested slopes can also be minimized by not operating logging equipment when soils are saturated, during wet weather, or when the ground is thawing.*

◆ *Select waterway opening sizes to minimize the risk of washout during the expected life of the structure. Opening size will vary depending on the drainage area of the watershed where the stream-crossing structure is to be placed.*

Additional management practice recommendations for the NIPF landowner

◆ *Locate property lines.*

The location of property lines might restrict the use of the best access locations. If significant environmental impact (e.g., erosion, water body sedimentation, numerous stream crossing) could be avoided by crossing adjacent property to provide access, consider negotiating or purchasing a right-of-way from the owner of the property.

The USDA Forest Service has produced a document titled A Landowner's Guide to Building Forest Access Roads (Wiest, 1998). This document, along with the assistance of a consulting forest engineer, provides support in road planning and location. To receive a copy of this document, contact the USDA Forest Service, Northeastern Area State and Private Forestry, in Radnor, Pennsylvania, (610) 975-4017, or order a copy from the web site at <http://www.na.fs.fed.us/spfo/pubs/stewardship/accessroads/accessroads.htm>.

◆ *Inventory the property.*

Managing timberland requires knowledge of what is on the property. Conduct an inventory to identify features of the land such as streams, steep slopes, eroding or erodible soils, roads and trails, and sensitive wildlife habitats. Aerial photos can be useful for an inventory, but if they are not available for the property, U.S. Geological Survey (USGS) quadrangle map(s) of the area can be used to locate these resources and create a permanent record of them on a map. USGS quadrangle maps show contour lines (steepness of the terrain), existing roads, waterbodies, springs, and buildings. They cost approximately $5 per map and are available for all of the United States.

◆ *Develop a forest management plan.*

Before harvesting operations begin, develop a forest management plan that contains goals, objectives, possible alternatives to harvesting, future planning, and the trade-offs that accompany altering the land. Contact the state department of forestry or cooperative extension service for information on forest harvesting BMPs and their implementation. A logging company is often the primary source of information regarding forestry and nonpoint source pollution control for NIPF owners, and only by first becoming familiar with the various BMPs can the NIPF landowner be assured that a contractor is choosing and implementing BMPs properly.

The use of a consulting forester or state forester is extremely helpful when developing a forest management plan. The forester can assist with all aspects of forest management and harvest, including the layout of roads and logging decks, BMP implementation, stream protection, and the proper use of chemical. The forester can also educate the NIPF owner about topics such as watershed protection and sustainable forest management.

Streamside Management Areas:

Below are listed some of the more important management practices for achieving the Management Measure for Streamside Management Areas. Complete discussions of these and other management practices for preharvest planning can be found in Section 3B.

- Minimize disturbances that would expose the mineral soil of the SMA forest floor. Do not operate skidders or other heavy machinery in the SMA.

- Locate all landings, portable sawmills, and roads outside the SMA.

- Restrict mechanical site preparation in the SMA, and encourage natural revegetation, seeding, and hand planting.

- Limit pesticide and fertilizer usage in the SMA. Establish buffers for pesticide application for all flowing streams.

- Directionally fell trees away from streams to prevent logging slash and organic debris from entering the water body. If slash and debris are in the stream as a result of harvesting practices, remove them immediately.

- Apply harvesting restrictions in the SMA to maintain its integrity.

Road Construction/Reconstruction:

Below are listed some of the more important management practices for achieving the Management Measure for Road Construction and Reconstruction. Complete discussions of these and other management practices for preharvest planning can be found in Section 3C.

Road Surface Construction Practices

◆ *Follow the design developed during preharvest planning to minimize erosion by properly timing and limiting ground disturbance operations.*

◆ *Properly dispose of organic debris generated during road construction.*

◆ *Prevent slash from entering streams and promptly remove slash that accidentally enters streams to prevent problems related to slash accumulation.*

Road Surface Drainage Practices

◆ *Install surface drainage controls at intervals that remove storm water from the roadbed before the flow gains enough volume and velocity to erode the surface. Route discharge from drainage structures onto the forest floor so that water will disperse and infiltrate. Methods of road surface drainage include the following:*

◆ *Install turnouts, wing ditches, and dips to disperse runoff and reduce the amount of road surface drainage that flows directly into watercourses.*

◆ *Install appropriate sediment control structures to trap suspended sediment transported by runoff and prevent its discharge into the aquatic environment.*

Road Slope Stabilization Practices

◆ *Use straw bales, straw mulch, grass-seeding, hydromulch, and other erosion control and revegetation techniques to complete the construction project. These methods are used to protect freshly disturbed soils until vegetation is established.*

◆ *Revegetate or stabilize disturbed areas, especially at stream crossings.*

Stream Crossing Practices

◆ *Construct stream crossings to minimize erosion and sedimentation.*

◆ *Install a stream crossing that is appropriate to the situation and conditions.*

Fish Passage Practices

◆ *On streams with important spawning areas, avoid construction during egg incubation periods.*

◆ *Design and construct stream crossings for fish passage according to site-specific information on stream characteristics and the fish populations in the stream where the passage will be installed.*

Road Management:

Below are listed some of the more important management practices for achieving the Management Measure for Road Management. Complete discussions of these and other management practices for preharvest planning can be found in Section 3D.

Road Maintenance Practices

◆ *Blade and reshape the road to conserve existing surface material; to retain the original, crowned, self-draining cross section; and to prevent or remove berms (except those designed for slope protection) and other irregularities that retard normal surface runoff.*

◆ *Maintain road surfaces by mowing, patching, or resurfacing as necessary.*

◆ *Clear road inlet and outlet ditches, catch basins, culverts, and road-crossing structures of obstructions as necessary.*

Wet and Winter Road Practices

◆ *Before winter, all permanent, seasonal, and temporary roads should be inspected and prepared for the winter months.*

Stream Crossing and Drainage Structure Practices

◆ *When temporary stream crossings are no longer needed, and as soon as possible upon completion of operations, remove culverts and log crossings to maintain adequate streamflow.*

◆ *During and after logging activities, ensure that all culverts and ditches are open and functional.*

◆ *Revegetate disturbed surfaces to provide erosion control and stabilize the road surface and banks.*

Timber Harvesting:

Section 319 requires states to assess nonpoint source pollution and implement management programs, and it authorizes EPA to provide grants to assist state nonpoint source pollution control programs.

Below are listed some of the more important management practices for achieving the Management Measure for Timber Harvesting. Complete discussions of these and other management practices for preharvest planning can be found in Section 3E. Additional management practices that are particularly applicable to the NIPF landowner follow this listing.

Harvesting Practices

◆ *Fell trees away from watercourses whenever possible, keeping logging debris from the channel, except where debris placement is specifically prescribed for fish or wildlife habitat.*

◆ *Immediately remove any tree accidentally felled in a waterway.*

◆ *Remove slash from the water body and place it outside the SMA.*

Practices for Landings

◆ *Landings should be no larger than necessary to safely and efficiently store logs and load trucks.*

◆ *Upon completion of a harvest, clean up, regrade, and revegetate the landing.*

Ground Skidding Practices

◆ *Skid uphill to log landings whenever possible. Skid with ends of logs raised to reduce rutting and gouging.*

◆ *Skid perpendicular to the slope (along the contour), and avoid skidding on slopes greater than 40 percent.*

Cable Yarding Practices

◆ *Use cabling systems or other systems when ground skidding would expose excess mineral soil and induce erosion and sedimentation.*

◆ *Avoid cable yarding in or across watercourses.*

Petroleum Management Practices

◆ *Service equipment at a location where any spilled fuel or oil will not reach watercourses, and drain all petroleum products and radiator water into containers.*

◆ *Dispose of wastes and containers in accordance with proper waste disposal procedures.*

◆ *Take precautions to prevent leakage and spills.*

Additional management practice recommendations for the NIPF landowner

◆ *Participate actively in the timber harvest.*

It is important that the NIPF landowner be an active participant in the timber harvest process. Working with the harvesting contractor and state forester, verify that road layout, stream protection, landing locations, skid trail layout, and drainage BMPs all follow the plan developed in the preharvest planning phase. Review the management measures in this guidance prior to developing a plan, note those measures and BMPs particularly relevant to your situation, discuss them with a state forester, and then participate in the harvest to be certain that it is conducted in a manner compatible with the sustainability of your property.

Site Preparation and Forest Regeneration:

Below are listed some of the more important management practices for achieving the Management Measure for Site Preparation and Forest Regeneration. Complete discussions of these and other management practices for preharvest planning can be found in Section 3F.

Site Preparation Practices

◆ *Mechanical site preparation should not be conducted on slopes greater than 30 percent.*

◆ *Do not conduct mechanical site preparation in SMAs.*

Forest Regeneration Practices

◆ *Order seedlings well in advance of planting time to ensure their availability.*

◆ *Hand plant highly erodible sites, steep slopes, and lands adjacent to stream channels (SMAs).*

Fire Management:

Below are listed some of the more important management practices for achieving the Management Measure for Fire Management. Complete discussions of these and other management practices for preharvest planning can be found in Section 3G. Additional management practices that are particularly applicable to the NIPF landowner follow this listing.

Prescribed Fire Practices

◆ *Carefully plan burning to take into account weather, time of year, and fuel conditions so that these help achieve the desired results and minimize impacts on water quality.*

◆ *Intense prescribed fire for site preparation should not be conducted in the SMA.*

◆ *Execute the burn with a trained crew and avoid intense burning.*

Additional management practice recommendations for the NIPF landowner

◆ *Contact a state forester before any prescribed burning.*

Prescribed burning poses many potential hazards, and the NIPF landowner must be aware of these. Before using fire as a management tool, consult with a professional forester to obtain information on permits, burning times and procedures, equipment, current fire conditions, and safety precautions.

◆ *Notify adjacent landowners.*

Before burning, notify adjacent landowners, the local county sheriff, and local fire departments to let them know the date of the burn. A permit might be required for the burn, and it might specify a time period during which the burn must occur. If the burn is not done during the specified period, a new permit must be obtained. Letting all potentially affected parties know that a burn will take place will lessen the likelihood that the fire department will be called to put out the fire. The date of the prescribed burn is always subject to change due to changing weather and fire hazard conditions, and if the date does change, inform the previously notified parties of the new date.

◆ *Hire a professional.*

A landowner who is not proficient in prescribed burning should hire a contractor to perform the burn. Investigate the background and record of any contractor contacted and ask the contractor to provide testimonies of his or her work. Ask the local forestry department, cooperative extension service, or fire department if they have knowledge of the contractor as well. Remember that having a contractor perform the burn does not release the landowner of obligations to notify potentially affected parties, obtain legal information and permits, and ensure that the burn is conducted within the conditions of the permit or recommendations made by the fire or forestry department with respect to time of day, safety precautions, and so forth.

Revegetation of Disturbed Areas:

Below are listed some of the more important management practices for achieving the Management Measure for Revegetation of Disturbed Areas. Complete discussions of these and other management practices for preharvest planning can be found in Section 3H.

◆ *Use mixtures of seeds adapted to the site, and avoid the use of exotic species. Species should consist primarily of annuals to allow natural revegetation of native under-story plants, and they should have adequate soil-binding properties.*

◆ *Seed during optimum periods for establishment, preferably just before fall rains or whenever the optimum period might be for the region.*

◆ *Fertilize according to site-specific conditions.*

◆ *Inspect all seeded areas for failures, and make necessary repairs and reseed within the planting season.*

◆ *During non-growing seasons, apply interim surface stabilization methods to control surface erosion.*

Forest Chemical Management:

Below are listed some of the more important management practices for achieving the Management Measure for Forest Chemical Management. Complete discussions of these and other management practices for preharvest planning can be found in Section 3I. Additional management practices that are particularly applicable to the NIPF landowner follow this listing.

◆ *Apply pesticides and fertilizers during favorable atmospheric conditions.*

◆ *Apply slow-release fertilizers when possible.*

◆ *Apply fertilizers during maximum plant uptake periods to minimize leaching.*

◆ *Consider the use of pesticides as only one part of an overall program to control pest problems.*

Additional management practice recommendations for the NIPF landowner

◆ *Contact a state forester.*

Forest landowners who intend to apply chemicals to manage their timber stands should first contact a local forester. The forester will be able to provide information on approved

---

pesticides and fertilizers, application guidelines or requirements, and a list of licensed applicators. It might be possible to hire state foresters to apply chemicals, or they might be willing to act as a foreman on the site to ensure that proper application procedures are followed and hire a licensed contractor to perform the work. Information on such arrangements, for which the landowner pays only part of the total cost, should be available from the state department of forestry or the local cooperative extension service.

Wetlands Forest Management:

Below are listed some of the more important management practices for achieving the Management Measure for Wetlands Forest Management. Complete discussions of these and other management practices for preharvest planning can be found in Section 3J. Additional management practices that are particularly applicable to the NIPF landowner follow this listing.

◆ *Select the harvesting method to minimize soil disturbance and hydrologic impacts on the wetland.*

Additional management practice recommendations for the NIPF landowner

◆ *Contact a state forester or soil scientist to identify forested wetlands.*

Forested wetlands can be difficult to identify. They can occupy very small areas or large areas, can be of any shape, and need not be permanently flooded. Delineation of an area as a wetland requires that three criteria be met:

- Hydrology—a degree of flooding or soil saturation
- Hydrophytic vegetation (vegetation specific to wetlands)
- Hydric soils

These three components can be very site-specific. Differentiating a forested wetland from a non-wetland forest can be difficult. Wetland areas on a property need not be contiguous, and it is possible for a property to have several wetland areas. Some wetlands might be large and easily identified, whereas others might be small and very inconspicuous (Mitsch et al., 1993). Furthermore, different plant species are adapted to the various conditions that wetlands can occupy, so the absence of wetland plants identified in one wetland area from other areas does not mean that other wetlands do not exist on the property. Because of the complexity of wetland identification, a person licensed in wetland delineation should be consulted if there is any doubt as to whether wetlands exist on a property.

An initial assessment of the existence of wetlands on a property can be done by walking the property and asking some simple questions (Maryland DNR, undated):

- Is the ground moist underfoot?
- Are there springs in the area? (Look at a USGS quadrangle map.)
- Are the tree species considered hydrophytic vegetation? (Use a wetlands tree guide.)
- Are there high-water marks or silt deposits on tree trunks?
- Is water ponded anywhere?

- Do your feet sink into the soil when you walk?

- Dig a hole about a foot deep. Is the soil mostly gray?

- Does the soil in the hole smell like sulphur or rotten eggs?

- Does the hole fill up with water? Does water leak into the hole?

- Is there lush vegetation in some areas and not in others?

To help answer some of the questions, it is useful to have field guides to identify wetland species. Field guides provide descriptions of trees and other wetland vegetation and information on their ranges and habitats.

Contact the local office of the Soil Conservation District to determine whether there are hydric soils on the property. The office will be able to provide a map of the soil series of the property.

Water Quality Protection During Invasive Species Control

Invasive species are gaining a foothold in many parts of the United States, and they can cause extensive damage to a forest. Introduced insects, diseases, and plants can all cause problems for the forest landowner, and the means of control include mechanical, chemical, and biological. Mechanical and chemical control methods, in particular, have the potential to affect water quality. Prior to attempting control of an invasive species, consider using the practices below for the protection of water quality during invasive species control activities. The U.S. Department of Agriculture, the U.S. Forest Service, state forestry agencies, cooperative extension agencies, and local or state universities can provide additional assistance with the identification of invasive species, the problems they cause, and appropriate control methods. Even if you do not believe that you have an invasive species problem, or that your problem is not serious enough to do anything about, it is advised to find out what the invasive species in your area are and what their signs are. Knowing what the problems are can help prevent them or help you identify them before the problem becomes insurmountable and your losses significant.

◆ *Consult a state forester before using mechanical control methods.*

The control of invasive species usually requires the implementation of either chemical or mechanical means of control. To ensure that water quality is not compromised when these practices are used, consult with the local county forester before taking any action.

Mechanical control methods used to eradicate an invasive plant, insect, or disease can potentially impair water quality. Some mechanical methods of invasive species removal are cutting, girdling, hand pulling, burning, and grubbing. Some species that can be managed through mechanical control are kudzu (*Pueraria lobata*), tree of heaven (*Ailanthus altissima*), leafy spurge (*Euphorbia esula*), mistletoe (*Phorandendron serotinum*), purple loosestrife (*Lythrum salicaria*), scotch broom (*Cytisus scoparius*), saltcedar (*Tamarix ramosissima*), spruce bark beetle (*Dendroctonus rufipennis*), Douglas fir beetle (*Dendroctonus pseudotsugae*), fusiform rust (the fungus *Cronartium fusiforme*), and pine pitch canker (the fungus *Fusarium subglutinans*). The cooperative extension service should be able to provide information on invasive species in your area and appropriate control methods. The following guidelines apply to water quality protection during invasive species control activities:

- Remove invasive species from the SMA only if water quality will not be compromised.

- Do not burn SMAs to eradicate an invasive species.

- Avoid removing infected trees during wet weather periods. This will help reduce erosion potential at the site of removal and on haul roads.

Chemical control of invasive species involves the application of herbicides, pesticides, or fungicides to remove unwanted pests. Review the guidelines for chemical applications in this guidance and provided by your state forestry department before using chemicals for invasive species control.

Additional Resources for the NIPF Landowner:

Landowner's Guide to Building Forest Access Roads, by Richard L. Wiest, is a designed for landowners in the northeastern United States who will use a tractor and ordinary earth moving equipment to build the simplest access roads on their property, or who will contract for these services. Recommendations cover basic planning, construction, drainage, maintenance and closure of such forest roads. Also covers special situations involving water that require individual consideration. Describes geotextiles to be used during temporary road construction. The guide is published by the U.S. Department of Agriculture, Forest Service, Northeastern Area, State and Private Forestry Division. (1998; 47 p.; order online at http://www.na.fs.fed.us/spfo/pubs/stewardship/accessroads/accessroads.htm; first copy free, other copies $8 ea.).

APPENDIX E: STATE AND PRIVATE FORESTRY PROGRAMS

Education and Training

Education and training are vital to effective BMP implementation. Educating and training loggers and landowners about the importance and use of BMPs is an effective way to reduce water quality effects from forest operations because harvesters and landowners are responsible for forest harvesting and decisions concerning the management of much of the forested land in the Nation. A logger education program that has been adopted in various forms and under numerous names in many states is the Logger Education to Advance Professionalism (LEAP) program (APA, 1995). It is modeled after Vermont's very successful Silviculture Education for Loggers Project and began as a national pilot program of the USDA Extension Service to promote responsible forest BMPs and to teach forest ecology and silviculture to loggers. These programs are based on the premise that it is important to teach forest ecology and silviculture to loggers because professional foresters supervise less than a third of all the acres harvested in the United States while loggers are involved in all of the harvests. Before these programs, few people employed in logging had training in forestry and silviculture, and the logger education programs are changing that situation. To accomplish its goal, logger training emphasizes five areas— safety and first aid, business management, harvesting operations, professionalism, and forest ecology and silviculture.

Currently there are nearly 500 million acres of non-federal forests in the United States. More than 50 percent of these acres are privately owned (USDA Forest Service).

A USDA Natural Resources Conservation Service (NRCS) program, *Soil and Water Conservation Assistance* (SWCA), provides cost share and incentive payments to farmers and ranchers to voluntarily address threats to soil, water, and related natural resources, including forest land, grazing land, wetlands, and wildlife habitat. SWCA can help landowners comply with federal and state environmental laws and make beneficial, cost-effective changes their land management practices. Through the nearly 3,000 Soil and Water Conservation Districts nationwide with 2,500 field offices, nearly a million private landowners are assisted annually with land management decisions.

NRCS also administers the Forestry Incentives Program (FIP), which supports good forest management practices on privately owned, nonindustrial forest lands nationwide. FIP is designed to benefit the environment while meeting future demands for wood products. Eligible practices are tree planting, timber stand improvement, site preparation for natural regeneration, and other related activities. FIP is a nationwide program available in counties designated on the basis of a Forest Service survey of total eligible private timber acreage that is potentially suitable for production of timber products. Federal cost-share money is available—with a limit of $10,000 per person per year with the stipulation that no more than 65 percent of the cost may be paid. A local USDA office, state forester, conservation district, or Cooperative Extension office can provide information on whether a particular county participates in FIP.

Numerous non-governmental organizations, such as the Forest Stewards Guild (http://www.foreststewardsguild.org/) and National Network of Forest Practitioners (http://www.nnfp.org/) are also available to be contacted for assistance in sustainable management of forest land.

Cooperative Forestry Programs

Cooperative Forestry is a nationwide program funded through Congress and administered nationally by the USDA Forest Service. Since 1978, the USDA has connected rural, urban, and nonindustrial private forest (NIPF) landowners with resources and ideas to assist with the care of their forests. The Cooperative Forestry program provides technical and financial assistance through partnerships with the state and private forestry organizations (USDA Forest Service, 1999). The Cooperative Forestry program was created under section 2101 of Title 16 of the United States Code, in which it is stated that it is the policy of Congress that the Secretary of Agriculture work through and in cooperation with state foresters, or equivalent state officials, nongovernmental organizations, and the private sector in implementing federal programs affecting non-federal forestlands. The landowner assistance programs covered under Cooperative Forestry are the Forest Legacy Program, the Forest Stewardship Program, and the Forest Land Enhancement Program. The Forest Service's Web site for Forestry Landowner Assistance, http://www.fs.fed.us/ spf/coop/, provides further information about the programs discussed below.

- *Forest Legacy Program.* The Forest Legacy Program (FLP), a federal program in partnership with states, supports state efforts to protect environmentally sensitive forest lands. Designed to encourage the protection of privately owned forest lands, FLP is an entirely voluntary program. To maximize the public benefits it achieves, the program focuses on the acquisition of partial interests in privately owned forest lands. FLP helps the states develop and carry out their forest conservation plans. It encourages and supports acquisition of conservation easements, legally binding agreements transferring a negotiated set of property rights from one party to another, without removing the property from private ownership. Most FLP conservation easements restrict development, require sustainable forestry practices, and protect other values.

- *Forest Stewardship Program.* This program helps private forest landowners develop plans for the sustainable management of their forests. This is accomplished through active forest management for present and future landowners, increasing the economic value of the timber along with providing environmental benefits. The Forest Service also provides public outreach programs to assist NIPF landowners with information regarding seedling production and tree stand improvements.

The 2002 Farm Bill incorporates the following cooperative forestry assistance programs:

- Forest Land Enhancement Program: The Forest Land Enhancement Program (FLEP) is established to provide financial, technical, educational and related assistance to state foresters to assist private landowners in actively managing their land. Note that the FLEP replaces the Stewardship Incentives Program (SIP) and the Forestry Incentives Program (FIP). To be eligible for cost-share assistance under the FLEP on up to 1,000 acres, a landowner must agree to develop and implement for not less than 10 years a management plan that has been approved by the state forester.

Cost share payments will be available to landowners for up to 75 percent of the total cost of implementing the plan.

- Enhanced Community Fire Protection: Recognizing the significant federal interest in enhancing community protection from wildfire, the Department of Agriculture will cooperate with state foresters to manage lands to (1) focus the federal role in promoting optimal firefighting efficiency at the federal, state and local levels; (2) expand outreach and education programs to homeowners and communities about fire protection; and (3) establish space around homes and property that is defensible against wildfire.

Congress passed the Healthy Forests Restoration Act of 2003 (P.L. 108-148) on December 3, 2003, based on legislation proposed by the Bush Administration. The law provides critical tools needed to fully implement the Healthy Forests Initiative and the funding necessary to reduce wildfire risks and improve forest and rangeland health (USDOI, USDA, 2004). The Healthy Forests Restoration Act establishes procedures to expedite forest and rangeland restoration projects on Forest Service and BLM lands. It focuses on lands (1) near communities in the wildland urban interface, (2) in high risk municipal watersheds, (3) that provide important habitat for threatened and endangered species where catastrophic wildfire threatens the survival of the species, and (4) where insects or disease are destroying the forest and increasing the threat of catastrophic wildfire. The law:

- Helps communities use wood, brush, and other plant materials removed in forest health projects as a fuel supply for biomass energy.

- Authorizes a program to support community-based watershed forestry partnerships that address critical forest stewardship and watershed protection and restoration needs at the state and local level.

- Directs research focused on the early detection and containment of insect and disease infestations.

- Establishes a private forestland easement program focused on recovering forest ecosystem types and protecting valuable wildlife habitat.

The Watershed Forestry Assistance Program, created by the law, enacts the Watershed Forestry Cost-Share Program. The cost-share program provides up to 75 percent of project funding to communities, nonprofit groups, and NIPF landowners for watershed forestry projects that:

- Use trees as solutions to water quality problems in urban and rural areas.

- Employ community-based planning, involvement, and action through State, local and nonprofit partnerships.

- Apply and disseminate monitoring information on forestry best management practices relating to watershed forestry.

- Implement watershed-scale forest management activities and conservation planning.

- Restore wetland and stream-side forests and establish riparian vegetative buffers.

Forest Land Ownership

Nonindustrial private forest land (NIPF) owners in the United States own 58 percent of all timberland. Of this, 29 percent is owned by farmers who can benefit from the numerous provisions of the 2002 Farm Bill that involve land management. The rest of the timberland in the United States is owned by the federal government (20 percent), the forest industry (14 percent), state government (6 percent), and counties and municipalities (2 percent). Because of the large percentage of timberland owned by nonindustrial private forest land owners, an important part of protecting forests and water quality during forest harvest is educating those landowners about forest management and proper timber harvesting techniques to protect water quality (Powell et al., 1994). Birch (1996a) reports that private forest land owners (including industrial owners) have diverse reasons for owning their land, including "… it's just part of the land" (40 percent), a private source for forest products (8 percent), recreation and aesthetic enjoyment (23 percent), investment (9 percent), and timber production (3 percent). The last group, those who hold their land for timber production, represents 29 percent of private forest land ownership. It is estimated (Birch, 1996a) that 5 percent of private forest land owners have a written management plan and these owners control 39 percent of private forest land.

With so much land owned and controlled by private forest land owners, and specifically NIPF owners, it is crucial that the importance of protecting water quality be considered as part of NIPF harvesting. Some private landowners may not place an emphasis on water quality protection when planning a harvest because it appears to provide benefits only for downstream users, not for the harvesting landowner. Other management measures–such as site preparation to improve regeneration–provide direct benefits to landowners and are therefore more likely to be part of the landowner's harvest plan (Alden et al., 1996).

Forest Program Administration and BMP Effectiveness

A survey to compare the attitudes of persons involved with forestry program administration and implementation about the effectiveness of various approaches to protecting water quality and forests in general rated methods for protecting water quality from most effective to least effective as follows (Ellefson et al., 1995): technical assistance, fiscal incentives, educational programs, voluntary programs, regulatory programs, and tax incentives (Figure E-1).

In this survey, forestry program administrators were asked to rate specifically the effectiveness of educational programs for protecting water quality: 19 were neutral about their effectiveness, 17 said that they thought they were effective, and 12 thought that they were ineffective. The results for a similar rating of the effectiveness of technical assistance programs for protecting water quality showed that 26 administrators thought they were effective, 17 were neutral about their effectiveness, and 6 thought them to be ineffective.

The importance of education in forest harvesting and forest stewardship can be judged from the fact that many state departments of forestry have BMP guidebooks and education programs geared not only to loggers and industrial owners but also to the landowners who are not trained in forest management and harvesting. A review of some states' educational programs is provided below, and this review represents the variety of

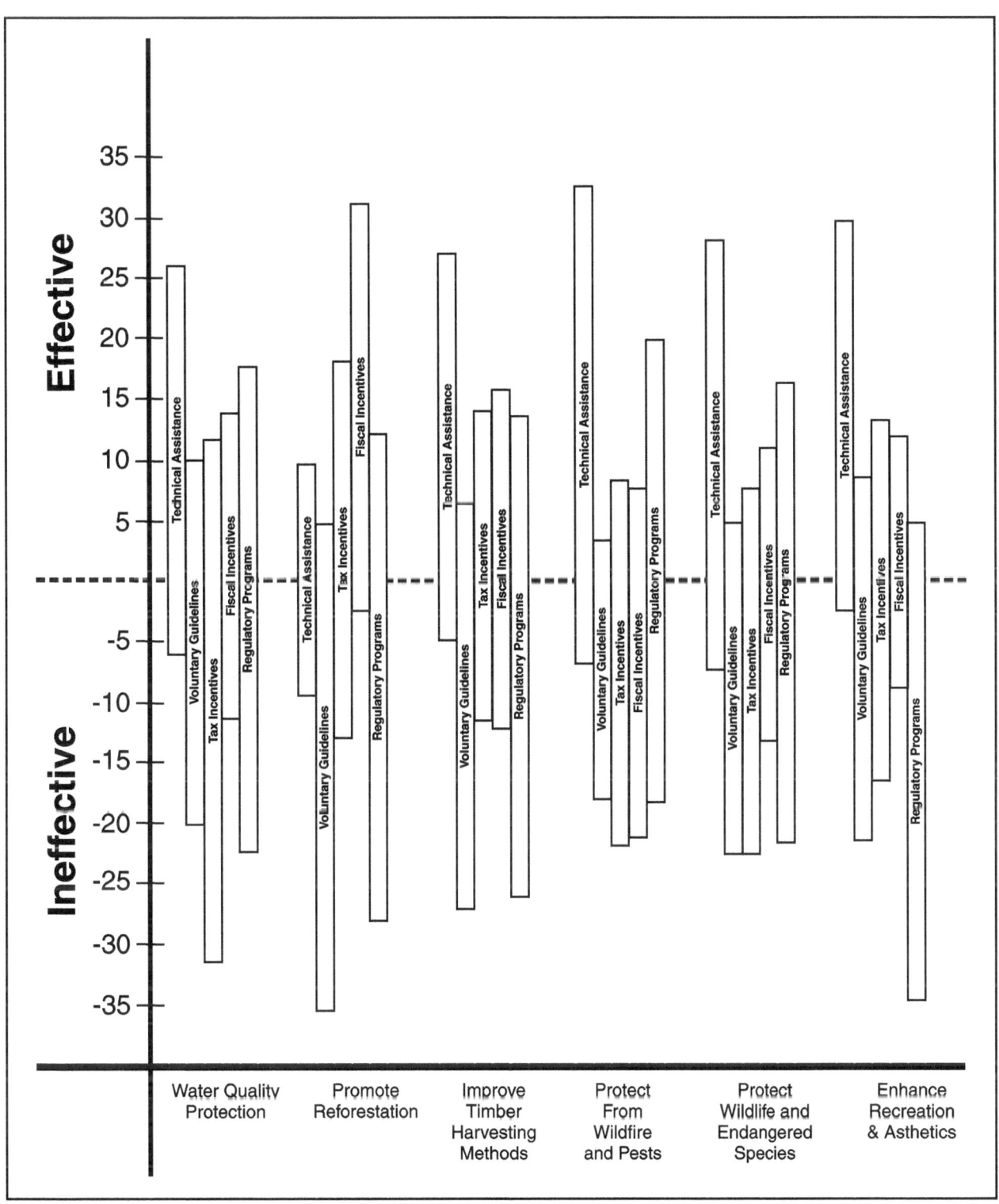

Figure E-1. Ratings of the effectiveness of various types of programs for accomplishing specific forestry objectives. Height of line above or below the center line indicates the number of state program administrators who rated the program type as effective or ineffective, respectively, for accomplishing the specific objective (Ellefson et al., 1995).

educational and technical assistance programs offered by states and the importance states place on education.

Examples of State Forestry Assistance Programs

Provided below are some examples of state programs for forestry assistance and education. Links to information on state forest protection and education programs can be found at the Web site www.usabmp.net.

Washington State

In 1999, Washington State created a Forestry Riparian Easement Program to be managed by a Small Forest Landowner Office within DNR. Responding to the federal Endangered Species Act by listing several salmon species and authorizing the Forest Practices Board to adopt rules for salmon recovery, the size of riparian buffers was increased and further measures were created to protect water quality and restore salmon habitat. Recognizing that these rules would have a disproportionate impact on small forest landowners, the easement program under the Forestry Riparian Easement Program acknowledges the importance of small forest landowners and the contributions they make to protect wildlife habitat. The program is also intended to help small forest landowners keep their land in forestry.

DNR's Forestry Riparian Easement Program partially compensates eligible small forest landowners in exchange for a 50-year easement on "qualifying timber." This is the timber the landowner is required to leave unharvested as a result of new forest practices rules protecting Washington's forests and fish. Landowners cannot cut or remove the qualifying timber during the easement period. The landowner still owns the property and retains full access, but has "leased" the trees and their associated riparian function to the state.

Washington's Backyard Forest Stewardship Program is especially designed for owners of small-forested parcels (from a "forested lot" up to ten acres) and anyone who owns a home in a forested environment. Guidelines for forest protection are provided on a DNR Web site (http://www.wa.gov/dnr/htdocs/rp/stewardship/bfs/) and can be obtained in print as well. Landowners who implement the guidelines relevant to their property can apply for recognition under the program from the state.

> The Federal Coastal Nonpoint Pollution Control Program (6217) is designed to enhance state and local efforts to manage land use activities that degrade coastal habitats and waters.

Virginia

The Virginia Department of Forestry (DOF) reports that surveys show most landowners sell timber and make other forest management decisions without professional advice. These same studies have demonstrated that landowners who sell timber with the assistance of a professional forester receive 50 percent more for their timber (Virginia DOF, 1998). Since professional foresters are knowledgeable of water protection BMPs, having a landowner contact a professional benefits both the landowner and the environment (Virginia Department of Forestry, 1998).

The Virginia DOF inspects harvesting sites for compliance with the Seed Tree Law and The Silvicultural Water Quality Law. During an inspection, compliance with other state and federal laws is observed so the landowner and logger can be informed and kept in compliance with applicable regulations. Other laws that landowners need to be aware of and in compliance with include, depending on their particular location and situation, the Chesapeake Bay Act, the Virginia Marine Resources Law, and the Federal Clean Water Act. The logger, consultant forester, industry forester, and/or the landowner are contacted by Virginia DOF during logging operations concerning BMP installation. The landowner is contacted concerning needs for forest renewal and future management.

Regardless of the origin of the request, if the landowner wishes to reforest an area or implement other recommended management practices, Virginia DOF will provide them with the names of consultants or contractors who can implement the recommended practices, and will inform them of any cost share assistance for which they might be eligible.

The Virginia DOF has the responsibility to administer and give technical approval for cost-share programs. A reforestation cost-share examination must be completed along with application forms and other paperwork for cost-share programs. For cost-share assistance, the area must be inspected for needs determination before the practice is started and after the practice is completed to determine if the practice was completed correctly. Again, required compliance with all applicable state and federal laws and regulations are checked.

Tennessee

Forestry assistance in Tennessee is handled by the Tennessee Department of Agriculture (DOA), Forestry Division. The Forestry Division trains loggers and others involved in land management in the use of logging techniques to prevent erosion and leave streams unharmed. Tennessee DOA has also developed a number of training aids for water quality, including a video, printed material, and a number of forest management demonstration sites. One of the Forestry Division's primary services is offering advice to landowners, often in person on the individual's property. A forest land owner can contact a local Area Forester to discuss management objectives for the property. The Area Forester will work through a sequence of steps to help meet the objectives. A local forestry office can also provide information on what landowner options are for managing their land. The DOA Forestry Division web site provides *A Practical Introduction to Forestry for Landowners* that gives information on a variety of forest management options and has references and links to other sources of information.

The Tennessee Reforestation Incentive Program (TRIP) was created in mid-1997 to provide financial assistance to landowners for planting trees on marginal and highly erodible crop and pasture lands. Money provided by the State Agricultural Nonpoint Water Pollution Control Fund administered by the Department of Agriculture is used to share the cost of planting trees to stabilize eroding lands and improve water quality.

Another training program available to loggers is the Master Logger Program. The mission of the Master Logger Program is "to enhance the professionalism of the Tennessee logger" through a complete educational program designed to improve the health and well-being of the logging industry and the forest resource. The Master Logger curriculum consists of five 1-day courses, one of which is on forest ecology and BMPs. Loggers attend individual sessions of the program 1 day every 2 weeks, and it takes 10 weeks to complete the workshop. Master Loggers must continue their education to retain Master Logger status. Many other states provide programs similar to the Master Logging

Program under various names, and all of the programs stem from the original pilot program of the USDA Extension Service, the LEAP program.

the number 10 years ago. The largest number of operations occur on small private forests where the landowners are typically not as familiar with the state's forest practice rules as are large industrial landowners. The state therefore puts a great deal of energy into providing information, training, and resources to landowners and operators (Oregon DOF, 1997).

The Oregon Department of Forestry's Forest Practices Program involves more than 150 people in the department's main offices and in field offices who provide face-to-face information and guidance to landowners. Program staff work with industry and environmental representatives to develop programs and incentives for encouraging sound stewardship of forest resources.

Small woodland owners in Oregon can request on-site assistance from their local service forester, who can provide information and guidance on insect and disease issues, reforestation and young growth management, financial incentives, and other forest related topics and resources. Private forest consultants are available throughout the state to provide comprehensive assistance to landowners. Consultants provide services that are beyond the scope of public agency assistance programs, such as the development of Forest Stewardship Plans.

The Oregon Forest Resource Trust provides monies for the direct cost payments of site preparation, tree planting, seedling protection, and competitive release activities. The program encourages landowners to establish and maintain healthy forests on underproducing forestlands—lands capable of growing forests but that are in brush, cropland, pasture, or that are very poorly stocked. The landowner commits to establishing a healthy "free-to-grow" forest stand and takes responsibility for seeing that the work gets done. The service forester provides technical assistance on how to complete the reforestation project and is available to provide direction with respect to the landowner's project management responsibility. If timber is harvested from the forests created with trust monies, participating landowners repay the trust (up to set amounts) with a portion of the profits. Eligible underproducing land must be at least 10 contiguous acres, zoned for forest or farm use, located in Oregon, and part of a private forestland ownership of no more than 5,000 acres. The trust can fund 100 percent of the reforestation cost up to $100,000 every two years.

The Oregon 50% Tax Credit, the "Underproductive Forest Land Conversion Tax Credit," encourages landowners to establish and maintain healthy and productive forests. Fifty percent of the cost of establishing a stand of trees on underproductive forestland may be applied as a credit against Oregon state taxes. The 50 percent tax credit applies on brushland, grassland, or on very poorly stocked forestland.

South Carolina

The South Carolina Forestry Commission provides timber management assistance to forest landowners in the state. Forestry Commission foresters will examine forestland and potential forestland at the request of a landowner. A written plan and map are prepared for the landowner, giving forest management recommendations that best meet the owner's needs and objectives, provided that they are compatible with good forest BMPs (South Carolina Forestry Commission, 1998). When conditions warrant, such as a request for a detailed plan on a large tract, the Forestry Commission forester can recommend consultants or industry foresters who can be of assistance.

Two-thirds of the state's forestlands are under private ownership, and the South Carolina Forestry Commission provides assistance to these landowners geared toward educating them so that they can take an active role in managing their forests. A South Carolina Forestry Commission staff member will help the landowner put together a multiple-resource Stewardship Management Plan (SMP) that provides detailed recommendations for timber management activities designed to help prevent soil erosion and protect water quality and might also provide details on wildlife habitat improvement. Anyone who owns at least 10 acres of forestland can qualify for assistance under the SMP program.

Ohio

The Ohio Department of Natural Resources Division of Forestry participates in the Service Forestry Program, the mission of which is to develop better stewardship of the forest resources on private lands in Ohio through on-site technical assistance and the dissemination of information to landowners. There are twenty-five Service Foresters statewide that work one on one with the woodland owners. The Service Foresters are available to provide landowners with current information for the long term management of their woodlands. The Service Foresters can provide management plans and advice on how to accomplish the plan's objectives. The Service Foresters also provide landowners with technical assistance and information on tree planting projects, woodland improvement activities and timber marketing assistance. The Service Foresters also direct landowners to other education participation programs in the state.

The Ohio Forestry Association maintains a Safety Training and Certification Program for logging contractors and their employees. It is the Ohio equivalent of a LEAP program. One of the requirements for certification as a Certified Logging Company is to have employees trained to use BMPs to reduce soil erosion and improve the appearance of timber harvesting activities (Ohio Forestry Association, 1999).

California

The California Department of Forestry & Fire Protection (CDF) administers several state and federal forestry assistance programs with the goal of reducing wildland fuel loads and improving the health and productivity of private forest lands. California's Forest Improvement Program (CFIP) and other federal programs that CDF administers, offer cost-share opportunities to assist individual landowners with land management planning, conservation practices to enhance wildlife habitat, and practices to enhance the productivity of the land.

The CFIP provides technical assistance to private forest landowners, forest operators, wood processors, and public agencies. Cost share assistance is provided to private forest landowners, Resource Conservation Districts, and nonprofit watershed groups. Cost-shared activities include management planning, site preparation, tree purchase and planting, timber stand improvement, fish and wildlife habitat improvement, and land conservation practices for ownerships containing up to 5,000 acres of forest land.

A Forest Legacy Program (FLP) protects environmentally important forestland threatened with conversion to non-forest uses, such as subdivision for residential or commercial development by promoting the use of permanent conservation easements.

Maine

The Forest Policy and Management Division of the Maine Department of Conservation, Forest Service provides technical assistance, information, and educational services to forest landowners. Part of the Division's implementation of the Forest Practices Act is providing educational workshops, field demonstrations, and media presentations, and contacting landowners personally to discuss forest management issues (Maine DOC, 1998).

North Dakota

The majority of North Dakota's rural forests are privately owned. Forest resource management in the state focuses on education and assisting nonindustrial private landowners to better manage, protect, and use their natural resources. This is accomplished through the development of a forest stewardship plan and direct financial assistance for forest improvement practices. Rural forestry services are delivered through an agreement with North Dakota's local Soil Conservation Districts (NDSU, 1998).

The Environmental Quality Incentives Program (EQIP) and the Wildlife Habitat Incentives Program (WHIP) offer up to 75 percent cost-share assistance to landowners for accomplishing forest stewardship projects such as tree planting, forest stand improvement, soil and water protection, riparian protection, windbreak renovation and wildlife habitat enhancement. Eligible landowners may sign up at their local FSA office for WHIP or EQIP practices.

Technical forestry assistance is provided to more than 600 rural landowners each year in North Dakota. Since 1991, 1,405 forest stewardship plans have been requested and completed for 71,777 acres of privately-owned native and planted woodlands and 456 forest improvement practices were awarded $548,887 in Stewardship Incentive Program cost-share funds. A total of 587 landowners enrolled 39,384 acres in the Forest Stewardship Tax Law.

Missouri

The vast majority of land in Missouri is under direct ownership and influence of private landowners. Private individuals own more than 93 percent of all land and 85 percent of forest land. The Department offers two levels of assistance based upon the landowner's need and interest in long term forest management. The two levels are Advisory Service and Management Service. Advisory Service is available to all landowners, including urban residents. This service includes group training sessions, publications, film and video loan, office consultation, insect and disease identification and analysis, referrals to consultants, on-site visits under certain conditions, and help with evaluating and choosing land management options.

Management Service is available to landowners interested in the long term management of their forest land. Those who receive management services agree to develop and carry out a management program for the immediate and long term stewardship of their property. Management plan implementation activities include guidance in soil and watershed protection, erosion control, wildlife habitat improvement, and forest road location and construction. A visit to the landowner's property is part of MDC's assistance in management plan development (Missouri DOC, 2000).

The Society of American Foresters' Certified Forester Program

The Society of American Foresters (SAF), a nonprofit, scientific, and educational organization, established the Certified Forester (CF) program in 1994. The term *Certified Forester* is registered with the U.S. Patent and Trademark Office and may only be used by individuals who meet SAF's certification requirements. The CF program is voluntary, nongovernmental, and open to qualified SAF members and nonmembers. A Certified Forester agrees to abide by current CF program requirements and procedures for certification and recertification; to maintain continuing professional development; and to conduct all forestry practices in a responsible, professional manner consistent with state and federal regulations governing environmental quality and forest BMPs.

Through the CF program and other activities, SAF advocates wise stewardship in forest resources management. The CF program provides a consistent, national credential. Certification constitutes recognition by SAF that, to the best of SAF's knowledge, a Certified Forester meets and adheres to certain minimum standards of academic preparation, professional experience, continuing education, and professionalism. No individual is eligible to receive or to maintain Certified Forester status or recertification unless the individual meets and continues to adhere to all requirements for eligibility. Some of the requirements that must be met by all CF applicants can be found in Appendix C.

Effectiveness of Education and Technical Assistance

Researchers with the U.S. Forest Service reviewed state BMP implementation and monitoring programs and the results from those programs in 1994. At the time, 21 states were assessing BMP effectiveness. The U.S. Forest Service found that the states had generally concluded that carefully developed and applied BMPs can prevent serious deterioration of water quality and that the availability of well-qualified personnel at the field level is probably the most cost-effective approach to meeting water quality standards. Most water quality problems, they found, were associated with poor BMP implementation, and trained field personnel could help correct problems with implementation (Greene and Siegel, 1994).

The researchers also concluded that an iterative self-education process at the state level was important for BMP improvement. Water quality monitoring is essential to understanding the relationship between land disturbance and water quality, they found, and it leads to improved understanding of the interaction of soils and topography with BMP implementation. This understanding was considered essential to continually reassessing BMP guidelines to make them more cost effective. BMPs need to be specified, used, monitored, and fine tuned to provide cost-effective water quality protection.

Ellefson and others (1995) reviewed forest practice programs in many states, and one aspect of their review involved asking program managers what they thought were the most effective means to protect water quality. State program managers rated the following in program effectiveness, from most effective to least effective: technical assistance, fiscal incentives, educational programs, voluntary programs, regulatory programs, and tax incentives. For promoting reforestation and improving timber harvesting methods, technical assistance and fiscal incentives were rated as the most effective means and regulatory programs and voluntary guidelines were rated as the two least effective.

When the Vermont Agency of Natural Resources (ANR) studied BMP implementation and effectiveness, ANR personnel accompanied harvesters in the field during harvests. During the harvests monitored, logging personnel appeared to become much more aware of the water quality issues related to their activities and the intent of the BMPs. By the end of the project, the loggers were extremely conscientious in their efforts to protect water quality. Vermont ANR personnel felt that without the oversight of the forestry agency, it was likely that water quality problems would have been more severe, particularly in the early phase of the project. After the assistance provided by the personnel, managers for the logging companies were fully capable of implementing appropriate BMPs with little or no oversight.

www.ingramcontent.com/pod-product-compliance
Lightning Source LLC
Chambersburg PA
CBHW080635180526

45168CB00008B/3179